V. V. Beletsky

Essays on the Motion of Celestial Bodies

Translated from the Russian
by Andrei Iacob

Springer Basel AG

Author:

V. V. Beletsky
Keldysh Institute of Applied Mathematics
Russian Academy of Sciences (KIAM RAS)
Miusskaja Sq. 4
125047 Moscow
Russia
e-mail: beletsky@keldysh.ru

Originally published in Russian by Nauka, Moscow

2000 Mathematics Subject Classification 70Fxx

A CIP catalogue record for this book is available from the
Library of Congress, Washington D.C., USA

Deutsche Bibliothek Cataloging-in-Publication Data
Beleckij, Vladimir V.
Essays on the motion of celestial bodies / V. V. Beletsky. Transl. from the Russ. by
Andrei Iacob. - Basel ; Berlin ; Boston : Birkhäuser, 2001
ISBN 978-3-0348-9533-0 ISBN 978-3-0348-8360-3 (eBook)
DOI 10.1007/978-3-0348-8360-3

ISBN 978-3-0348-9533-0

Originally published by Birkhäuser Verlag in 2001
Softcover reprint of the hardcover 1st edition 2001
Member of the BertelsmannSpringer Publishing Group
Cover design: Micha Lotrovsky, CH-4106 Therwil, Switzerland
Printed on acid-free paper produced of chlorine-free pulp. TCF ∞

www.birkhasuer-science.com

ISBN 978-3-0348-9533-0

9 8 7 6 5 4 3 2 1

The Russian edition of this book was awarded the 1999 F. A. Zander Prize of the Russian Academy of Sciences

The Beginning of a New Style in the Scientific Literature

(from the review of the 1st edition of V. V. Beletsky's "Essays" by V. I. Arnold and Ya. B. Zeldovich, *Priroda,* No. 10, 1973, 115–117.)

..."Dear Fagot, show us something simple for the start" – it is with this epigraph from M. Bulgakov's *The Master and Margarita* that the first essay of the book under review begins. And the reader is not being lied to – the book begins with some well known, and for that reason simple, classical results. Appreciating the style of the exposition, which is rather unusual in the scientific literature, from the very first pages (more precisely, from the brilliant dedication to Private Ferri Bren, a soldier in the Austro-Hungarian army, later a fighter in the International Red Guard), the reader will only gradually begin to be captivated by the unexpected novelty of the results. And after turning the last page, he will certainly agree with these reviewers that the appearance of books such as this, and of writers such as its author, is a reflection of the amazing revival that celestial mechanics is currently experiencing in response to the demands of the theory of space flight.

Here the rules encountered more than once in other fields of knowledge repeat themselves. One would have thought that the problem of the motion of a body in the gravitational field of the Earth has been thoroughly studied. However, the practical implementation of the process has necessarily led to demands for more detailed and precise information, for the development of optimal strategies. At first glance, when one tries to answer those demands, the clarity and beauty of the simplified theory are lost. But in the next stage new points of view as well as theories that describe the results of more detailed calculations and new beauty arise. It is precisely this stage that is reflected in V. V. Beletsky's book.

As the author rightly notes, the names of the specialists working in the applied areas, whose successes in the development of space technology and the conquest of cosmic space have stimulated the progress of many exact sciences, are not commonly found in the pages of scientific journals; these people, as a rule, do not write books and only rarely defend dissertations. But without their efforts the development of mechanics, as reflected in many dissertations, would have been unthinkable, and many books would have remained unwritten, among them the present "Essays on the Motion of Celestial Bodies." For that reason, from the mouth of its author, a well-known specialist in the mechanics of space flight, the ordinary-sounding words "*This book is a modest attempt to pay tribute to my professors, friends, and colleagues*" are filled with a special importance.

... For a solid scientific monograph, V. V. Beletsky's book is out of ordinary in many respects. Without exaggeration one can say that it marks the affirmation of a new style in the scientific literature. The author explains in a frank and detailed manner the reasons behind each calculation, its difficulties, and the psychological side of the research. The book contains no attempts to inflate the importance of results or to give results while hiding the methods used to obtain them. The books is adorned by humorous illustrations by I. V. Novozhilov, Doctor in Physico-Mathematical Sciences.... The general impression that the "*Essays*" make is not that this is a boring lesson, but rather a discussion with a brilliant, knowledgeable and wise interlocutor. Even people with little interests in space problems will go through the book with satisfaction, perhaps omitting the calculations.

... The famous XIXth century Frenchman Camille Flammarion had published in succession an *Astronomy*, an *Astronomy for All*, and an *Astronomy for Ladies*. Shouldn't V. V. Beletsky and the "Nauka" publishing house follow the same path in publishing new editions of the "*Essays*"?

Contents

Dedication

Private Ferri Bren, a soldier in the Austro-Hungarian army, was finally exhausted. Late at night, he came out of the woods and onto a country road. A transport was passing. He could barely see the silhouettes of the drivers in the dark; mud splashed under the hooves of the horses. Without thinking twice, Ferri Bren climbed up into a driverless cart, covered himself with a tarp, and fell asleep.

A push in the side woke him at dawn. Drivers – soldiers of the Russian army – stood around him. Cursing, but without anger, laughing and poking him with their fists, they confiscated his weapons and pack. Slightly frightened, the Slovenian German Ferri Bren climbed down from the cart onto the muddy autumn ground – the ground of Russia. Not yet knowing that this was forever.

That waiting for him were battles in the International Red Guard. And half a century of life and work in a new homeland. And a large new family, that he would raise and feed with his rare skill as a theatrical tailor. And a distant grave in a quiet cemetery in Irkutsk...

And that ahead was this book, dedicated to his memory – to Fyodor Iosifovich Bren, my grandfather.

Vladimir Beletsky

A Word from the Author

> Infinitely breathes the Universe,
> Rockets rush through it, like blobs of Sun.
> These are your dreams and visions,
> Your knowledge. Your insomnias.
>
> R. Rozhdestvenskiĭ

This book is the story of some interesting theoretical investigations in the mechanics of space flight, i.e., in the theory of motion of spacecraft. Some new problems of celestial mechanics are discussed as well.

This is not a textbook, and we did not seek to present a well-composed survey or a systematic exposition of the mechanics of space flight. As a matter of fact, many interesting results of utmost importance are not covered here. The book has a different purpose. In the process of its impetuous development the mechanics of space flight brought to life a whole series of fascinating and novel problems. Our goal was to present to the reader some interesting and at times unexpected achievements that have allowed some old problems of mechanics to be reconsidered in a new light; to reveal connections between classical problems and new results of the mechanics of space flight; and to explore some new formulations and solutions that have arisen in the dynamics of space flight.

At the same time, our story about the mechanics of space flight is also a story about methods of investigation. The mechanics of space flight uses methods and results of theoretical mechanics, celestial mechanics, certain technological sciences, and many fields of mathematics. As is now fashionable to state, the mechanics of space flight was born at the junction of the aforementioned branches of science.

The problems that will be described in the book vary in importance. The solutions of some of them already lie at the foundation of all known achievements (flights of satellites and lunar stations), the solutions of the others will find their application in a not so distant future (space flights with low-thrust engines); and it may be that some of the questions dealt with in the book are altogether void of current importance. But all of the problems considered here have one major feature in common: they are interesting. This is what we have sought to demonstrate to the readers to whom this book is primarily addressed: students specializing in mechanics who are beginning to think about narrowing down the field in which they wish to work. Our goal is to help them make that choice.

If this book will help its readers become aware, even to a small extent, of how astonishing and rich in events and phenomena the mechanics of space flight is, then we will be satisfied that our goal was accomplished.

In this kind of book one cannot always avoid long calculations; but here again, we have done our best to reveal, whenever possible, the beauty of the research process leading to this or that result. The emphasis is put on the analysis of results that can be carried out to the stage of graphs and drawings, and sometimes to numbers. Whenever possible, the investigation relies on maximally intuitive, elegant geometric tools. A typical example is the analysis of first integrals of the equations of motion.

Sometimes we will "digress" and present some background notions and results from mechanics and mathematics that are necessary for reading the book, but this material should not detract from the informal tone of the essays. In the course of the exposition, we were sometimes unable to resist the pleasure of resorting to less serious statements or unexpected analogies, borrowing tools from the science-fiction literature. However, the main mathematical objects treated in the book are systems of ordinary differential equations, and therefore the exposition cannot be regarded as accessible to an unprepared reader.

The basic material of the book should be accessible to students in the third and fourth (and perhaps also second) years of college programs in mechanics-mathematics, physics-mathematics, and technology, and some points can even be understood by diligent high-school students. If some passages seem difficult or even incomprehensible at first, one should by no means despair. Rather, those passages should be worked out with pencil in hand. However, we made every effort to lighten the exposition and to purposefully avoid a dry "academic" style. If on occasion we were not successful, the reader will have to be indulgent and understand that at times tradition is stronger than the desire for simplicity.

We hope that the book will provide some moments of satisfaction to other groups of potential readers as well: graduate students, teachers of theoretical mechanics, engineers and scientists working in the mechanics of space flight and related specialties. In fact, we have already tested the contents and style of the offered essays on these categories of listeners in survey lectures delivered at various universities.

The book uses results published by contemporary scientists – specialists in mechanics and mathematicians; a major place is also occupied by our own research.

As a whole the book addresses only a small part of the questions and problems that arise in the modern mechanics of space flight and celestial mechanics. In some of the problems described in the book substantial progress toward a solution was achieved only very recently. But one should remember that in the face of Nature our successes are never sufficiently important. There have always been and there will always be more unsolved (as well as not yet formulated) problems than solutions. As Isaac Newton put it:

> "I don't know what I may seem to the world, but, as to myself, I seem to have been only like a boy playing on the sea shore, and diverting myself in now and then finding a smoother pebble or a prettier shell than ordinary, whilst the great ocean of truth lay all undiscovered before me."

We were fortunate enough over the course of many years to learn from prominent scientists and teachers[1], share the same classroom with now-well-known researchers, and collaborate with first-rate specialists. This book is a modest attempt to pay tribute to my professors, friends, and colleagues.

To begin with, we wish to expresses our gratitude to Dmitry Evgen′evich Okhotsimskiĭ for supporting the idea of writing this book; to Vladimir Grigor′evich Demin for undertaking the painful task of editing it; to Natal′ya Samuilovna Konikov and Evgenia Alekseevna Stepanov for preparing the manuscript; and to Nikolaĭ Alekseevich Parusnikov and Georgiĭ Rafailovich Sazykin for their support and enthusiasm.

This book would have not come to light without the tireless care of Nina Aleksandrovich Beletsky. We are also thankful to Timur Magometovich Eneev, whose his kind advice was used in the preparation of the manuscript.

Special thanks are due to Igor′ Vasil′evich Novozhilov for the beautiful illustrations that he devised for this book.

We also want to note the role played by specialists in applied fields, whose successes in developing space technologies and conquering the cosmos have stimulated the progress of many exact branches of science. The names of these specialists are not familiar items in the pages of scientific journals; as a rule, they do not write books, and they rarely defend dissertations. But without their work advances that have been made in mechanics, as reflected in many dissertations, would have not occurred, and many books would have remained unwritten, among them the present book.

The first edition (in Russian) of the *Essays* was favorably met by its readers; several journals (*Zemlya i Vselennaya*, *Priroda*, *Uspekhi Fizicheskikh Nauk*, the Polish journal *Urania*) responded with kind reviews.

We are grateful to the referees of the book, V. I. Arnold, Ya. B. Zel′dovich, K. Ziolkovskiĭ, and A. L. Kunitsyn, for their analysis of the text and observations. To the extent it was possible, these observations were taken into account in the present edition. Errors and misprint discovered in the first edition were corrected. As a whole, the book retains the plan and style of exposition of the first edition, whose material was only minimally rewritten. However, the reader familiar with the first edition will discover a number of additions: on the resonant motions of celestial bodies and the generalized Cassini laws; on the evolution of galaxies and Poincaré's recurrence theorem; on a new point of view concerning the phenomenon of lunar-solar precession of the Earth's axis and on the magnetic stabilization of satellites; on the development of the concept of a gravity flier in the science, science-fiction, and fiction literature; finally, on the "miracle of closed orbits" in a Newtonian central gravity field.

[1]Here let us mention especially Vladimir Vasil′evich Golubev, Nikolaĭ Gur′evich Chetaev and Arkadiĭ Aleksandrovich Kosmodem′yanskiĭ.

I. V. Novozhilov has provided a number of new illustrations for the second edition.

Also, for the present translation into English we have added a number of comments at the end of some of the essays, and additional references. We wish to thank the translator for completing this not always easy to accomplish project.[2]

[2]In his turn, the translator wishes to thank Maryse Brouwers for her invaluable help in polishing difficult passages, to Ilya Bernstein, Prof. Assya Hummesky, Prof. Peter Read and Suzanne Zeitman for helping with the translation of the poetry epigraphs, to Alexandra Berland, Ward Bouwsma, Olga Luppov and Smilka Zdravskovska for help with tricky words, and last but not least to Patrick Ion for teaching me how to tame – at least partially – the TEX beast. I am also grateful to Thomas Hintermann and Sylvia Lotrovsky for entrusting me with this translation project – it was a real joy to read the book – and to Micha Lotrovsky for his artwork and efforts to rejuvenate the large number of figures and illustrations from the original edition. Perhaps some of the author's vivid style was lost in the translation, for which I apologize to his readers, as I do for any graphic or linguistic quality problems in the translated typeset manuscript.

First Essay

On the Unperturbed and Perturbed Motion of a Satellite, with a Digression on Asymptotic Methods of Nonlinear Mechanics

...“Dear Fagot, show us something simple for the start.”

M. Bulgakov, *The Master and Margarita*

1. Lucky us!

We begin our story with a description of some classical results, perhaps a bit too well known to intrigue and stir the reader's interest, but absolutely necessary for our exposition.

The fundament on which celestial mechanics and the mechanics of space flight lie is *Newton's law of gravitation*: two point masses (particles), of mass m and M, attract each other with the force

$$F = \frac{fmM}{r^2}, \tag{1.1.1}$$

where r is the distance between the masses and f is the universal constant of gravitation, which has the same value throughout the Universe ($f = 6.67 \cdot 10^{-8}$ cm^3/g·s^2).

If one consider the motion of one of these particles (say m) relative to the other (M), then assuming that one can neglect all forces except for (1.1.1), the differential equations of motion take on the form [1.1]

$$\ddot{x} + \frac{\mu x}{r^3} = 0, \quad \ddot{y} + \frac{\mu y}{r^3} = 0, \quad \ddot{z} + \frac{\mu z}{r^3} = 0, \tag{1.1.2}$$

where $\mu = f(M+m)$, x, y, z are the coordinates of the particle m in a coordinate system in translational motion with the origin at M, and $r = \sqrt{x^2+y^2+z^2}$.

The equations (1.1.2) are easily integrated in terms of elementary functions. This circumstance is so important that some further discussion is in order. As a matter of fact, we have neglected “extraneous” forces that act on our particle m. But such forces are present, and if one takes them into account the equations of motion take the form

$$\ddot{x} + \frac{\mu x}{r^3} = f_x, \quad \ddot{y} + \frac{\mu y}{r^3} = f_y, \quad \ddot{z} + \frac{\mu z}{r^3} = f_z, \tag{1.1.3}$$

where f_x, f_y, f_z are the components of the supplementary accelerations. Equations (1.1.3), generally speaking, are no longer integrable. Unfortunately, almost all problems of mechanics are described by nonintegrable equations. And only a few problems are integrable, such as, for instance, the problem of the motion of two particles under the action of the force of mutual attraction (1.1.1).

Fate had it that the human race emerged and developed on a planet that belongs to the planetary system of a single star – the Sun, and where, moreover, the planets lie at large distances from one another and their masses are considerably smaller than the mass of the central heavenly body, the Sun. Under these circumstances, whenever one studies the motion of any of the planets one can with high accuracy work with the equations (1.1.2), which account only for the force of attraction exerted on that planet by the Sun. Indeed, the "exact" equations of type (1.1.3) will contain corrections f_x, f_y, f_z that under the indicated conditions are very small:

$$\sqrt{f_x^2 + f_y^2 + f_z^2} \ll \mu/r^2.$$

Therefore, in equations (1.1.3) one can neglect the right-hand sides, i.e., work with the equations (1.1.2). And these equations enjoy two very important properties:

(1) They are integrable.

(2) The result of the integration reveals that the motion is very simple.

In other words, it is the specific structure of our Solar system that is responsible for the facts that, first, the motion of each planet is relatively simple, and second, that this motion can be described by integrable differential equations. This in turn has allowed mankind, in a relatively short period of time,[1] to grasp the laws governing the motions of planets (Kepler), and then to explain them (Newton).

We should mention that a large number of stars in our galaxy are multiple stars – double, triple – rather than single ones, like our Sun. Assuming the existence, in a multi-star system, of planets with stable orbits (and such orbits may indeed exist), it is possible that some planet with such an orbit harbors a race of intelligent beings [1.2]. Intelligent beings emerging on a planet belonging to the system of, say, a double star would be placed by Nature in worse conditions, compared with the humans inhabiting Earth, when it comes to understanding its laws. Indeed, the motion of a planet in the system of a double star will be described by the equations (1.1.3) with non-negligible right-hand sides (the influences of the two stars are comparable in magnitude). Consequently, the trajectory of the planet will be very complicated; to discover the laws that govern such a motion one would require a comparatively longer time, and even then it would be quite difficult to realize that these laws do indeed originate in the simple law (1.1.1) of gravitational attraction by each of the two stars. Therefore, the local Keplers and Newtons of such a double-star system planet would be in a tough position. The validation of

[1] The few thousand years of existence of our civilization are less than one percent of the time passed since the emergence of the *Homo Sapiens* species.

the law (1.1.1) requires unequivocal agreement between the solutions of equations (1.1.3) and observation data; but the equations (1.1.3) are not integrable.

The path to knowledge of such a hypothetical intelligent race will be more tedious and slower (just try to imagine how the Ptolemaic epicycles in a multiple-star system would look!). And since the pace at which a civilization develops depends on the totality of the knowledge accumulated, the author would like to formulate, not without taking some risk, the following statement:

The development of a civilization on a planet belonging to a multiple-star system under otherwise identical conditions[2] will be slower than in the case of a single-star system.

We should therefore consider ourselves very lucky!

2. Keplerian motion

Since the mass of a planet [respectively, the mass of an artificial satellite] is negligibly small compared with the mass of the Sun [respectively, the mass of the Earth], one can consider that the center of mass of the system under study coincides with the center of mass of the larger body. In what follow, unless otherwise stipulated, it will be understood that the central body is the Earth and the body of negligible mass is an artificial satellite.

The satellite is subject to the acceleration generated by Newton's force of attraction

$$\mathbf{f}_{\mathrm{N}} = -\frac{\mu}{r^2}\,\mathbf{e}_r, \tag{1.2.1}$$

directed toward the center of the Earth. Here $\mathbf{e}_r$ denotes the unit vector directed from the center of the Earth to the satellite (the latter being regarded as a particle). To this acceleration there corresponds the *force function* of the Newtonian central force field

$$U = \frac{\mu}{r}, \tag{1.2.2}$$

so that the components of the acceleration on the x-, y-, and z-axes of the fixed coordinate system with origin at the center of the Earth are equal to

$$f_x = \frac{\partial U}{\partial x} = -\frac{\mu x}{r^3}, \quad f_y = \frac{\partial U}{\partial y} = -\frac{\mu y}{r^3}, \quad f_z = \frac{\partial U}{\partial z} = -\frac{\mu z}{r^3}, \tag{1.2.3}$$

and the equations of motion (1.1.2) are integrable. The orbits satisfying these equations are called *Keplerian orbits.*

Above we were interested in the integrability of the equations (1.1.2), rather than in the integration process. The latter is treated in all textbooks on theoretical

[2] I do not exclude the possibility of a faster path to knowledge which does not make use of such a tool – in essence imperfect – as differential equations, or even the existence of civilizations whose qualities are not comparable to ours. This is how the phrase "under otherwise identical conditions" should be understood.

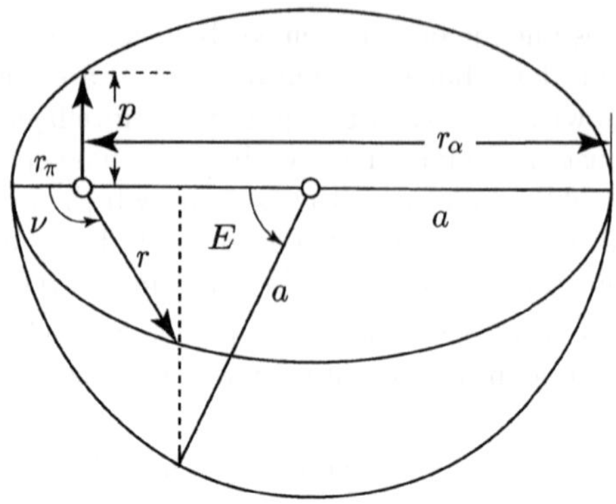

FIGURE 1.1. Elliptic Keplerian orbit

and celestial mechanics (see, e.g., [1.1], [1.3]), and so there is no point in discussing it here. We will only describe the final results of the integration. As it turns out, the satellite moves along an elliptic, parabolic, or hyperbolic orbit, such that the center of the Earth coincides with a focus of the ellipse, parabola, of hyperbola.

We will be interested mainly in the elliptic case. Figure 1.1 shows such an orbit. In polar coordinates r, ν the equation of the ellipse reads

$$r = \frac{p}{1 + e\cos\nu}, \tag{1.2.4}$$

where the angle ν is measured from the direction r_π of the ray from the center of the Earth to the closest point of the orbit, called the *perigee.* The satellites reaches the maximal distance r_α from the Earth at the *apogee*, which corresponds to the value $\nu = 180°$. In equation (1.2.4) the quantities p and e are constant. p is called the *focal parameter* or *semilatus rectum*, and its geometric meaning is clear from Figure 1.1. The second quantity – the *eccentricity* of the orbit, e, – characterizes how elongated, or oblate the orbit is. When $e = 0$ the orbit is circular, and when $e \to 1$ the orbit tends to a parabolic one. The quantities p and e can be expressed through the apogee (r_α) and perigee (r_π) distances:

$$p = \frac{2r_\alpha r_\pi}{r_\alpha + r_\pi}, \quad e = \frac{r_\alpha - r_\pi}{r_\alpha + r_\pi}. \tag{1.2.5}$$

The largest dimension of the ellipse is characterized by its *semi-major axis* a; one has

$$a = \frac{r_\alpha + r_\pi}{2}, \tag{1.2.6}$$

and p, e, and a obey the relation

$$p = a(1 - e^2). \tag{1.2.7}$$

The angle ν in formula (1.2.4) is called the *true anomaly.*

The dependence of $\nu(t)$ on time gives the law of motion of the satellite on its orbit. In the theory of Keplerian orbits the most difficult step is finding the explicit expression of ν in terms of time t. The angular velocity of orbital motion, $d\nu/dt$, satisfies the so-called *area integral* or *law*:

$$r^2 \frac{d\nu}{dt} = \sqrt{\mu p}. \tag{1.2.8}$$

If here we substitute the expression of $r(\nu)$ from (1.2.4) and carry out the corresponding quadrature, we obtain the explicit expression of time in terms of ν: $t = t(\nu)$. The problem is now to solve this transcendental equation for ν . To this end one introduces a new variable E, called the *eccentric anomaly* (its meaning is seen in Figure 1.1), which is connected with ν by the relations

$$\cos\nu = \frac{\cos E - e}{1 - e\cos E}, \qquad \sin\nu = \frac{\sin E}{1 - e\cos E}\sqrt{1-e^2}; \tag{1.2.9}$$

moreover,

$$r = a(1 - e\cos E). \tag{1.2.10}$$

The relationship between the eccentric anomaly and the time t is expressed by the *Kepler equation*

$$E - e\sin E = n(t - \tau^*), \tag{1.2.11}$$

where $n = \sqrt{\mu/a^3}$ is the so-called *mean motion*; the constant τ^* denotes the *time of perigee passage.* From (1.2.11) it readily follows that the orbital period of revolution of the satellite is

$$T = 2\pi\sqrt{a^3/\mu}. \tag{1.2.12}$$

If one solves Kepler's equation (1.2.11) for E, i.e., one determines $E(t)$, then using (1.2.9) $\nu(t)$ will be also determined as an explicit function of t, and so will $r(t)$. The reader interested in methods for solving Kepler's equation and the results that they yield is referred to G. N. Duboshin's book [1.1] already mentioned above.

Finally, the magnitude V of the velocity vector of orbital motion satisfies the relation

$$V = \sqrt{\mu/p}\cdot\sqrt{1 + e^2 + 2e\cos\nu}, \tag{1.2.13}$$

its radial and transversal projections being respectively equal to

$$V_r = \sqrt{\mu/p}\, e\sin\nu, \quad V_\tau = \sqrt{\mu/p}\,(1 + e\cos\nu); \tag{1.2.13$'$}$$

hence, at perigee ($\nu = 0$) the velocity is maximal, and at apogee ($\nu = \pi$) it is minimal. Naturally, since the motion takes place in a conservative force field (1.2.2), the *law of conservation of energy* (the *energy integral*) holds:

$$\frac{V^2}{2} - \frac{\mu}{r} = h = \text{const}. \tag{1.2.14}$$

Substituting here the expression (1.2.13) and (1.2.4) and recalling relation (1.2.7) we conclude that

$$h = -\frac{\mu}{2a}. \tag{1.2.15}$$

Note also that instead of the area integral (1.28) one can write the vector area integral

$$\mathbf{r} \times \mathbf{V} = \mathbf{c}, \tag{1.2.16}$$

where $\mathbf{r}$ is the radius (position) vector and $\mathbf{V}$ is the velocity vector of the satellite. Then, as it turns out, the vector

$$\mathbf{l} = -\frac{\mu}{r}\mathbf{r} + \mathbf{V} \times \mathbf{c} \tag{1.2.17}$$

(called the *Laplace vector*) is constant, i.e., $\mathbf{l}$ is an integral of motion. This vector is directed along the semi-major axis toward the perigee and its length is

$$|\mathbf{l}| = \mu e. \tag{1.2.18}$$

These are the basic laws of Keplerian motion on elliptic orbits. However, to this point we have considered the motion in the plane of the orbit, and to completely determine the motion in three-dimensional space we need to know also the position of the orbit in space. This position is specified as follows (Figure 1.2). We project the orbital plane on the celestial sphere and consider a system of coordinates XYZ such that the Z-axis points to the North pole of the cosmos (which is a completely determined point on the celestial sphere, close to the North Star) and the X-axis points to the vernal equinox point (which is also completely determined – in the present era it is located in the Pisces constellation); the origin of coordinates is the Earth's center, and the XY-plane coincides with the Earth's equatorial plane. The line of intersection between the orbital plane and the equator is called the *line of nodes*; its traces on the celestial sphere are called *nodes*. More specifically, the point on the celestial sphere where the satellite intersects the equator moving from the Southern hemisphere to the Northern one is called the *ascending node* of the orbit, while the opposite node is called the *descending node*. The position of the orbital plane is completely determined by two constant angles: the angle between the direction from the origin to the vernal equinox point and the direction from the origin to the ascending node, denoted by Ω and called *longitude of the ascending node*, and the angle between the equatorial plane and the orbital plane, denoted by i and called *inclination*. Further, the perigee π of the orbit lies at a constant angular distance ω from the nodal line, where the angle ω is measured in the orbital plane from the line of nodes to the direction from the origin to the perigee; ω is called the *argument of perigee*. Often instead of the true anomaly ν one introduces another variable, the polar angle $u = \omega + \nu$, called the *argument of latitude*. This is the angle measured in the orbital plane from the line of nodes to the current radius vector of the satellite in orbit.

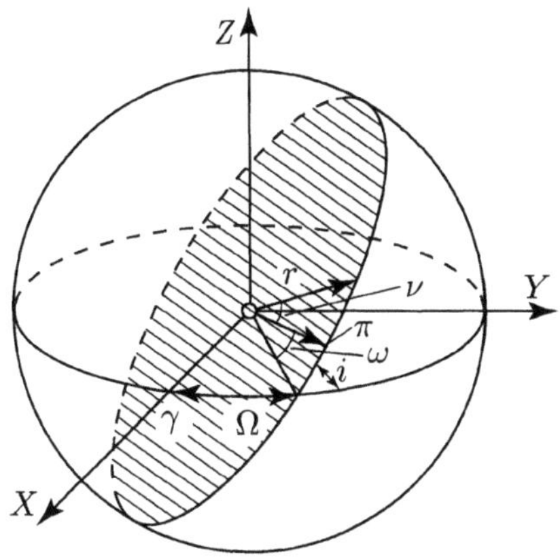

FIGURE 1.2. Position of the orbit in space

It is time to sum up. The orbit of the satellite is characterized by two independent constant parameters: p (or a) and e; the position of the orbit in space is determined by three independent angles: Ω, ω, and i; finally, the position of the satellite in orbit at each time is determined by the parameter τ^*. Thus, we have six independent constant parameters, which completely determine the position of the satellite in space (i.e., its coordinates and velocity at each time), for example,

$$a, \ e, \ \Omega, \ \omega, \ i, \ \tau^*. \tag{1.2.19}$$

The parameters (1.2.19) are referred to as *orbital elements* of the satellite.[3]

Let us point out the following circumstance. The original system of differential equations of motion (1.1.2) is of order six (three second-order equations); accordingly, the solution of system (1.1.2) must contain six arbitrary integration constants. Once the values of these constants are determined from the initial (or boundary) conditions, the concrete motion in question will be completely determined. The orbital elements (1.2.19) form precisely a set of such arbitrary integration constants (there are six of them, and they are independent). The initial conditions assign concrete numerical values to the orbital elements and in this manner determine a concrete motion.

3. Perturbed motion. Osculating elements

The Keplerian motion considered in the preceding section (which satisfies equations (1.1.2)) is also referred to as the *unperturbed motion* of the satellite. Indeed, the true motion of the satellite is governed by some equations of the type (1.1.3),

[3]Orbital elements can be defined in alternative ways. For instance, the position of the orbital plane can be specified relative to the ecliptic plane instead of the equatorial plane; instead of the parameters e and ω one can consider two independent combinations of them, and so on.

because in addition to Newton's force of attraction, which imparts the acceleration (1.2.3), the satellite is acted upon by a large number of different forces: atmospheric resistance, the deviation of the Earth's force of attraction from the Newtonian force, the attraction of the Moon and of the Sun, and so on. All the forces enumerated above (and many others) are small compared with the Newtonian force of attraction, i.e., obey the condition $\sqrt{f_x^2 + f_y^2 + f_z^2} \ll \mu/r^2$. Consequently, taking into account these forces results only in small corrections in the equations of motion, more precisely, the presence of the components f_x, f_y, f_z in the right-hand sides of equations (1.1.3).

The motion of the satellite under the influence of the Newtonian force of attraction to the center of the Earth and additional forces is called *perturbed motion*; the perturbed motion is governed by the equations (1.1.3). The concrete form of the components f_x, f_y, f_z of the perturbing acceleration depends on the nature of the perturbing forces (atmospheric resistance, or gravitational perturbations, or radiation pressure, etc., or all of them, depending on the character of the problem under study).

Equations (1.1.3) are not integrable, and so the question of whether the perturbed motion lends itself to investigation is of crucial importance. Here the following argument comes to help. Since the perturbing forces are small compared with the main force of gravitational attraction, one expects that the perturbed motion will differ only slightly, in some sense or another, from Keplerian motion. Hence, the solution of equations (1.1.3) should be sought in a form that is close to the solution of equations (1.1.2). Specifically, we will assume that the perturbed motion takes place along some "elliptic" orbit with the elements (1.2.19), which now, however, are not constant, but change with time:

$$a(t),\ e(t),\ \Omega(t),\ \omega(t),\ i(t),\ \tau^*(t). \tag{1.3.1}$$

Then the problem reduces to searching for the explicit time dependencies (1.3.1) (by virtue of the equations (1.1.3) of perturbed motion).

The "ellipse" with variable elements (1.3.1) is called *osculating ellipse* and the variable elements themselves are called *osculating elements.* This terminology reflects the tight (or – as N. D. Moiseev put it – "intimate")[4] closeness between the perturbed and unperturbed orbits.[5] If we suddenly "freeze" the values of the parameters (1.3.1) at some moment of time t^*, then the motion will continue along a Keplerian orbit and for values of time in a neighborhood of t^* will remain very close to the motion (on the same neighborhood of t^*) with non-frozen parameters (1.3.1).

[4]Professor Nikolaĭ Dmitrevich Moiseev (1902–1955) was the founder of the Moscow school of celestial mechanics; his numerous publications and pedagogical activities have exerted a decisive influence on the contemporary generation of specialists in celestial mechanics.

[5]Osculating elements were invented by the great French scientist Laplace. In Latin *osculari* refers to the act of kissing (see, e.g., Webster's New Universal Unabridged Dictionary, Deluxe Second Edition, Dorset & Baber, 1979, p. 1265).

To derive the differential equations for the osculating elements (1.3.1) we need to pass from the variables x, y, z, $\dot{x}$, $\dot{y}$, $\dot{z}$ to the variable (1.3.1), substitute them in equations (1.1.3), and solve the resulting equations with respect to the derivatives dp/dt, de/dt, $d\Omega/dt$, $d\omega/dt$, di/dt, $d\tau^*/dt$ of the osculating elements. The method of osculating elements is readily recognized as the method of variation of the constants of the unperturbed motion by virtue of the equations of perturbed motion. Let us mention the main law of osculatory motion: the coordinates x, y, z and the velocity components $\dot{x}$, $\dot{y}$, $\dot{z}$ are expressed through the osculating elements in the same manner for the perturbed and for the unperturbed motions; the only difference, as we wish to emphasize once more, is that in the unperturbed motion the elements are constant, whereas in the perturbed motion they are functions of time.

4. Osculating orbit of an equatorial satellite

Let us examine a relatively simple example in which one can observe a number of effects specific to the perturbed motion.

Consider a satellite that moves in the equatorial plane of the Earth [1.16], [1.17]. The Earth is slightly oblate (squeezed toward its equator), and consequently the force acting on the satellite will slightly differ from the Newtonian force. However, if the satellite moves in the equatorial plane this force remains central, i.e., permanently directed toward the center of the Earth. The central acceleration f that this force imparts to the satellite is given by the formula

$$f = -\frac{\mu}{r^2} - \varepsilon\,\frac{\mu R_0^2}{r^4}. \tag{1.4.1}$$

Here the first term is the ordinary Newtonian central acceleration and the second term is the perturbing acceleration produced by Earth's oblateness. Also, r is the distance from the satellite to the center of the Earth, R_0 the equatorial radius of the Earth, and ε is a dimensionless constant that depends on the oblatneness of the Earth; specifically, for our Earth one can take $\varepsilon = 0.0016$. The acceleration $f(r)$ corresponds to the force function

$$U(r) = \int f(r)\,dr = \frac{\mu}{r} + \varepsilon\,\frac{\mu R_0^2}{3r^3}.$$

Since our motion derives from a force function $U(r)$ that does not depend on time, then, as is known from mechanics, the total energy of motion is conserved:

$$\frac{1}{2}V^2 - \frac{\mu}{r} - \varepsilon\,\frac{\mu R_0^2}{3r^3} = H_0. \tag{1.4.2$'$}$$

Furthermore, since the acting force is central, the equations of motion necessarily admit an additional first integral – the *area integral* (which holds for all central forces):

$$r^2\frac{d\varphi}{dt} = c. \tag{1.4.3}$$

Here φ is the polar angle in the polar coordinate system r, φ.

Now let us try to derive the equations in osculating elements for our problem. In the unperturbed motion the area integral has the form (1.2.8). In both the perturbed and the unperturbed motion the right-hand side of the area integral can be written as rV_n, where V_n is the transversal projection of the velocity of the satellite. But, by the basic rule of osculatory motion, the coordinates (r) and the velocity (V_n) are expressed in terms of elements in identical manner for both the perturbed and the unperturbed motion. Hence, the constant c in the area integral (1.4.3) of our perturbed motion will have the same expression as the right-hand side of (1.2.8). Consequently, $c = \sqrt{\mu p}$, whence $p = \text{const}$, or, if one prefers,

$$\frac{dp}{dt} = 0. \tag{1.4.4}$$

This is precisely the first of the equations in osculating elements for our motion: it says that the focal parameter of the osculating orbit remains constant.

Further, recalling expressions (1.2.14), (1.2.15), and (1.2.7), we can write the energy integral (1.4.2) of the perturbed motion in the form

$$\frac{\mu(e^2-1)}{2p} - \frac{\varepsilon\mu R_0^2}{3r^3} = H_0. \tag{1.4.2$'$}$$

Differentiating with respect to time we get

$$e\,\frac{de}{dt} = -\varepsilon\,\frac{pR_0^2}{r^4}\,\dot{r}\,.$$

Recalling once again the basic rule of osculatory motion, we conclude that the expression of the radial velocity $\dot{r} = V_r$ must have the same form (1.2.13$'$) as for the unperturbed motion. Thus, the last equation takes the final form

$$\frac{de}{dt} = -\varepsilon\,\frac{\sqrt{\mu p}R_0^2}{r^4}\,\sin\nu. \tag{1.4.5}$$

Here it is understood that r is given by formula (1.2.4). Now let us return to the area integral (1.4.3). Note that $\varphi = \nu + \omega$, and so the area integral can be recast as

$$\frac{d\omega}{dt} = \frac{\sqrt{\mu p}}{r^2} - \frac{d\nu}{dt}. \tag{1.4.6}$$

Therefore, in order to write the differential equation for the third osculating element ω we need to find the expression of $d\nu/dt$ for the perturbed motion.

To that end, we take the expression for the radial velocity provided by (1.2.13$'$) and write it in the form $e\sin\nu = \sqrt{p/\mu}V_r$. Differentiating this equation with respect to time and recalling that in the osculatory motion $p = \text{const}$, we obtain

$$\frac{de}{dt}\,\sin\nu + \frac{d\nu}{dt}\,e\,\cos\nu = \sqrt{\frac{p}{\mu}}\,\dot{V}_r.$$

Now recall from kinematics that $\dot{V}_r = \ddot{r} = f_r + r\dot{\varphi}^2$, where f_r is the total radial acceleration. This yields formula (1.4.1). Using the area integral (1.4.3) to express $\dot{\varphi}$ in terms of r and substituting the result in $\dot{V}_r$, we obtain

$$\dot{V}_r\sqrt{\frac{p}{\mu}} = \frac{\sqrt{p\mu}}{r^2}\left(\frac{p}{r} - 1\right) - \frac{\varepsilon\sqrt{p\mu}\,R_0^2}{r^4}.$$

Since $p/r - 1 = e\cos\nu$, using the explicit expression (1.4.5) for de/dt we finally obtain

$$\frac{d\nu}{dt} = \frac{\sqrt{p\mu}}{r^2} - \frac{\varepsilon\sqrt{p\mu}\,R_0^2}{er^4}\cos\nu. \tag{1.4.7}$$

Now from (1.4.5) it readily follows that

$$\frac{d\omega}{dt} = \frac{\varepsilon}{e}\frac{\sqrt{p\mu}\,R_0^2}{r^4}\cos\nu, \tag{1.4.8}$$

and the system of differential equation (1.4.4), (1.4.5), (1.4.7), (1.4.8) is closed with respect to p, e, ω, ν. Rigorously speaking, we should have added an equation for τ^*, but we did omit it because the equations already listed are sufficient for carrying out an analysis of the motion.

First of all, let us point out that if one neglects the oblateness of the Earth (i.e., set $\varepsilon = 0$), then it follows right away that not only $p = p_0$, but also $e = e_0$, $\omega = \omega_0$ (as it must be for Keplerian motion). However, when $\varepsilon \neq 0$ the eccentricity e and the argument of perigee ω are functions of time, which is precisely what distinguishes the perturbed motion from the unperturbed one.

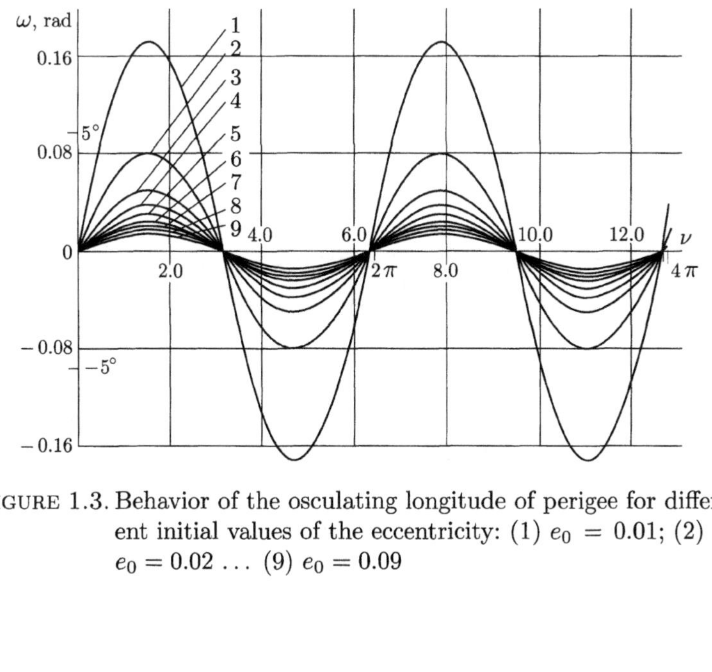

FIGURE 1.3. Behavior of the osculating longitude of perigee for different initial values of the eccentricity: (1) $e_0 = 0.01$; (2) $e_0 = 0.02$ ... (9) $e_0 = 0.09$

Figure 1.3 shows the graphs of the dependence $\omega(\nu)$, while Figure 1.4 shows the graphs of the dependence $e(t)$ in the osculatory motion for various initial data. These graphs were obtained by numerical integration of the equations (1.4.4), (1.4.5), (1.4.7), (1.4.8). We see that the eccentricity of the orbit changes periodically in time with a small amplitude and does not undergo systematic (in the customary terminology – *secular*) variations, which in time could lead to a considerable deviation of the eccentricity from its initial value. The amplitude of the oscillations of the eccentricity can be estimated by writing the energy integral (1.4.2) in the form

$$\frac{3}{2}(e^2 - e_0^2) = \varepsilon \frac{R_0^2}{p^2}\left[(1 + e\cos\nu)^3 - (1 + e_0)^3\right].$$

From this it follows that at the osculating perigee ($\nu = 0$) we have $e = e_{\max} = e_0$. At the osculating apogee, $\nu = \pi$; solving approximately the above cubic equation for e, we obtain, to first order in ε,

$$e = e_{\min} \approx e_0 - 2\varepsilon\left(\frac{R_0}{p}\right)^2,$$

provided only that $e_0 > 2\varepsilon\left(\frac{R_0}{p}\right)^2$.

The fact that the osculating eccentricity is smaller at apogee than at perigee means that the curvature of the orbit at apogee is smaller than the curvature at perigee.

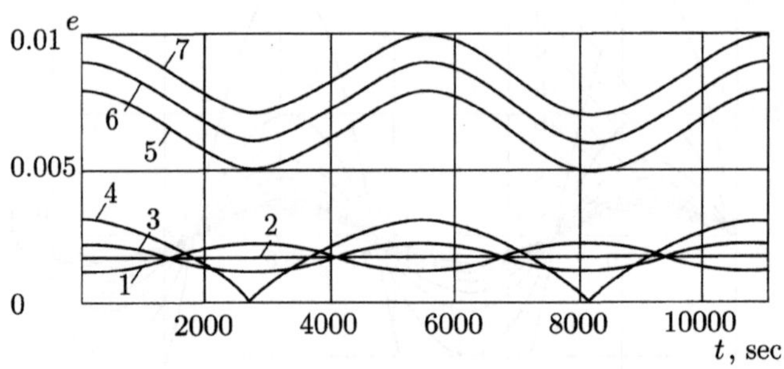

FIGURE 1.4. Behavior of the osculating eccentricity for different initial values: (1) $e_0 = 0.001$; (2) $e_0 = 0.0016$; (3) $e_0 = 0.002$; (4) $e_0 = 0.003$; (5) $e_0 = 0.008$; (6) $e_0 = 0.009$; (7) $e_0 = 0.01$

Examining the graphs $\omega(\nu)$ (Figure 1.3) we see that in addition to undergoing periodic changes, the position of the perigee is also subject to a systematic, secular drift, which accumulates with each revolution of the satellite. The magnitude of this drift per one revolution is equal to

$$\begin{aligned}\Delta\omega &= \int_0^{2\pi} \frac{d\omega}{dt} \cdot \frac{dt}{d\nu}\, d\nu \\ &= \int_0^{2\pi} \frac{\varepsilon R^2 p^{-2}(1 + e\cos\nu)^2 \cos\nu \, d\nu}{e - \varepsilon R^2 p^{-2}(1 + e\cos\nu)^2 \cos\nu}.\end{aligned}$$

If $e_0 > 2\varepsilon\left(R_0/p\right)^2$, then this integral can be computed approximately to first order in ε. Setting $e \approx e_0$, we obtain

$$\Delta\omega = 2\pi\varepsilon\left(\frac{R_0}{p}\right)^2. \tag{1.4.9}$$

Therefore, the orbit of the satellite can be represented as an oval curve of constant size ($p =$const), "sharper" at the perigee and "less sharp" at the apogee, which rotates in its plane by the angle $\Delta\omega$ given by (1.4.9) during one revolution of the satellite in its orbit. The orbit of an equatorial satellite is schematically shown in Figure 1.5.

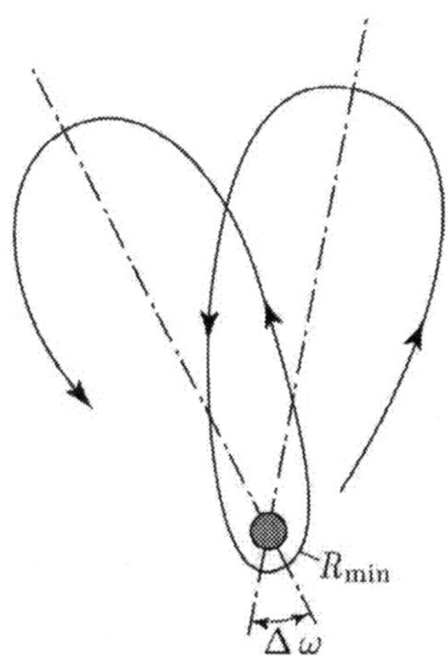

FIGURE 1.5. Schematic representation of the orbit of an equatorial satellite of Earth

If $e \sim \varepsilon$ or $e \ll \varepsilon$, then the character of the orbit does not change, but the behavior of the osculating elements changes qualitatively and the osculating orbit may differ qualitatively from the actual orbit in a rather unexpected manner.

This is particularly transparent in the following example. Among the orbits of an equatorial satellite there is a strictly circular orbit on which the satellite's velocity of revolution is constant. Is this orbit described in osculating elements by a circular orbit as well?

The answer turns out to be negative. Indeed, to a circular orbit of radius r_0 there corresponds the following exact solution of the equations (1.4.4), (1.4.5), (1.4.7), (1.4.8):

$$\nu = 0,$$

$$e = e^* = \varepsilon \left(\frac{R_0}{r_0} \right)^2 ,$$

$$p = r_0 \left(1 + \varepsilon \frac{R_0^2}{p^2} \right) ,$$

$$\omega = \frac{\sqrt{\mu p_0}}{r_0^2} (t - t_0) + \omega_0 .$$

Thus, a circular orbit is described in osculating elements by an ellipse that rotates with angular velocity $\dot{\omega} = \sqrt{\mu p_0}/r_0^2$, with the satellite always located at the perigee of this ellipse ($\nu \equiv 0$). We see that for arbitrarily small perturbations ($\varepsilon \neq 0$) in the perturbed motion it may happen that $d\nu/dt \equiv 0$, which is not possible in the unperturbed motion.

We conclude the analysis of our example – the osculating motion of an equatorial satellite – by listing the main properties of this motion that were revealed by the above analysis:

As a rule (but not always), the osculating ellipse reflects qualitatively the properties of the actual motion; the perturbations in osculating elements can have a periodic and secular character. The secular perturbations, which describe the evolution of the motion, play the dominant role, and they are the first that one should single out.

5. The equations in osculating elements. Delaunay elements

All the assertions made in the conclusion of Section 4 remain valid in more complex cases, i.e., arbitrary (but small) perturbing forces and motion in three-dimensional space. Naturally, for these cases the form of the equations in osculating elements is also more complex. The reader interested in a detailed derivation of these equations is referred to textbooks on celestial mechanics (for example, the lecture notes of G. N. Duboshin [1.1]). An elegant and brief derivation of the equations in osculating elements was given by A. I. Lur′e [1.4]. For reference purposes, we write below the complete set of equations in osculating elements.

These equations have the following form:

$$\left.\begin{aligned}
&1.\quad \frac{dp}{dt} = 2r\widetilde{T},\\
&2.\quad \frac{de}{dt} = \widetilde{S}\sin\nu + \left[\cos\nu + (e+\cos\nu)\frac{r}{p}\right]\widetilde{T},\\
&3.\quad \frac{d\omega}{dt} = -\frac{\cos\nu}{e}\,\widetilde{S} + \frac{\sin\nu}{e}\left(1+\frac{r}{p}\right)\widetilde{T} - \frac{r}{p}\sin u\,\cot i\,\widetilde{W},\\
&4.\quad \frac{di}{dt} = \frac{r}{p}\cos u\,\widetilde{W},\\
&5.\quad \frac{d\Omega}{dt} = \frac{r}{p}\frac{\sin u}{\sin i}\widetilde{W},\\
&6.\quad \frac{d\tau^*}{dt} = \frac{p}{e}\sqrt{\frac{p}{\mu}}\left[(eN\sin\nu - \cos\nu)\widetilde{S} + \frac{p}{r}N\widetilde{T}\right]\frac{r^2}{p^2}.
\end{aligned}\right\} \qquad (1.5.1)$$

Here

$$N = 2\frac{p^2}{r^2}\int_0^\nu \frac{\cos\nu\,d\nu}{(1+e\cos\nu)^3}, \qquad u = \nu + \omega.$$

Instead of the equation for the focal parameter p one can use the equation for the semi-major axis a:

$$\frac{da}{dt} = \frac{2a^2 e\sin\nu}{p}\widetilde{S} + \frac{2a^2}{r}\widetilde{T}. \qquad (1.5.2)$$

Then, of course, in the right-hand side of the equations we need to express p through a and e via (1.2.7). In (1.5.1) and (1.5.2) we used the notation

$$\widetilde{S} = \sqrt{\frac{p}{\mu}}\,S, \qquad \widetilde{T} = \sqrt{\frac{p}{\mu}}\,T, \qquad \widetilde{W} = \sqrt{\frac{p}{\mu}}\,W,$$

where S, T, and W are respectively the radial, transversal, and normal perturbing accelerations (Figure 1.6). Instead of r one takes the expression (1.2.4). It is assumed that the explicit rules of dependence of S, T, W on the osculating elements and time are known. It often turns out that S, T, and W depend on time not explicitly, but through the true anomaly ν and the argument of latitude u, which appear in S, T, and W explicitly.

The area conservation law (1.2.8) does not hold for the perturbed motion. The analogous formula for the perturbed motion is

$$r^2\frac{d\nu}{dt} = \sqrt{\mu p} + \sqrt{\frac{p}{\mu}}\left[\frac{\cos\nu}{e}S - \frac{\sin\nu}{e}\left(1+\frac{r}{p}\right)T\right]. \qquad (1.5.3)$$

The equations (1.5.1) in osculating elements are in no way simpler than the equations (1.1.3) in coordinates. As a matter of fact, equations (1.1.3) are “nicer”

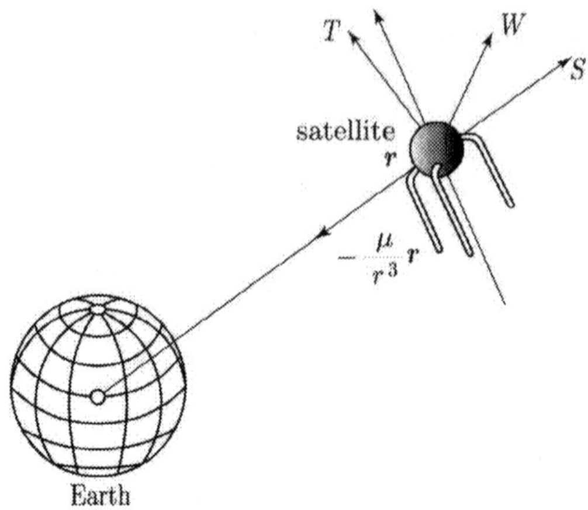

FIGURE 1.6. The components of the perturbing acceleration

– more symmetric and compact. If the equations (1.1.3) are not integrable, neither are the equations (1.5.1) in osculating elements. However, equations (1.5.1) enjoy the following advantage: they allow us right away to understand how the perturbed trajectory differs from the unperturbed one. Indeed, the right-hand sides of these equations contain the perturbing acceleration and become equal to zero if the perturbing acceleration vanishes. In this last case equations (1.5.1) immediately give the solution

$$p = \text{const}, \quad e = \text{const}, \quad \omega = \text{const},$$
$$i = \text{const}, \quad \Omega = \text{const}, \quad \tau^* = \text{const},$$

i.e., a Keplerian orbit.

If now S, T, W are different from zero, then, as is usually the case, they are small quantities. It is therefore natural to expect that the osculating elements will remain close, in some sense or another, to their initial values. In this case we can use for equations (1.5.1) the well-developed methods for approximate solution of differential equations containing a small parameter. The application of such methods to equations (1.5.1) yields a wealth of results. Particularly effective in the study of such problems are the asymptotic methods of nonlinear mechanics.

An outline of the asymptotic methods will be given below. They enable us to replace a nonintegrable problem by an integrable problem whose solution approximates with a high order of accuracy the unknown solution of the original nonintegrable problem.[6]

[6]The elaboration of asymptotic methods was in fact initiated by the classics of celestial mechanics, and in the famous treatise of H. Poincaré [1.5] such a method was already widely used to investigate motions. The further development, justification, and wide dissemination of asymptotic methods in the theory of oscillations is due to N. M. Krylov, N. N. Bogolyubov, Yu. A. Mitropol′skiĭ, V.M. Volosov, and other scientists from the former Soviet Union [1.6], [1.7], [1.8].

We have introduced the osculating elements under the assumption that the perturbed motion differs only slightly from the unperturbed one thanks to the smallness of the perturbing accelerations. However, small perturbations can accumulate over time. The perturbed motion is close to the unperturbed one for a short duration, but over a sufficiently long time interval the accumulating small perturbations can substantially distort the motion. If we now "freeze" the osculating elements at the values they have attained, we obtain a Keplerian motion that differs quite substantially from the original one. These long-period changes of the motion – or, in the usual terminology – this evolution or drift of the motion, is the most important issue in the investigation of perturbed motion. The notions introduced above are illustrated by Figure 1.7, which depicts schematically the change of some osculating element η with time.

The asymptotic methods allow us, based on the equations (1.5.1) in osculating elements, to obtain approximate equations which describe only the evolution of the motion (curve 3 in Figure 1.7); this description is sufficiently accurate on a very large (but, in general, finite) time interval. It is important that such approximate *evolution equations* are often integrable, since this enables us to carry out a detailed analysis of the qualitative and quantitative portrait of the evolution of motion. Whenever desired, the asymptotic methods can be used to construct higher-oder approximations of the perturbed motion, which describe even the small-scale variations of curve 2 in Figure 1.7.

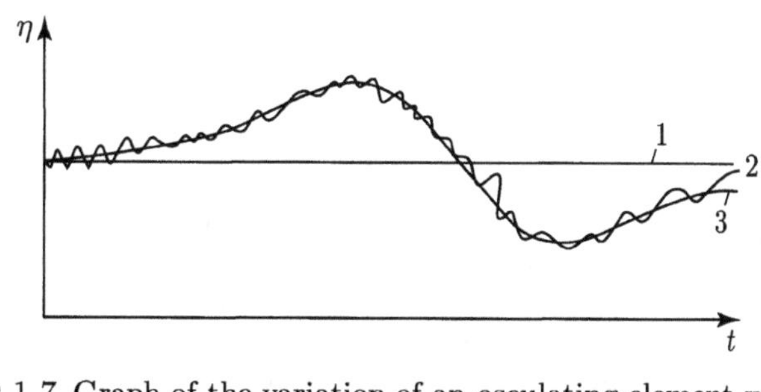

FIGURE 1.7. Graph of the variation of an osculating element with time: 1 – unperturbed motion; 2 – perturbed motion; 3 – evolution of motion

In celestial mechanics, and also in the mechanics of space flight, instead of the classical osculating elements it is sometimes convenient to use certain combinations of them. If the perturbing forces derive from a force function U, then the equations of the perturbed motion acquire a particularly symmetric form when written in the *Delaunay variables*

$$L,\ G,\ H,\ l,\ g,\ h. \tag{1.5.4}$$

These elements are introduced in terms of the osculating elements (1.3.1) by means of the following recipe:

$$\left.\begin{aligned} L &= \sqrt{\mu a}, & l &= n(t-\tau^*),\\ G &= \sqrt{\mu a(1-e^2)}, & g &= \omega,\\ H &= \sqrt{\mu a(1-e^2)}\cos i, & h &= \Omega. \end{aligned}\right\} \tag{1.5.5}$$

One assumes that the force function U is explicitly expressed in terms of the Delaunay elements and one introduces the Hamiltonian

$$F = \frac{\mu}{2L^2} + U(L, G, H, l, g, h). \tag{1.5.6}$$

Then the equations of the perturbed motion take on the canonical form

$$\left.\begin{aligned} \frac{dL}{dt} &= \frac{\partial F}{\partial l}, & \frac{dl}{dt} &= -\frac{\partial F}{\partial L},\\ \frac{dG}{dt} &= \frac{\partial F}{\partial g}, & \frac{dg}{dt} &= -\frac{\partial F}{\partial G},\\ \frac{dH}{dt} &= \frac{\partial F}{\partial h}, & \frac{dh}{dt} &= -\frac{\partial F}{\partial H}. \end{aligned}\right\} \tag{1.5.7}$$

This form of the equations of motion is particularly convenient for the application of asymptotic methods, as we shall see a little later.[7]

Let us emphasize the following important circumstance. In the unperturbed motion the elements L, G, H, g and h are constant. When small perturbations are present, these elements undergo a slow evolution, and for this reason they are referred to as *slow variables*. The Delaunay element l is not constant in the unperturbed motion: it changes linearly in time, with the rate $\dot{l} = n = \sqrt{\mu/a^3}$. In the perturbed motion, l retains approximately the same fast rate of variation, experiencing only a small distortion under the action of the small perturbations. For this reason, l is referred to as a *fast variable*. This separation of variables into slow and fast is typical for the problems to which asymptotic methods of investigation apply.

[7] For a derivation of the equations of motion in Delaunay elements see, e.g., G. N. Duboshin's book [1.1].

6. Digression on asymptotic methods of nonlinear mechanics. Oscillations of a satellite about its center of mass. Averaging of canonical equations

It is convenient to begin our exposition of asymptotic methods with some typical example, and only then move on to more general algorithms. Accordingly, we will consider first the particular problem of the oscillations of a satellite about its center of mass.

The satellite, whose center of mass moves in orbit, is acted upon by the torque of gravitational forces, which stabilizes the satellite with respect to its radius vector. This will discussed in more detail in one of the subsequent essays. Here, where our aim is to illustrate the asymptotic methods of the theory of oscillations, we consider the equation governing the plane oscillations of a satellite in an elliptic orbit [1.17], [1.18]:

$$(1 + e\cos\nu)\delta'' - 2e\sin\nu\,\delta' + n^2\sin\delta = 4e\sin\nu, \tag{1.6.1}$$

where $\delta = 2\theta$ and $n^2 = 3(A - C)/B$. Here (see Figure 1.8) ν is the true anomaly, e is the eccentricity of the orbit, θ is the angle between the axis of the satellite, which remains permanently in the orbital plane, and the current radius vector, and A, B, C are the principal central moments of inertia of the satellite. For definiteness, we will assume that $A > B > C$ and that the axis corresponding to the moment of inertia B is normal to the orbital plane. Finally, the primes in (1.6.1) denote differentiation with respect to ν.

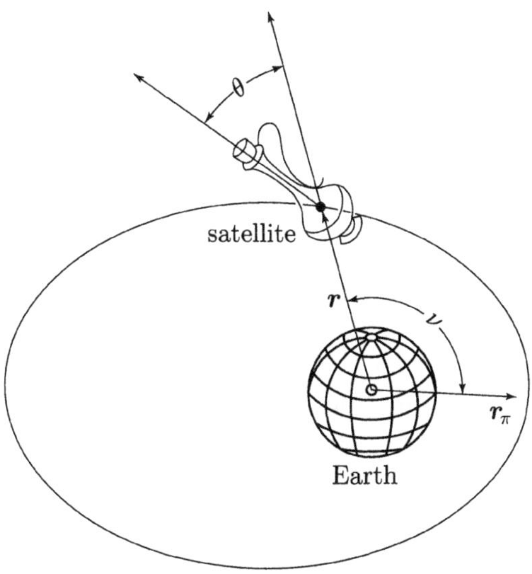

FIGURE 1.8. Satellite in orbit

If the orbit were circular ($e = 0$), equation (1.6.1) would admit the solution $\delta = 0$, which corresponds to the relative equilibrium of the satellite (the axis of the moment of inertia C permanently coincides with the radius vector for all times). For $e \neq 0$ there is no relative equilibrium – the satellite oscillates about the direction of the radius vector. These oscillations are not easy to investigate, because equation (1.6.1) is not integrable. However, we can expect that for small values of the eccentricity e there exists small oscillations, described by the linear equation obtained from (1.6.1) if we replace $\sin\delta$ by δ. We begin our analysis of small oscillations precisely with this linear equation.

Changing the variables by means of the rule $\delta/2 \equiv \theta = z/(1 + e\cos\nu)$, we obtain the new equation

$$z'' + \left(\frac{n^2 + e\cos\nu}{1 + e\cos\nu}\right) z = 2e\sin\nu,$$

which, despite its linearity, it is still not integrable because of the presence of variable coefficients. Since the eccentricity e is small by assumption, the last equation can be replaced, to first order in e, by

$$z'' + n^2 z = e[2\sin\nu + (n^2 - 1)z\cos\nu]. \tag{1.6.2}$$

Note that when $e = 0$ equation (1.6.2) has the solution

$$z = a\cos\psi, \quad z' = -an\sin\psi, \tag{1.6.3}$$

where $\psi = n\nu + \psi_0$. If we now regard (1.6.3) simply as a change of variables (i.e., if we introduce the new variables a, ψ instead of z, z') in equations (1.6.2), then the equation is transformed into the equivalent system

$$\left.\begin{aligned} \frac{da}{d\nu} &= -\frac{1}{n} f(\nu, a\cos\psi)\sin\psi, \\ \frac{d\psi}{d\nu} &= n - \frac{1}{na} f(\nu, a\cos\psi)\cos\psi, \\ \frac{d\nu}{d\nu} &= 1, \end{aligned}\right\} \tag{1.6.4}$$

where

$$f(\nu, z) = e[2\sin\nu + (n^2 - 1)z\cos\nu].$$

Since f is small together with e, it follows that the system (1.6.4) has one slow variable, a, and two fast ones, ψ and ν. For $e = 0$ we have $f = 0$ and the system (1.6.4) admits the solution $a = a_0$, $\psi = n\nu + \psi_0$, which in view of (1.6.3) describes small oscillations in a circular orbit. For $e \neq 0$ we will seek the solution of (1.6.4) in the form

$$\left.\begin{aligned} a &= \bar{a} + e\alpha_1(\bar{\psi}, \bar{a}, \nu) + e^2\alpha_2(\bar{\psi}, \bar{a}, \nu) + \cdots, \\ \psi &= \bar{\psi} + e\beta_1(\bar{\psi}, \bar{a}, \nu) + e^2\beta_2(\bar{\psi}, \bar{a}, \nu) + \cdots. \end{aligned}\right\} \tag{1.6.5}$$

Here $\bar{a}(\nu)$, $\bar{\psi}(\nu)$, $\alpha_i(\bar{\psi},\bar{a},\nu)$, $\beta_i(\bar{\psi},\bar{a},\nu)$ are unknown functions subject to determination; $\bar{a}(\nu)$ and $\bar{\psi}(\nu)$ must satisfy the system of equations

$$\left.\begin{aligned} \frac{d\bar{a}}{d\nu} &= e\widetilde{A}_1(\bar{a}) + e^2\widetilde{A}_2(\bar{a}) + \cdots , \\ \frac{d\bar{\psi}}{d\nu} &= n + e\widetilde{B}_1(\bar{a}) + e^2\widetilde{B}_2(\bar{a}) + \cdots , \end{aligned}\right\} \tag{1.6.6}$$

where, again, $\widetilde{A}_i(\bar{a})$, $\widetilde{B}_i(\bar{a})$ are functions of only the slow variable $\bar{a}$, which need to be determined. A solution of the original system (1.6.4) in the form of a series (1.6.5), (1.6.6) in powers of the small parameter e is called an *asymptotic solution*, in the sense that as $e \to 0$ it tends to the exact solution of the system obtained from the original system by setting $e = 0$ (here the functions α_i and β_i, which need to be determined, are required to be bounded). The accuracy of the asymptotic solution depends on the number of terms retained in the series expansion. One can assert that if $\bar{a}_n$ is the solution of the nth approximation in the asymptotic sense, then $|a - \bar{a}_n| \sim e^n$ over a time interval $t \sim 1/e$. In other words, the asymptotic method yields a sufficiently accurate solution on a time interval which in general is bounded (and sometimes even on an infinite time interval).

Let us remark that already when we wrote the original differential equation (1.6.2) of our problem we did retain only the terms of first order in e, and therefore in (1.6.5) and (1.6.6) it makes sense to retain only such terms. The rationale for the change of variables that takes us to (1.6.5) and (1.6.6) is that the system (1.6.6) is considerably simpler than the original system (1.6.4) (and in the present case it is even integrable), and the functions α_i and β_i can be found by simple calculations. Thus, one can find in a simple manner an approximate solution of a system whose exact solution is not known.

Instead of solution (1.6.5)–(1.6.6) of the system (1.6.4) one can seek a solution of an equivalent problem, posed directly for the original equation (1.6.2). Up to terms of first order in e this problem is formulated as follows:

Find a solution of the equation (1.6.2) of the form

$$z = a\cos\psi + eu_1(\psi, a, \nu), \tag{1.6.7}$$

where u_1 is bounded and where a and ψ satisfy the equations

$$\frac{da}{d\nu} = eA_1(a), \quad \frac{d\psi}{d\nu} = n + eB_1(a); \tag{1.6.8}$$

here A_1 and B_1 need to be determined.

Substituting (1.6.7) in (1.6.2), using (1.6.8), and collecting the terms of order no higher than one in e we obtain, after some manipulations, the equation

$$\begin{aligned} &- 2en(A_1\sin\psi + B_1 a\cos\psi) + e\left(\frac{\partial^2 u_1}{\partial\psi^2}n^2 + 2\frac{\partial^2 u_1}{\partial\psi\partial\nu}n + \frac{\partial^2 u_1}{\partial\nu^2}\right) + n^2 eu_1 \\ &= e\left(2\sin\nu + \frac{a(n^2-1)}{2}[\cos(\nu+\psi) + \cos(\nu-\psi)]\right). \end{aligned} \tag{1.6.9}$$

We can divide both sides of this equation by e. Since the right-hand side does not contain the harmonics $\sin\psi$ and $\cos\psi$, it follows that $A_1 = 0$ and B_1, whence

$$a = a_0, \quad \frac{d\psi}{d\nu} = n. \tag{1.6.10}$$

Note that the same result is obtained if we *average the right-hand sides of the system* (1.6.4) *independently with respect to the phases* ν *and* ψ. Further, under the conditions $A_1 = B_1 = 0$ the partial differential equation (1.6.9) has the solution

$$\begin{aligned} u_1 = &\frac{2}{n^2-1}\sin\nu + \\ &+\frac{1}{2}a(n^2-1)\left[\frac{1}{2n-1}\cos(\nu-\psi) - \frac{1}{2n+1}\cos(\nu+\psi)\right]. \end{aligned} \tag{1.6.11}$$

Formulas (1.6.10) and (1.6.11) determine completely the sought-for solution (1.6.7) of equation (1.6.2). In the first approximation the small oscillations in a elliptic orbit differ only insignificantly from the small oscillations in a circular orbit, and one can work with the approximate solution

$$z = a_0\cos(n\psi + \psi_0).$$

However, this solution looses meaning when the frequency n of the proper oscillations and the frequency "1" of the orbital motion are commensurable: according to (1.6.11), when $n \to 1$ the amplitude of the oscillations grows without limit: a resonance takes place. (Another resonance, termed *parametric*, takes place for $n = 1/2$). But unbounded oscillations are not compatible with the setting of our problem (the linear equation (1.6.2) describes only small oscillations). For this reason near a resonance it is necessary to investigate a nonlinear equation.

To this end let us rewrite the original equation (1.6.1) in the form

$$\delta'' + n^2\delta = f(\nu, \delta, \delta', \delta'') \tag{1.6.12}$$

where $f = e[4\sin\nu + 2\delta'\sin\nu - \delta''\cos\nu] + n^2(\delta - \sin\delta)$, and regard the quantity f as small; this in turn requires that both the eccentricity e and the difference $\sin\delta - \delta$ be small. In addition, n^2 is close to 1 by assumption (we are analyzing the vicinity of a resonance). For $f = 0$ the solution has the form (1.6.3), in which at resonance we must also set $n = 1$. We shall seek the solution of equation (1.6.12) in the form

$$\delta = a\cos\psi, \quad \psi = \nu + \varkappa, \tag{1.6.13}$$

where a and $\varkappa$ are new variables. As in the no-resonance case, equation (1.6.12) is equivalent to a system of type (1.6.4), which now takes on the form

$$\left.\begin{aligned} \frac{da}{d\nu} &= -\frac{1}{n}f(\psi-\varkappa, a\cos\psi, -an\sin\psi, an^2\cos\psi)\sin\psi, \\ \frac{d\varkappa}{d\nu} &= n - 1 - \frac{1}{na}f(\psi-\varkappa, a\cos\psi, -an\sin\psi, an^2\cos\psi)\cos\psi, \\ \frac{d\psi}{d\nu} &= 1 + \frac{d\varkappa}{d\nu}. \end{aligned}\right\} \tag{1.6.14}$$

Since f is small and n is close to 1, our system now has *two* slow variables (a and $\varkappa$) and *one* fast variable (ψ). There is no fundamental difference between the systems (1.6.4) and (1.6.14), and the solution of the latter can also be sought in a form similar to (1.6.5)–(1.6.6), where in the first approximation we must set

$$\delta = a\cos\psi, \qquad \psi = \nu + \varkappa,$$
$$\frac{da}{d\nu} = A_1(a,\varkappa), \qquad \frac{d\varkappa}{d\nu} = n - 1 + B_1(a,\varkappa).$$

As in the case examined above, the functions $A_1(a,\varkappa)$ and $n-1+B_1(a,\varkappa)$ of the slow variables a, $\varkappa$ are obtained in the process of implementing the algorithm by *averaging over the fast variable* ψ *the right-hand sides of equations* (1.6.14) *for the slow variables*:

$$A_1 = \frac{1}{2\pi}\int_0^{2\pi}\left(-\frac{1}{n} f\sin\psi\right) d\psi,$$

and similarly for B_1. This yields the following explicit formulas (see [1.17]):

$$\left.\begin{aligned}
\frac{da}{d\nu} &= -\frac{4e}{n+1}\cos\varkappa \equiv -\frac{1}{a}\frac{\partial\Phi}{\partial\varkappa},\\
\frac{d\varkappa}{d\nu} &= \left[\frac{n}{2} + \frac{n}{a}J_1(a)\right] - 1 + \frac{4e}{a(n+1)}\sin\varkappa \equiv \frac{1}{a}\frac{\partial\Phi}{\partial a},\\
\Phi(a,\varkappa) &= \frac{4e}{n+1}a\sin\varkappa + n\left[\frac{a^2}{4} - (J_0(a)-1)\right] - \frac{a^2}{2}.
\end{aligned}\right\} \qquad (1.6.15)$$

Here $J_0(a)$ and $J_1(a)$ are the Bessel functions of order zero and one, respectively.[8] Differentiating Φ with respect to ν and using (1.6.15) we find that

$$\frac{d\Phi}{d\nu} = \frac{\partial\Phi}{\partial a}\frac{da}{d\nu} + \frac{\partial\Phi}{\partial\varkappa}\frac{d\varkappa}{d\nu} \equiv 0$$

Hence, equations (1.6.15) possess the first integral

$$\Phi(a,\varkappa) = C_0 = \text{const},$$

which provides "phase-amplitude" curves (the dependence of the amplitude a on the phase $\varkappa$ of near-resonance trajectories). An example of such curves is shown

[8] The only property of Bessel functions that we need here is that the first terms of their expansions in powers of a are

$$J_1(a) = \frac{a}{2} - \frac{a^3}{4} + \cdots, \qquad J_0(a) = 1 - \frac{a^2}{2^2} + \frac{a^4}{2^6} - \cdots.$$

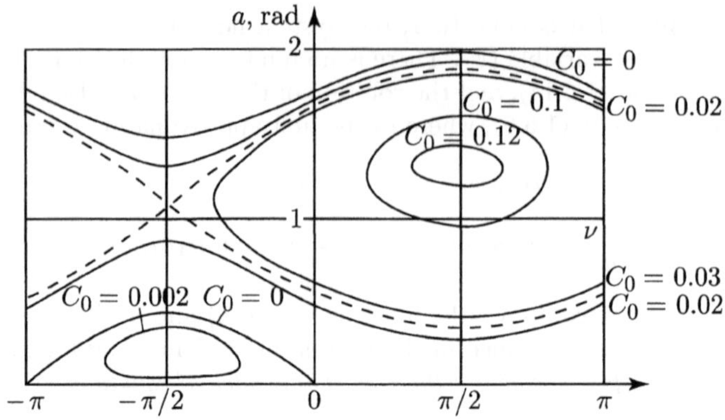

FIGURE 1.9. Phase-amplitude curves $(n^2 = 1.2;\ e = 0.01)$

in Figure 1.9 for different values of C_0. When $a(\nu)$ and $\varkappa(\nu)$ vary (slowly) with ν, the representing point in the $(\varkappa, a)$-plane traces one of the integral curves (Figure 1.9).

The fixed points in this figure ($\varkappa = \pm\pi/2$, $a = a_*$) correspond to periodic oscillations (with period 2π in ν and constant amplitude). In all the remaining situations the amplitude oscillates slowly between its maximal and minimal values. From (1.6.5) one can deduce that the constant amplitude a_* of the periodic regime is related to the eccentricity e and the parameter n^2 by the formula

$$n^2 = \frac{a_* \pm 4e}{2J_1(a)}.$$

The dependence $a_*(n^2)/2$ for fixed e is shown in Figure 1.10. We see that for $e \approx 0.01$ and values of n^2 far from resonance, the satellite oscillates about its radius vector with an amplitude of order 0.5°, whereas at resonance (i.e, when $n^2 \sim 1$) the amplitude reaches 30°! (Generally, $a_* \sim e$ for $n^2 \gg 1$ and $n^2 \ll 1$, and $a_* \sim \sqrt[3]{e}$ for $n^2 \sim 1$.) And one last fact: there exist one periodic solution for $e > \sqrt{\frac{2}{27}}\,(n^2-1)^{3/2}/n$, and three periodic solutions for $e < \sqrt{\frac{2}{27}}\,(n^2-1)^{3/2}/n$. All this follows from (1.6.15). To sum up, the application of asymptotic methods enable us to gain rich information about the character of the oscillations of the satellite.

The example analyzed above demonstrates the following:

(i) The asymptotic method applies whenever, thanks to the presence of a small parameter in the system of equations, the variables can be divided into "slow" and "fast."

(ii) The first-approximation equations can be obtained by averaging the original equation (reduced to a certain form) independently with respect to each of the fast variables; this kind of averaging is possible if there are no resonance relations among the fast variables.

(iii) If resonances are present, one needs to transform the system of equations in the neighborhood of a resonance; as a result, the number of slow variables increases, while the numbers of fast variables decreases accordingly.

(iv) Whereas the original equations of motion are not integrable, the equations of the first approximation of the asymptotic methods (the averaged system) may turn out to be integrable, in which case one can obtain a rich information on the behavior of solutions.

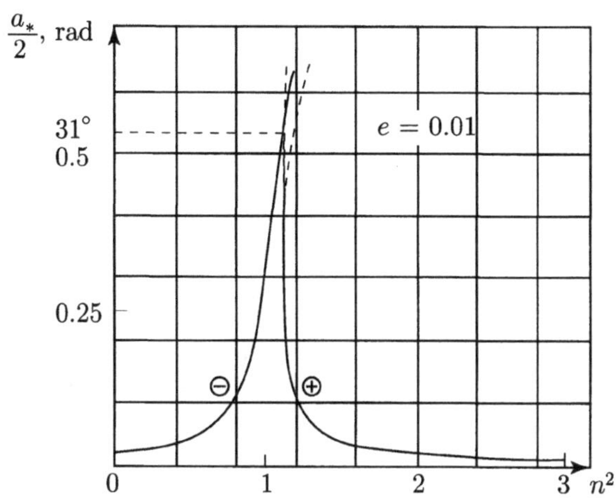

FIGURE 1.10. Phase-amplitude characteristic $(e = 0.01)$

It is useful to convince ourselves of the validity of all these assertions by following the basic ideas of the asymptotic method at work in a somewhat more general situation.[9] Namely, let us observe that the systems (1.6.4), (1.6.14), as well as the system (1.5.7) of differential equations of perturbed motion have the following structure:

$$\left.\begin{aligned} \dot{x} &= \varepsilon X(x, y, \varepsilon), \\ \dot{y} &= \omega(x) + \varepsilon Y(x, y, \varepsilon), \end{aligned}\right\} \tag{1.6.16}$$

where x is the vector of slow variables, y is the vector of fast variables, and ε is the small parameter. In the concrete case of equations (1.5.7) there is only one fast variable, l, but in general there can be more, and this is precisely the kind

[9]For a first acquaintance with the present status of asymptotic methods of the theory of nonlinear oscillations the reader will find useful consulting the book of N. N. Moiseev on this subject [1.8], where many problems of the dynamics of space flight are worked out as examples of application of asymptotic methods.

of generalization we have in mind. The fact that the perturbations affecting the motion are small is emphasized by the explicit presence of the small parameter $\varepsilon \ll 1$. If $\varepsilon = 0$, then $x = \text{const} = x_0$, $y = \omega(x_0)t + y_0$. We shall assume that X and Y are 2π-periodic in the fast variables. We shall seek the solution of the system (1.6.16) in the form

$$\left.\begin{aligned} x &= \bar{x} + \varepsilon u_1(\bar{x},\bar{y}) + \varepsilon^2 u_2(\bar{x},\bar{y}) + \cdots, \\ y &= \bar{x} + \varepsilon v_1(\bar{x},\bar{y}) + \varepsilon^2 v_2(\bar{x},\bar{y}) + \cdots, \end{aligned}\right\} \tag{1.6.17}$$

where $\bar{x}$, $\bar{y}$ are new variables, and where u_i, v_i need to be determined, and must be chosen so that the variables $\bar{x}$, $\bar{y}$ will satisfy the new system of differential equations

$$\left.\begin{aligned} \dot{\bar{x}} &= \varepsilon A_1(\bar{x}) + \varepsilon^2 A_2(\bar{x}) + \cdots, \\ \dot{\bar{y}} &= \omega(\bar{x}) + \varepsilon B_1(\bar{x}) + \varepsilon^2 B_2(\bar{x}) + \cdots. \end{aligned}\right\} \tag{1.6.18}$$

Retaining some concrete number of terms in (1.6.17), we obtain a solution of a certain order of accuracy. The system (1.6.18) is manifestly considerably simpler than (1.6.16), since in it the fast and slow variables are separated. The essence of asymptotic methods is precisely to achieve this separation.

We require that the functions u_i, v_i be bounded (because only in this case each successive term in (1.6.17) will be small compared with the preceding term). Substituting the expansions (1.6.17) in (1.6.16), using the system (1.6.18), and grouping the like terms in ε, we arrive at the following system of partial differential equations with known right-hand sides:

$$\sum_{i=1}^{k} \frac{\partial u_1}{\partial \bar{y}_i}\,\omega_i(\bar{x}) = X(\bar{x},\bar{y},0) - A_1(\bar{x}),$$

$$\sum_{i=1}^{k} \frac{\partial v_1}{\partial \bar{y}_i}\,\omega_i(\bar{x}) = Y(\bar{x},\bar{y},0) + \sum_{\alpha=1}^{m} \frac{\partial \omega}{\partial x_\alpha}\, u_1^\alpha(\bar{x},\bar{y}) - B_1(\bar{x}),$$

....

$$\sum_{i=1}^{k} \frac{\partial u_n}{\partial \bar{y}_i}\,\omega_i(\bar{x}) = g_n - A_n,$$

$$\sum_{i=1}^{k} \frac{\partial v_n}{\partial \bar{y}_i}\,\omega_i(\bar{x}) = h_n - B_n.$$

Here the first [resp., last] two equations correspond to the terms of order 1 [resp., n] in ε, u_j, v_j, A_j, B_j are the unknown vector functions, k is the number of fast variables (and the dimension of the vectors v_j, B_j, ω_j, Y), and m is the number of slow variables (and the dimension of the vectors u_j, A_j, X); u_1^α is the "α-component" of the vector u_1, and the functions h_n, g_n, A_n, B_n are explicitly expressible in terms of the functions found from the preceding approximations. Since all the partial differential equations appearing above are of the same type, it suffices to solve the first of them – the remaining ones are solved in much the same manner.

The vector function X is by definition periodic in each of the variables y, and as such it admits a Fourier series expansion

$$X(\bar{x},\bar{y}) = \sum_{n_1,\dots,n_k} a^{(x)}_{n_1,\dots,n_k}(\bar{x})e^{i(n_1\bar{y}_1+\dots+n_k\bar{y}_k)}.$$

This allows us to seek $u_1(\bar{x},\bar{y})$ in the form

$$u_1(\bar{x},\bar{y}) = \sum_{n_1,\dots,n_k} b_{n_1,\dots,n_k}(\bar{x})e^{i(n_1\bar{y}_1+\dots+n_k\bar{y}_k)} + \sum_{i=1}^{k} c_i(\bar{x})\bar{y}_i. \qquad (1.6.19)$$

Substituting this expression of u_1 in the first of the partial differential equations written above, we get

$$\left.\begin{aligned} b_{n_1,\dots,n_k} &= \frac{a^{(x)}_{n_1,\dots,n_k}(\bar{x})}{i(n_1\omega_1+\dots+n_k\omega_k)}, \\ \sum_{i=1}^{k} c_i(\bar{x})\omega_i(x) &= a^{(x)}_{0,\dots,0}(\bar{x}) - A_1(\bar{x}). \end{aligned}\right\} \qquad (1.6.20)$$

In the first of these formulas at least one of the integers n_i is different from zero. Since we assumed that u_1 is bounded, all functions c_i must be equal to zero (otherwise, as seen from (1.6.19), u_1 will grow unboundedly with $\bar{y}$). But if $c_i = 0$ for all i, then from (1.6.20) one immediately concludes that $A_1(\bar{x}) = a_{0,\dots,0}(\bar{x})$. Since the "0th" term of a Fourier series is equal to the mean value of the function being expanded, we see that

$$\left.\begin{aligned} A_1(\bar{x}) &= \frac{1}{(2\pi)^k}\underbrace{\int_0^{2\pi}\cdots\int_0^{2\pi}}_{k\text{ times}} X(\bar{x},\bar{y},0)\,dy_1\dots dy_k = \bar{X}, \\ u_1 &= \sum_{n_i\neq 0} \frac{a^{(x)}_{n_1,\dots,n_k}(\bar{x})}{i(n_1\omega_1+\dots+n_k\omega_k)} e^{i(n_1\bar{y}_1+\dots+n_k\bar{y}_k)} + b^{(x)}_{0,\dots,0}(\bar{x}). \end{aligned}\right\} \qquad (1.6.21)$$

The function $b_{0,\dots,0}(\bar{x})$ is arbitrary and can be determined from other considerations.

Similarly,

$$B_1 = \overline{Y + \sum_{\alpha=1}^{m} \frac{\partial\omega}{\partial x_\alpha} u_1^\alpha},$$

$$v_1 = \sum_{n_i\neq 0} \frac{a^{(y)}_{n_1,\dots,n_k}(\bar{x})}{i(n_1\omega_1+\dots+n_k\omega_k)} e^{i(n_1\bar{y}_1+\dots+n_k\bar{y}_k)} + b^{(y)}_{0,\dots,0}(\bar{x}),$$

and so on.

Needless to say, all these calculations are meaningful if, and only if, there are no resonances in the system investigated, i.e.,

$$n_1\omega_1 + \cdots + n_k\omega_k \neq 0$$

for all nonzero n-tuples of integers $(n_1, \ldots, n_k)$. In the opposite case the functions u_1 and v_1 are not defined (are infinite).

If the nonresonace condition is satisfied, then upon truncating the series (1.6.17) at the terms ε^{n-1}, the slow [resp., fast] variables are determined on a time interval $t \sim 1/\varepsilon$ with an accuracy of order ε^n [resp., ε^{n-1}], and so

$$|x - \bar{x}| \sim \varepsilon^n, \quad |y - \bar{y}| \sim \varepsilon^{n-1}.$$

Therefore, the terms of order higher that ε^{n-1} in the fast variables can be neglected. In particular, in the *first approximation* the system (1.6.16) is replaced by the new equations

$$\left.\begin{aligned} x &= \bar{x}, & y &= \bar{y}, \\ \frac{d\bar{x}}{dt} &= \varepsilon A_1(\bar{x}), & \frac{d\bar{y}}{dt} &= \omega(\bar{x}), \end{aligned}\right\} \tag{1.6.22}$$

where $A_1(\bar{x})$ is given by formula (1.6.21).

Thus, the first approximation in the asymptotic method is the averaging method. Solving equations (1.6.22) on the time interval $t \sim 1/\varepsilon$ yields the accuracy $|x - \bar{x}| \sim \varepsilon$, $|y - \bar{y}| \sim 1$. Since the parameter ε is usually small (because the perturbations are assumed to be small), the time interval over which the first approximation is valid is sufficiently large to allow us to understand the tendencies of the evolution of the system.

The fast variables are determined with less accuracy, but this is not important, since usually one is interested mainly in the evolution of the slow variables, and those variables are asymptotically (in ε, as $\varepsilon \to 0$) close to their exact values (as we already remarked, this is the meaning of the term "asymptotic method.")

It is remarkable that the equations of the first approximation can be derived in such a simple manner, by averaging independently with respect to each of the fast variables. Recall though that this is possible only in the absence of resonances. If there exist a set of integers n_i° such that $\sum_{i=1}^k n_i^\circ \omega_i = 0$, i.e., the frequencies of the motion are commensurable, then in the vicinity of the resonance the quantity

$$n_1^\circ \theta = n_1^\circ \bar{y}_1 + \cdots + n_k^\circ \bar{y}_k$$

will change slowly, because its derivative

$$n_1^\circ \dot{\theta} = \sum_{i=1}^k n_i^\circ \omega_i = \varepsilon h(x)$$

is close to zero. Hence, it is no longer possible to average over each $\bar{y}_i$ independently: one averages with respect to the slow variable θ! However, θ may be regarded as a new independent variable. Denoting the vector of fast variables by $z = (y_2, \dots, y_k)$, the system (1.6.16) can be reduced to the form

$$\begin{aligned}
\dot{x} &= \varepsilon X^*(x, \theta, z, \varepsilon), \\
\dot{\theta} &= \varepsilon \Theta^*(x, \theta, z, \varepsilon), \\
\dot{z} &= \omega(x) + \varepsilon Z^*(x, \theta, z, \varepsilon).
\end{aligned}$$

This system has the same structure as (1.6.16), the only difference being that it contains one more slow variable and one less fast variable. The problem is thus reduced to the previous one, since the actual number of slow and fast variables is of no consequence. We shall consider next averaging algorithms for resonant and nonresonant motions in the case when the equations of perturbed motion are in canonical form, which is of special interest here.

Averaging canonical equations. As we have seen, in the asymptotic method the differential equations of the first approximation are obtained by averaging the right-hand sides of the equations for the slow variables with respect to the fast variables, where the unperturbed values of the variables are used in the averaging. Let us examine how the averaging process looks when the equations are in canonical form (of type (1.5.7)). Let η_i be a slow variable satisfying the equation

$$\frac{d\eta_i}{dt} = \pm \frac{\partial F}{\partial \eta_j}, \tag{1.6.23}$$

and let the Hamiltonian F be periodic in the fast variable. In our case there is a single fast variable, namely l (in the general case there can be more). In problems of space flight the function F usually depends explicitly on the true anomaly $\nu(l)$, and not on l, and so one can write

$$F = f(\nu(l)). \tag{1.6.24}$$

In the present case the fast period is the period $T = 2\pi/n$ of revolution of the satellite in its orbit, and the averaged right-hand side of equation (1.6.23) has the form

$$\pm \frac{1}{2\pi} \int_0^{2\pi} \left. \frac{\partial F}{\partial \eta_j} \right|_{\eta_k = \bar{\eta}_k} dl = \pm \frac{\partial}{\partial \eta_j} \frac{1}{2\pi} \int_0^{2\pi} F \, dl = \pm \frac{\partial}{\partial \eta_j} \widetilde{F}. \tag{1.6.25}$$

Here the notation $\eta_k = \bar{\eta}_k$ indicates that the averaging is carried out for fixed values of the parameters η_k.

Let us write the expression of $\widetilde{F}$ is more detail:

$$\begin{aligned}
\widetilde{F} = \frac{1}{2\pi} \int_0^{2\pi} F(\nu) \frac{dl}{d\nu} \, d\nu = \\
= \frac{1}{2\pi} \int_0^{2\pi} F(\nu) \frac{dl}{dt} \frac{dt}{d\nu} \, d\nu = \frac{1}{2\pi} \int_0^{2\pi} F(\nu) n \frac{dt}{d\nu} \, d\nu.
\end{aligned} \tag{1.6.26}$$

Here one needs to express $dt/d\nu$ explicitly through ν and the orbital elements. Hence, averaging the function F with respect to the fast variables l is equivalent to averaging the same F with respect the true anomaly ν, on which it explicitly depends, with the weight $ndt/d\nu$. It is convenient to introduce the dimensionless time $\tau = nt$ and the new mean value $\overline{F} = \widetilde{F}/n$ of F. This finally yields

$$\left.\begin{aligned} \frac{d\eta_i}{d\tau} &= \pm\frac{\partial \overline{F}}{\partial \eta_j}, \\ \overline{F} &= \frac{1}{2\pi}\int_0^{2\pi} F\,\frac{dt}{d\nu}\,d\nu. \end{aligned}\right\} \tag{1.6.27}$$

The evolution (averaged) equations (1.6.27) enjoy the following remarkable properties:

1°. Averaging the original exact equations of motion is equivalent to averaging the force function. This means that in order to obtain the evolution system *it is not necessary* to first write the full system of equations; rather, it suffices to write the characteristic function F, average it and, once $\overline{F}$ is known, write directly the evolution equations.

2°. The right-hand sides of system (1.6.27) do not depend on ν (indeed, we averaged over ν), and hence neither do they depend on l, i.e., $\partial\overline{F}/\partial l = 0$. Then the first of the Delaunay equations yields

$$L = l_0, \tag{1.6.28}$$

or, equivalently,

$$a = a_0. \tag{1.6.29}$$

Therefore, *if the force perturbing the motion of the satellite is conservative, then the semi-major axis of the satellite's orbit does not evolve*, i.e., it remains constant in the mean. This remains true also when F depends on several fast variables $l_1, l_2, \dots, l_k$. This fact is a reflection of the "Laplace theorem," well known in celestial mechanics, which asserts that the semi-major axes of the planets have no secular variations (i.e., variations that accumulate in time); for more details, see G. N. Duboshin's book [1.1]. "Laplace's theorem" was actually proved only recently by V. I. Arnold, and only after that the domain of applicability of this theorem became clear.[10]

3°. Since in the first approximation the fast variable l varies in exactly the same way in the perturbed motion and in the unperturbed motion, while the slow variable L does not evolve at all, instead of six equations the evolution system will contain

[10] See Arnold's works [1.9], [1.10], and also the essay on the stability of the Solar system in the present book and the excellent popular-science text of V. G. Demin [1.11].

only four:

$$\left.\begin{aligned} \frac{dG}{d\tau} &= \frac{\partial \overline{F}}{\partial g}, \quad \frac{dg}{d\tau} = -\frac{\partial \overline{F}}{\partial G}, \\ \frac{dH}{d\tau} &= \frac{\partial \overline{F}}{\partial h}, \quad \frac{dh}{d\tau} = -\frac{\partial \overline{F}}{\partial H}. \end{aligned}\right\} \tag{1.6.30}$$

4°. Since the function F does not depend on τ, the evolution equations (1.6.30) admit the first integral

$$\overline{F} = \text{const}. \tag{1.6.31}$$

Therefore, the full system of equations governing the evolution motion has two first integrals, (1.6.28) and (1.6.31).

To this point we have assumed that our equations contain only one fast variable, l. However, the same equations in the Delaunay variables (1.5.7) may contain in the expression of F other fast variables. For example, when we investigate the influence of the Moon and of the Sun on a Earth satellite, the function F unavoidably contains the coordinates of the Moon and of the Sun, which under certain conditions can also be regarded as fast variables.

The above conclusions **1°–4°** remain valid in this case, too, provided that the mean value $\overline{F}$ is understood as the result of averaging F over all fast variables, independently with respect to each of them. However, in the case of more than one fast variable one may encounter a qualitatively new situation: commensurability of the frequencies of the fast variables, or, in the customary terminology, resonance. The independent averaging with respect to the fast variables is feasible (and hence the previous conclusions about the evolution motion are valid) only in the absence of resonances. If a resonance is present, then a special study is needed and the evolution equations take a different form. To understand the special qualitative features of resonance cases is suffices to examine the case of two fast variables.

Thus, let us assume that in the equations written in Delaunay elements the function F has the form

$$F = f(\eta_j, l, l_1), \tag{1.6.32}$$

where η_j are slow variables and l, l_1 are fast variables.

For the unperturbed motion we set, slightly changing the notation,

$$\frac{dl}{dt} = \omega = \text{const}, \tag{1.6.33}$$

$$\frac{dl_1}{dt} = \omega_1 = \text{const}. \tag{1.6.34}$$

Here it will be more convenient to consider that F depends explicitly on l and l_1 (and not on some function of these arguments). The second relation (1.6.34) remains valid in the perturbed motion as well (which, incidentally, is not essential).

Further, suppose that one can find (positive or negative) integers n_0, m_0 such that

$$n_0\omega + m_0\omega_1 = 0. \tag{1.6.35}$$

Then we say that a *resonance* holds. The sum

$$|n_0| + |m_0| \tag{1.6.36}$$

is called the *order of the resonance.* If, on the contrary,

$$n_0\omega + m_0\omega_1 \neq 0 \tag{1.6.37}$$

for all positive or negative integers n_0, m_0, then the nonresonance case holds. In practice, a resonance (1.6.35) needs to be analyzed together with a small neighborhood, so that one takes $n_0\omega + m_0\omega_1 = \varepsilon$, where ε is small. If the motion is nonresonant, then upon introducing

$$\overline{F} = \frac{1}{(2\pi)^2}\int_0^{2\pi}\int_0^{2\pi} F(l, l_1)\, dl\, dl_1 \tag{1.6.38}$$

the evolution equations for the slow variables η_j reduce to the form

$$\frac{d\eta_i}{dt} = \pm\frac{\partial\overline{F}}{\partial\eta_j}. \tag{1.6.39}$$

Here the time t is a dimensional variable. But these last equations enjoy all the properties **1°–4°** of the evolution equations; in particular, this is true for "Laplace's theorem" stating that there is no evolution of the semi-major axes of orbits. These properties can be generalized to the case of an arbitrary number of fast variables, provided that there are no resonances between the frequencies of the fast variables.

Now let turn our attention to the resonance case. Since the function $F(l, l_1)$ is periodic in the variables l, l_1, it can be expanded in a Fourier series:

$$f(l, l_1) = \overline{F} + \sum_{n=1}^{\infty}\sum_{m=1}^{\infty}\left[a_{nm}\sin(nl + ml_1) + b_{nm}\cos(nl + ml_1)\right]. \tag{1.6.40}$$

Here $\overline{F}$ is given by formula (1.6.38), and as is well known, the Fourier coefficients a_{nm}, b_{nm} are easily calculated. Since we are interested in a neighborhood of the resonance (1.6.35), the variables $\varkappa_{nm} = nl + ml_1$ are all *fast*, provided that $n \neq kn_0$ and $m \neq km_0$ simultaneously ($k = 1, 2, \dots$). Consequently, the corresponding terms can be *averaged* and it turns out that their mean values are equal to zero. In formula (1.6.40) we must retain, in addition to $\overline{F}$, only the terms for which $n = kn_0$ and $m = km_0$, with the same k . The variable $\varkappa = n_0 l + m_0 l_1$ is *slow*, because near the resonance (1.6.35) we have

$$\frac{d\varkappa}{dt} = n_0\omega + m_0\omega_1 = \varepsilon,$$

i.e., $\varkappa$ changes slowly. Therefore, we can replace F by the following evolution terms:

$$\left.\begin{aligned} \widetilde{F}(\eta_i, l, l_1) &= \overline{F}(\eta_i) + \sum_{k=1}^{\infty}(a_k \sin k\varkappa + b_k \cos k\varkappa), \\ \overline{F}(\eta_i) &= \frac{1}{(2\pi)^2}\int_0^{2\pi}\int_0^{2\pi} F\, dl\, dl_1, \\ a_k &= a_{kn_0,km_0}(\eta_i), \quad b_k = b_{kn_0,km_0}(\eta_i). \end{aligned}\right\} \tag{1.6.41}$$

Here η_i denote the five slow variables L, G, H, g, h.

Let us derive the evolution equation for the new variable $\varkappa$. We have

$$\frac{d\varkappa}{dt} = n_0 \frac{dl}{dt} + m_0\omega_1 = -n_0 \frac{\partial \widetilde{F}}{\partial L} + m_0\omega_1.$$

Introducing the function

$$\Phi(\eta_i, \varkappa) = n_0 \widetilde{F} - m_0\omega_1 L, \tag{1.6.42}$$

we obtain

$$\frac{d\varkappa}{dt} = -\frac{\partial \Phi}{\partial L}, \quad \frac{dL}{dt} = \frac{\partial \widetilde{F}}{\partial l} = \frac{\partial \widetilde{F}}{\partial \varkappa} n_0 = \frac{\partial \Phi}{\partial \varkappa}.$$

Finally, the complete evolution system has the form

$$\left.\begin{aligned} \frac{dL}{dt} &= \frac{\partial \Phi}{\partial \varkappa}, & \frac{d\varkappa}{dt} &= -\frac{\partial \Phi}{\partial L}, \\ \frac{dG}{dt} &= \frac{1}{n_0}\frac{\partial \Phi}{\partial g}, & \frac{dg}{dt} &= -\frac{1}{n_0}\frac{\partial \Phi}{\partial G}, \\ \frac{dH}{dt} &= \frac{1}{n_0}\frac{\partial \Phi}{\partial h}, & \frac{dh}{dt} &= -\frac{1}{n_0}\frac{\partial \Phi}{\partial H}. \end{aligned}\right\} \tag{1.6.43}$$

Here all the variables are slow and $\Phi = \Phi(L, G, H, \varkappa, g, h)$ does not depend explicitly on time; consequently, as is readily verified, the system admits the first integral $\Phi =$const, or, in expanded form,

$$\begin{aligned} \{\overline{F}(L, G, H, g, h) + \sum_{k=1}^{\infty}[a_k(L, G, H, g, h) \sin k\varkappa + b_k(L, G, H, g, h) \cos k\varkappa]\}n_0 - \\ - m_0\omega_1 L = \text{const}. \end{aligned} \tag{1.6.44}$$

The above analysis suffices to demonstrate how different the resonance case is from the nonresonance one. In the function $\widetilde{F}$, in addition to the term $\overline{F}$, which is present in the nonresonance case, it is necessary to consider the correction term $\sum(a_k \sin k\varkappa + b_k \cos k\varkappa)$, which depends on the new slow variable $\varkappa$, and which is an exclusive manifestation of the resonance phenomenon. Due to the presence of this correction term, the evolution system contains six equations, and not four, as in the resonance case. As a consequence, the investigation of the evolution system is more difficult than in the nonresonance case.

In the resonance case *"Laplace's theorem" on the conservation of the magnitude of the semi-major axis of an orbit does not hold*: the semi-major axis evolves. In the resonance case, like in the nonresonance case, the evolution system has a first integral expressing the constancy of the new Hamiltonian. However, the structure of this integral is considerably more complex than the structure of the analogous integral in the nonresonant motion.

7. Satellite in the gravitational field of Earth

The analysis of the motion of a satellite can be carried out by means of the averaging method just described.

The Newtonian force of attraction (1.2.1) and the corresponding force function (1.2.2) describe only approximately the force of attraction by which the real Earth acts on the satellite. For an Earth that is perfectly spherical and homogeneous, or consists of homogeneous concentric layers, formulas (1.2.1) and (1.2.2) would be correct. But this is not the case: the Earth is slightly oblate in the direction of its poles (and to a very small extent even "sideways"), and is not perfectly symmetric and homogeneous. As a consequence, the force field generated by the Earth has a rather complex structure. As a better approximation, compared with (1.2.2), to the real force field one takes a force function of the following form:

$$U = \frac{\mu}{r}\left\{1 + \sum_{k=2}^{\infty} I_k \left(\frac{R}{r}\right)^k P_k(\sin\varphi)\right\}, \quad \mu = fM. \tag{1.7.1}$$

Here M is the mass of the Earth, f the constant of gravitation, R the equatorial radius of the Earth, and φ the geographic latitude of a point (at distance r from the center of the Earth). The coefficients I_k have fixed dimensionless values. Finally, the functions P_k are the *Legendre polynomials*, defined as follows:

$$\left.\begin{aligned} P_0(x) &= 1,\\ P_1(x) &= x,\\ P_2(x) &= \frac{1}{2}(3x^2 - 1),\\ P_3(x) &= \frac{1}{2}(5x^3 - 3x),\\ &\cdots\cdots\cdots\cdots\\ P_n(x) &= \frac{1}{2^n \cdot n!}\frac{d^n}{dx^n}(x^2 - 1)^n. \end{aligned}\right\} \tag{1.7.2}$$

Thus, the force function (1.7.1) depends not only on the distance to the center of attraction, like the force function (1.2.5), but also on the latitude of the point.

The form (1.7.1) assumes that the force field is axisymmetric (with respect to the Earth's polar axis). A more precise expression for U should take into account also the "sideways oblateness" of the force field, its asymmetry. Then U will depend also on the longitude of the point. However, for our purposes the expression (1.7.1) will suffice. Note that (1.7.1) is the sum of the Newtonian potential (1.2.5) and correction terms which, by the definition of the concept of "perturbations," must be small compared with the principal (first) term. Indeed, according to the most recent observation data, the first few coefficients I_k in (1.7.1) have the following values:

$$\left.\begin{aligned} I_2 = -1082.2 \cdot 10^{-6}, \quad I_3 = 2.3 \cdot 10^{-6}, \\ I_4 = 2.1 \cdot 10^{-6}, \dots , \end{aligned}\right\} \tag{1.7.3}$$

so that even the largest of these coefficients, I_2, gives a correction of the order of a tenth of 1%, while the remaining coefficients are several orders smaller. The coefficient I_2 characterizes the most essential deviation of the gravitational field of the real Earth from that of an ideally spherical "Earth," deviation caused by the notable oblateness of the Earth in the polar direction.

Now let us examine the effect of the main perturbation on the motion of a satellite in the Earth's gravitational field using the information on asymptotic methods given above. The problem of the motion of a satellite in the gravitational field of an "oblate" Earth provides a beautiful illustration of the application of asymptotic methods of nonlinear mechanics. At the same time, it is one of the basic problems in the entire theory of motion of artificial satellites.

Thus, we will consider the perturbing force function

$$U = \frac{\mu}{r} I_2 \left(\frac{R}{r}\right)^2 \frac{1}{2}(3\sin^2\varphi - 1). \tag{1.7.4}$$

Recalling the definition of the argument of latitude $u = \omega + \nu$ and of the inclination i of the orbit to the equatorial plane we can write (see Figure 1.11)

$$\sin\varphi = \sin i \sin u. \tag{1.7.5}$$

The Hamiltonian F differs from U by a term that depends only on the variable L and, as we have seen in Section 6 of the present essay, this variable does evolve. Hence, in the four evolution equations (1.6.30) instead of the mean value $\overline{F}$ it suffices to consider the mean value $\overline{U}$. According to formulas (1.7.4), (1.7.5), and (1.7.26),

$$\overline{U} = \frac{\mu R^2 I_2}{2} \frac{1}{2\pi} \int_0^{2\pi} [3\sin^2(\omega+\nu)\sin^2 i - 1] \frac{1}{r^3} \frac{r^2}{\sqrt{\mu p}}\, d\nu. \tag{1.7.6}$$

Substituting here $r = p/(1 + e\cos\nu)$ and carrying out the integration, we get

$$\overline{U} = \frac{\sqrt{\mu} I_2 R^2}{4p^{3/2}} (1 - 3\cos^2 i), \tag{1.7.7}$$

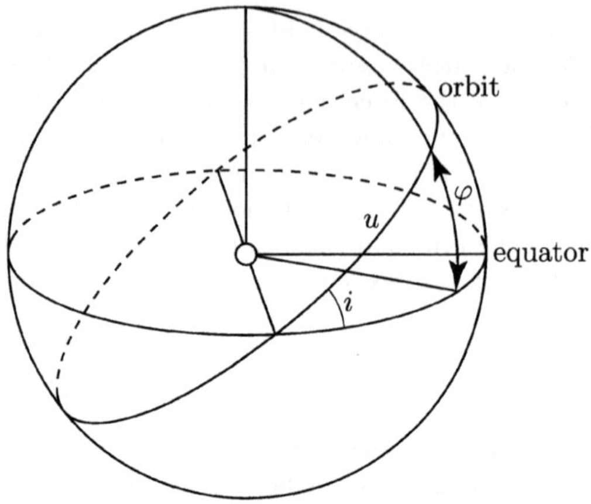

FIGURE 1.11. The orbit of a satellite in the equatorial coordinate system

or, upon replacing the osculating elements by the Delaunay elements (by means of formulas (1.5.5)),

$$\overline{U} = \frac{1}{4}\frac{\mu^2 I_2 R^2}{G^3}\left(1 - 3\frac{H^2}{G^2}\right). \tag{1.7.8}$$

We see that $\overline{U}$ does not involve the elements g, h. Consequently, in addition to the integral $L = l_0$ the equations of motion admit the additional first integrals

$$G = G_0, \quad H = H_0.$$

In the usual osculating elements this means that

$$a = \text{const}, \quad e = \text{const}, \quad i = \text{const}. \tag{1.7.9}$$

But then, of course, $p = a(1 - e^2)$=const as well. Hence, *the fact that the Earth is oblate does not modify the form of the orbit and its inclination to the equatorial plane* (we should mention here that we are speaking about the main characteristics of the motion described by the averaged, evolution equations). The effect of the evolution manifests solely as a variation of the elements g and h, i.e., as a variation of the longitude of ascending node Ω and the argument of perigee ω. These variations are readily calculated by substituting the force function (1.7.8) (instead of $\overline{F}$) in the remaining evolution equations for g and h and passing from the variables G, H back to the osculating elements. This yields

$$\frac{d\Omega}{d\tau} = \frac{3}{2} I_2 \left(\frac{R}{p}\right)^2 \cos i, \tag{1.7.10}$$

$$\frac{d\omega}{d\tau} = \frac{3}{4} I_2 \left(\frac{R}{p}\right)^2 (1 - 5\cos^2 i). \tag{1.7.11}$$

Thus, under the influence of the Earth's oblateness the orbital plane executes a precession with rate (1.7.10), while the orbit itself, without changing its shape, executes a precession with rate (1.7.11) in its plane. The motion of the orbital plane is retrograde ($I_2 < 0$, $\cos i > 0$ (Figure 1.12). For many (former) Soviet artificial satellites the inclination of the orbit to the equatorial plane is $i \approx 65°$. In this case the quantity $1 - 5\cos^2 i$ is small and the motion of the orbit's perigee is very slow compared with the motion of the orbit's node. The typical rate of motion of the node is $\sim 3°$ per day, whereas the rate of motion of the perigee is $\sim 0.5°$ per day.

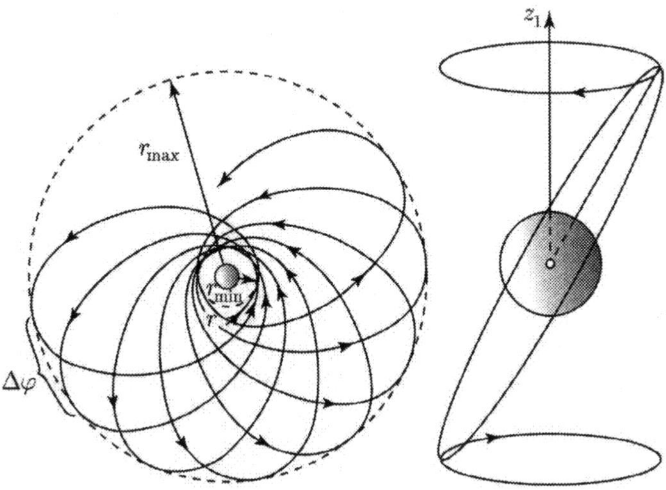

FIGURE 1.12. The perturbed orbit of a satellite in the gravitational field of the Earth: (a) motion in the orbital plane; (b) motion of the orbital plane itself

Formulas (1.7.10), (1.7.11) appear, for example, in a paper by D. E. Okhotsimkiĭ, T. M. Eneev, and G. P. Taratynova published in the first issue of the journal Uspekhi Fizicheskikh Nauk for 1957.[11] Their paper deals also with the perturbations of the motion of a satellite due to the braking action of the upper layers of the atmosphere. Although at the altitudes at which satellites fly the atmosphere is quite rarefied, it can nevertheless brake the motion, as a result of which it the satellite gradually approaches closer to the Earth (and finally falls on Earth). In this book we will not consider the influence of the atmosphere on orbits; the interested reader can find a detailed analysis of the unperturbed and perturbed motion

[11]In that issue, on the eve of the launching of the first Soviet artificial satellite (on the 14th of October, 1957), a group of specialists in mechanics published their already completed works on the dynamics of Earth artificial satellites and trajectories of lunar flights.

of a satellite in the books of D. E. Okhotsimskiĭ [1.13] and P. E. El′yasberg [1.14]. Excellent material on the mechanics of space flight is treated in the book of M. B. Balk [1.15] where, in particular, the reader will find the interesting paragraph "The satellite paradox" (page 287). In Balk's book this paradox is formulated as follows: "Due to the braking effect of atmospheric drag, the linear velocity of a satellite that moves on a nearly-circular orbit grows; in fact it turns out that the acceleration in the direction of motion is the same as it would be if the drag force were to change its direction and push the satellite forward" (! – inserted by the present author, V. B.).

If the reader (with pencil and paper in his hands) cannot figure out by himself why this paradox arises, he is well advised to read about it in Balk's book (see also our 8th essay).

A number of effects in the dynamics of artificial satellites, especially aerodynamic effects, are discussed in the survey paper by L. I. Sedov [1.19].

Additional comments for this translation

The equation (1.6.1) of planar oscillations (and rotations) of a satellite in an elliptic orbit, published by the author in 1959 [1.18] (see also [1.17]), proved to be a rewarding arena for studying the dynamical effects of the problem as well as for the application of new methods of analysis. The concrete form of this equation emerged first in 1956, in an internal report of an organization that is currently known as the M. V. Keldysh Institute of Applied Mathematics of the Russian Academy of Sciences (KIAM RAS).

Since that moment – more than 40 years have passed! – this equation has continued to attract the unabated attention of researchers. The first investigations, carried out by analytic and numerical methods, were published in [1.18′], [1.20]–[1.22], and were incorporated in the author's monograph [1.17], and partially also in the present book (6th essay). V. A. Sarychev's survey paper [1.23] covers results of these and other investigations of equation (1.6.1) up to 1978.

In the papers [1.24]–[1.26], V. A. Sarychev and his collaborators continued their studies, initiated in [1.21], of periodic rotations and oscillations of a satellite in an elliptic orbit and of the stability of these motions.

It is worth mentioning that the phase trajectories studied in the works listed above have a regular character (periodic, and sometimes quasi-periodic). In the beginning of the 1980s the analysis of planar motions of a satellite in an elliptic orbit reached a qualitatively new level. At that time the theory of dynamical systems and computational techniques were sufficiently developed to allow one to study not only isolated trajectories of the system, but also the whole structure of its phase space. It turned out that regardless of the deterministic character of the initial conditions, the typical situation is that of almost chaotic trajectories, which fill the phase space like a "sea;" in this chaotic sea there are "islands" of periodic and quasi-periodic motions. Such a structure of the phase space is intimately connected with the nonintegrability of the dynamical system.

The motions described by equation (1.6.1) also shared in these developments. Since the beginning of the 1980s numerous works investigating equation (1.6.1) from precisely the point of view mentioned above have been published in many countries. Results of these investigations are presented in [1.27]–[1.33]. In the case $e = 0$ (circular orbit) equation (1.6.1) is integrable and its phase space is filled by regular trajectories. For small values of e thin chaotic regions – layers – arise, which divide the phase space and which merge to form a chaotic sea when the magnitude of the orbit's eccentricity e increases (now and then in this process, phase portraits of amazing beauty are observed).

We should mention that the first calculations of fragments of phase portraits for equation (1.6.1) in which the ideas described above were pervasive were published in the earlier papers [1.21] and [1.41].

In recent years an even deeper level of study of equation (1.6.1) was attained in work of A. D. Bruno and his collaborators [1.34]–[1.40]. These studies are based on the new constructive methods of analysis of strongly nonlinear problems developed by A. D. Bruno [1.34]. As the authors write in [1.37]: "We decided first to apply this method to a concrete problem that is not too simple and at the same time not too complicated. As such a problem we took the problem of the oscillations of a satellite in the plane of its elliptic orbit." In particular, the method developed by A. D. Bruno allowed one to circumvent the difficulties connected with the fact that equation (1.6.1) is singular for values of e close to 1 (for $e = 1$ the coefficient of the highest derivative vanishes at the point $\nu = \pi$). Families of periodic solutions were studied in a generalized sense – for arbitrary values of the parameter n^2 (in the mechanical problem $|n^2| < 3$) and for all values of the eccentricity in the interval $0 < e < 1$. The stability and the asymptotics of these families for small and large values of n^2, and the rules governing their structure were investigated as well. The limits of these families for $e \to 1$ were constructed; one of these limit families coils into a fractal spiral.

Second Essay

On the Rebirth of an Old Problem, or what Happens if two Masses are Placed at a Purely Imaginary Distance from one Another

...and the more he looked at the bell-rope,
the more he felt that he had seen something
like it, somewhere else, sometime before.

A. A. Milne, *The World of Pooh*

"And – Afterwards?"...
"Then we all say '*Aha!*'"
"All three of us?"
"Yes."

A. A. Milne, *The World of Pooh*

1. From Euler to our days

In the first essay we did already speak about the motion of a satellite in the gravitational field of the Earth spheroid. We have seen that the well developed tools of equations in osculating elements and asymptotic methods of nonlinear mechanics allow us in a rather simply manner to establish the basic perturbations in the motion of a satellite caused by the oblateness of the Earth. The formulas obtained provide a transparent geometric picture of the motion of the satellite, revealing the qualitative and quantitative laws of motion. Needless to say, this is not sufficient for the exact calculation of orbits. An exact solution of the problem of the motion of a satellite in the gravitational field described by the force function (1.7.1) is not feasible because of the nonintegrability of the equations of motion. But if are willing to dwell deeper into the integrability problem (which we indeed intend to do), then amazing facts, and a new, remarkable point of view are brought to light.

It turns out that the problem at hand is in a certain sense equivalent to the old, classical problem of the motion of a particle in the gravitational field of two fixed Netwonian centers of attraction. As is known, this last problem was formulated and reduced to quadratures by Leonhard Euler. Since then, over a period of two hundred years, this problem was studied by many scientists, fascinated by its academic beauty as well as by its transparent resemblance with the famous "three-body problem," whose importance for celestial mechanics and the mechanics of space flight is enormous. Indeed, the so-called "restricted three-body" problem deals with the motion of a nonattracting particle in the gravitational field of two free Newtonian centers (which therefore move relative to one another according to Kepler's laws). Availability of a solution for the restricted three-body problem would allow one, for instance, to study directly the motion of a spaceship from the Earth to the Moon. The problem of the motion of a particle in the field of two fixed centers differs from the restricted three-body problem "merely" by the fact that the two attracting centers are fixed. It is not surprising that there were numerous attempts to apply the known solution of Euler's problem in the search for the unknown solution of the restricted three-body problem. But these attempts were crowned with no notable success. "The problem of two fixed centers" found no concrete applications in celestial mechanics and until very recently its investigations had an abstract mathematical character.

The emergence of artificial satellites stimulated rapid progress in mechanics and led, in particular, to the discovery of a far-reaching and unexpected analogy between the problem of two centers and the problem of the motion of a satellite in the Earth's gravitational field. Thus an old problem found a new and very important application in the theory of motion of artificial satellites. The first work that established this analogy [2.1] appeared in 1961 and belongs to three (then) young scientists, E. P. Aksenov, E. A. Grebenikov, and V. G. Demin. In subsequent studies based on this analogy, these authors have investigated in detail the motion of a satellite in the Earth's gravitational field.[1]

Let us embark in real earnest upon the consideration of this analogy.

2. The connection between the two problems

Suppose that two particles of mass m_1 and m_2, respectively, are placed in two fixed centers. Let us introduce a coordinate system $Oxyz$ such that the axis Oz passes through the points m_1 and m_2 (Figure 2.1). Denote the distance from O to the particle m_i by c_i, $i = 1, 2$. Further, suppose that in the resultant of the force fields of Newtonian attraction of the two particles m_1, m_2 there is a particle of mass m whose motion we are proposing to study. The motion of m is determined by the force function

$$U_{(2)} = f\left(\frac{m_1}{r_1} + \frac{m_2}{r_2}\right), \tag{2.2.1}$$

[1] In the 1961 book of D. Brouwer and G. M. Clemence [2.2] there is also a brief mention of such an analogy between the two problems. A detailed exposition of the problems at hand can be found in V. G. Demin's monograph [2.3] and in the book of G. N. Duboshin [2.4]. Apparently the idea of replacing the problem of the motion of a satellite in a real gravitational field by some close integrable problem was first advanced and implemented by M. D. Kislik in 1958–1959 [2.5]. In the United States a similar approach was put into practice in research by J. Vinti [2.6].

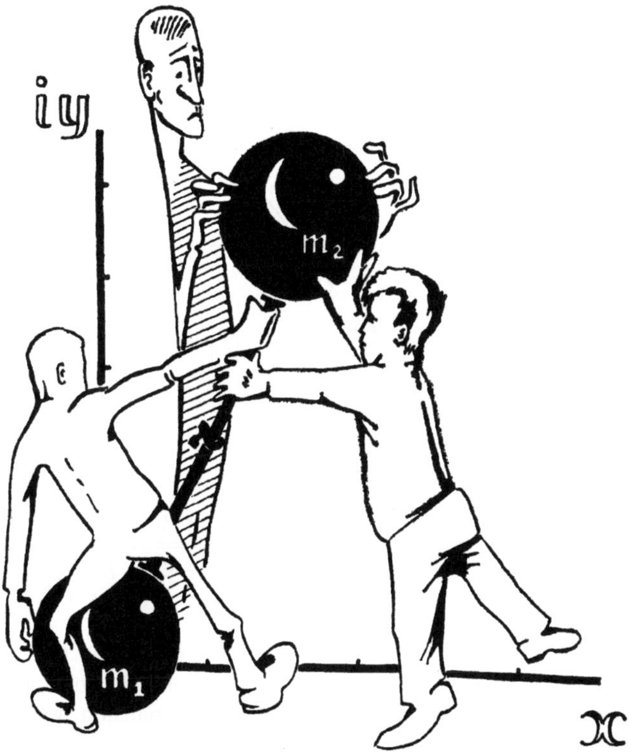
iy
m₂
m₁
x

i.e., the sum of the force functions of the two Newtonian centers of attraction (which is indicated by the subscript (2) on the left-hand side of formula (2.2.1)). Here f is the universal constant of gravitation and r_i is the distance from m_i to the moving particle m, $i = 1, 2$:

$$r_1 = \sqrt{x^2 + y^2 + (z - c_1)^2}, \quad r_2 = \sqrt{x^2 + y^2 + (z - c_2)^2}.$$

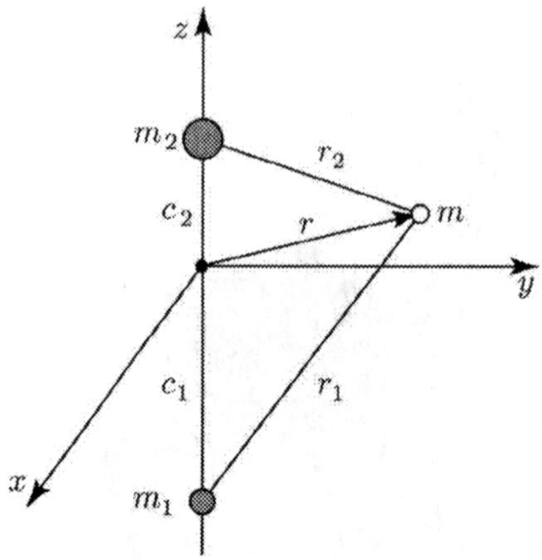

FIGURE 2.1. Problem setting

Consider, for example, the expression for $\frac{1}{r_2}$:

$$\frac{1}{r_2} = \frac{1}{\sqrt{r^2 - 2c_2 z + c_2^2}} = \frac{1}{r} \frac{1}{\sqrt{1 - 2\frac{z}{r}\frac{c_2}{r} + \left(\frac{c_2}{r}\right)^2}}, \tag{2.2.2}$$

where $r = \sqrt{x^2 + y^2 + z^2}$ is the distance from the moving particle to the origin. Denoting $z/r = \sin\varphi$ (where φ is the latitude of the moving point, measured from the plane Oxy) and $\alpha = c_2/r$, we use the following expansion, well known in the theory of Legendre polynomials:

$$\frac{1}{\sqrt{1 - 2\alpha \sin\varphi + \alpha^2}} = \sum_{n=0}^{\infty} \alpha^n P_n(\sin\varphi); \tag{2.2.3}$$

here P_n is the Legendre polynomial of order n. Combining relations (2.2.2) and (2.2.3), one can write

$$\frac{m_2}{r_2} = \frac{m_2}{r} \sum_{n=0}^{\infty} \left(\frac{c_2}{r}\right)^n P_n(\sin\varphi). \tag{2.2.4}$$

A similar expression can be written for m_1/r_1. Consequently, the force function (2.2.1) of the problem of two centers takes the form

$$U_{(2)} = \frac{f(m_1+m_2)}{r}\left\{1+\sum_{k=1}^{\infty}\frac{\gamma_k}{r^k}P_k(\sin\varphi)\right\}, \tag{2.2.5}$$

where the coefficients γ_k are found from the equations

$$(m_1+m_2)\gamma_k = m_1c_1^k + m_2c_2^k. \tag{2.2.6}$$

Here a comparison suggest itself between the force function $U_{(2)}$ given by (2.2.5) and the already familiar to us force function of the Earth spheroid (see formula (1.7.1)). The two force functions are almost identically in form; they would be identical if the constant coefficients in the corresponding terms of the expansions were the same. And since the problem with force function (2.2.5) is integrable, so would be the problem of the motion of a satellite. Of course, this is not really the case. The point is that the expansion (2.2.5) contains only *four* free parameters, m_1, m_2, c_1, and c_2, and so only *four* terms of the expansion (1.7.1) can be identified with four terms of the expansion (2.2.5). It is natural to make such an identification for the first terms of the expansion, i.e., to require that

$$\left.\begin{aligned} &1)\quad m_1+m_2=M,\\ &2)\quad m_1c_1+m_2c_2=0,\\ &3)\quad m_1c_1^2+m_2c_2^2=MI_2R^2,\\ &4)\quad M_1c_1^3+m_2c_2^3=MI_3R^3. \end{aligned}\right\} \tag{2.2.7}$$

Then the terms that follow these four in the expansions of U and $U_{(2)}$ cannot coincide, but this is not crucial because at any rate only the first few constants I_k are accurately known, whereas the remaining I_k are very small and their values are poorly known.

Thus, if we choose the parameters m_1, m_2, c_1, c_2 of the problem of two fixed centers so that relations (2.2.7) hold, then the force function $U_{(2)}$, while remaining the force function governing the problem, will nevertheless have the same *main* (or *leading*) *terms* as the force function U of the Earth's gravitational field. In other words, if the potential of the Earth's gravitational field is slightly "altered" by modifying its "tail" (i.e., by replacing the very small coefficients of the terms with $k \geq 4$ in the expansion by different, though also very small coefficients), then the problem of the motion of a satellite in the gravitational field of the Earth (or any other planet) *becomes integrable*. The motion of a particle under the action of such a "altered" potential will be very close to the motion under the action of the original potential because the effect of the main terms of the potential is taken into account exactly. Moreover, thanks to the integrability of the problem, the exact formulas obtained after integration should reflect all the effects of the

MAIN TERM
OF NEWTONIAN
POTENTIAL

influence of the main terms of the force function of the Earth's gravitational field (and not only the secular effects discussed above).

Thus, the integrable problem of two fixed centers approximates with a high degree of accuracy the problem of the motion of a particle in the gravitational field of the Earth spheroid provided that the parameters of the former problem are chosen in a special way, namely, that relations (2.2.7) are satisfied. The algebraic system (2.2.7) is solved with no difficulty. Its first two equations in (2.2.7) yield

$$m_1 = -\frac{Mc_2}{c_1 - c_2}, \quad m_2 = \frac{Mc_1}{c_1 - c_2}. \tag{2.2.8}$$

Substituting the expressions (2.2.8) in the last two equations (2.2.7), we obtain

$$c_1 c_2 = -I_2 R^2,$$
$$c_1 c_2 (c_1 + c_2) = -I_3 R^3,$$

or

$$c_1 c_2 = -I_2 R^2, \qquad c_1 + c_2 = \frac{I_3 R}{I_2}. \tag{2.2.9}$$

This shows that c_1 and c_2 are the roots of the following quadratic equations in c:

$$c^2 - \frac{I_3 R}{I_2} c - I_2 R^2 = 0. \tag{2.2.10}$$

Solving this equation we get

$$\left.\begin{aligned} c_1 &= \left(\frac{I_3}{2I_2} + \sqrt{\left(\frac{I_3}{I_2}\right)^2 + 4I_2} \right) \frac{R}{2}, \\ c_2 &= \left(\frac{I_3}{2I_2} - \sqrt{\left(\frac{I_3}{I_2}\right)^2 + 4I_2} \right) \frac{R}{2}. \end{aligned}\right\} \tag{2.2.11}$$

Now substituting the values (2.2.11) in (2.2.8) we complete the determination of the parameters m_1, m_2, c_1, c_2. Recalling the numerical values of I_2 and I_3 ($I_2 \sim -10^{-3}$, $I_3 \sim 10^{-5}$), we reach a surprising conclusion: in the problems of two fixed centers in which we are interested, the distances c_1 and c_2 and the masses m_1 and m_2 turn out to be complex numbers! Thus we arrive at a generalization of the classical problem to complex masses and distances. This should not frighten the reader – indeed, the force function remains real, and hence so does the motion.

The coefficient I_2 characterizes the oblateness of the Earth, while I_3 characterizes its asymmetry with respect to the equatorial plane. This asymmetry is rather small ($|I_3/I_2| \sim 10^{-2}$) and can be neglected. If we take into account only the oblateness of the Earth, i.e., put $I_3 = 0$, we obtain

$$m_1 = m_2 = \frac{M}{2}, \quad c_1 = iR\sqrt{|I_2|}, \quad c_2 = -iR\sqrt{|I_2|}. \tag{2.2.12}$$

Thus, the problem of the motion of a satellite in the gravitational field of an oblate Earth that is symmetric with respect to its equatorial plane can be interpreted as the (integrable) problem of the motion of a particle in the field of two fixed centers of equal masses, situated at a purely imaginary (!) distance from one another. (The distance would be real for an elongated Earth, instead of an oblate one). The analogy discovered by Aksenov, Grebenikov and Demin between the two problems allowed one to consider the motion of a satellite in the Earth's gravitational field in closed form, specifically, in elliptic functions (see below). One can say that the classical problem was considered mainly in the planar case, whereas the trajectories of a satellite are spatial in nature (i.e., they live in three-dimensional space), a fact that itself stimulated further investigations of the problem of two fixed centers.

It is important to emphasize that the analogy between the two problems took time to be discovered. It was gradually revealed by scientist from several countries. J. Vinti [2.6] in USA, M. D. Kislik [2.5] in the (former) Soviet Union, and others have reduced the problem of the motion of a satellite in the Earth's gravitational field to an integrable problem by means of a potential "truncated" in some or another manner. But only the research by Aksenov, Grebenikov and Demin in the setting indicated above did completely clarify the formulation of the problem, its character, and the role that it plays. The cases considered earlier are particular cases of their analysis. Presently the problem of two fixed centers of complex masses placed at a complex distance from one another is known as the *generalized problem of two fixed centers.* In this new form the problem continues to attract the attention of scientists (see work of V. M. Alekseev [2.7] and others).

3. Integration. Coordinate system

We are now ready to integrate our problem. We will consider the case (2.2.12). Denoting $c = R\sqrt{|I_2|}$, we have

$$U_{(2)} = f\frac{M}{2}\left(\frac{1}{\sqrt{x^2+y^2+(z+ic)^2}} + \frac{1}{\sqrt{x^2+y^2+(z-ic)^2}}\right). \tag{2.3.1}$$

Not every system of coordinates is suitable for carrying out the integration. Usually each concrete problem requires a clever choice of a coordinate system, in which the analysis is carried in the most convenient manner; making such a choice is not always easy, and it is something worth puzzling over, since it predetermines whether the study of the problem will be successful (or end in failure). In the problem at hand instead of the Cartesian coordinates x, y, z we will work with a system of curvilinear coordinates λ, μ, w such that

$$\left.\begin{aligned} x &= c\sqrt{(1+\lambda^2)(1-\mu^2)}\,\cos w,\\ y &= c\sqrt{(1+\lambda^2)(1-\mu^2)}\,\sin w,\\ z &= c\lambda\mu. \end{aligned}\right\} \tag{2.3.2}$$

Let us explain the meaning of the new variables. From (2.3.2) we derive the relations

$$\left.\begin{aligned} \frac{x^2+y^2}{c^2(1+\lambda^2)}+\frac{z^2}{c^2\lambda^2}&=1,\\ \frac{x^2+y^2}{c^2(1-\mu^2)}-\frac{z^2}{c^2\mu^2}&=1,\\ \frac{x}{y}&=\tan w. \end{aligned}\right\} \tag{2.3.3}$$

We see that the geometric locus of the points $\lambda = \text{const}$ $(0 \leq \lambda < +\infty)$ is an oblate ellipsoid of revolution whose symmetry axis coincides with the z-axis of the old coordinate system. The semi-minor [resp., semi-major] axis of the ellipsoid is equal to $c\lambda$ [resp., $c\sqrt{1+\lambda^2}$] (Figure 2. 2).

The geometric locus of the points $\mu = \text{const}$ is a one-sheeted hyperboloid of revolution (with the same symmetry axis z). (These hyperboloids fill up the entire space when μ^2 runs from 0 to 1.) Finally, $w = \text{const}$ $(-\infty < w < +\infty)$ is a plane containing the z-axis (the meridian plane). In these variables

$$r_1 = \sqrt{x^2+y^2+(z+ic)^2} = c(\lambda + i\mu),$$
$$r_2 = \sqrt{x^2+y^2+(z-ic)^2} = c(\lambda - i\mu),$$

and consequently

$$U_{(2)} = fM\,\frac{\lambda}{c(\lambda^2+\mu^2)}\,. \tag{2.3.4}$$

Since $U_{(2)} > 0$ everywhere, the range of the variable λ is $0 < \lambda < \infty$.

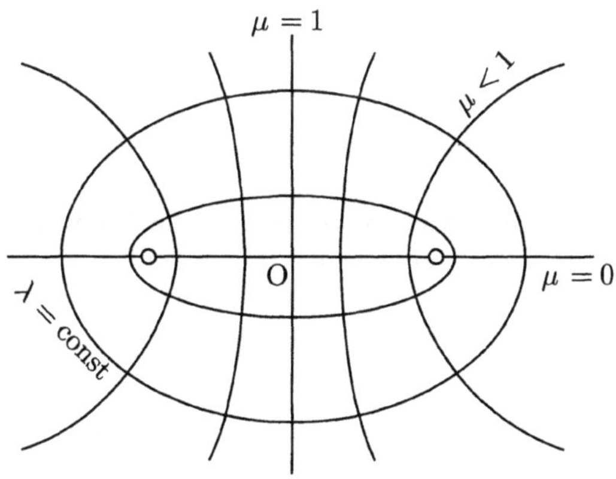

FIGURE 2.2. The elliptic coordinate system

Using (2.3.2) to write the expressions of the derivatives $\dot{x}$, $\dot{y}$, $\dot{z}$ in term of the derivatives $\dot{\lambda}$, $\dot{\mu}$, $\dot{w}$, we obtain the expression of the kinetic energy $T = \frac{1}{2}\left(\dot{x}^2 + \dot{y}^2 + \dot{z}^2\right)$ in terms of λ, μ, w, $\dot{\lambda}$, $\dot{\mu}$, $\dot{w}$:

$$T = \frac{c^2}{2}\left\{\frac{\lambda^2+\mu^2}{1+\lambda^2}\,\dot{\lambda}^2 + \frac{\lambda^2+\mu^2}{1-\mu^2}\,\dot{\mu}^2 + (1+\lambda^2)(1-\mu^2)\dot{w}^2\right\} \tag{2.3.5}$$

The total energy of the motion is

$$H = T - U_{(2)}. \tag{2.3.6}$$

4. The Hamilton-Jacobi method

Here and in the ensuing analysis we will need the Jacobi method of integrating the equations of dynamics. We will present this method without proofs, and in fact in the simplest form in which it will be used in this book, referring for details to textbooks on analytical or celestial mechanics (see, for instance, [2.8], [2.9]).

Let the kinetic energy depend on the *generalized* (curvilinear) *coordinates* q_1, q_2, q_3 and the corresponding (*generalized*) velocities $\dot{q}_1, \dot{q}_2, \dot{q}_3$:

$$T = T(q_1, q_2, q_3, \dot{q}_1, \dot{q}_2, \dot{q}_3). \tag{2.4.1}$$

Then the partial derivatives

$$p_1 = \frac{\partial T}{\partial \dot{q}_1}, \quad p_2 = \frac{\partial T}{\partial \dot{q}_2}, \quad p_3 = \frac{\partial T}{\partial \dot{q}_3} \tag{2.4.2}$$

are called *generalized momenta*. Let us use relations (2.4.2) to express the velocities $\dot{q}_1, \dot{q}_2, \dot{q}_3$ in terms of the generalized momenta p_1, p_2, p_3 (under the assumption that this is possible) and substitute the results in the expression (2.4.1) for the kinetic energy. This yields the transformed expression

$$\widetilde{T} = T(q_1, q_2, q_3, p_1, p_2, p_3).$$

Let $U(q_1, q_2, q_3)$ be the force function of the force field in which the motion takes place. We assume that U does not depend explicitly on time. Then the total energy of the motion,

$$H = \widetilde{T} - U,$$

is conserved:

$$H(q_1, q_2, q_3, p_1, p_2, p_3) = h\,(= \text{const}). \tag{2.4.3}$$

The function $H(q_1, q_2, q_3, p_1, p_2, p_3)$ is called the *Hamilton function* or the *Hamiltonian* of the system.

The equations of motion in the present case have the *canonical form*

$$\frac{dq_i}{dt} = \frac{\partial H}{\partial p_i}, \quad \frac{dp_i}{dt} = -\frac{\partial H}{\partial q_i}, \quad i = 1, 2, 3, \tag{2.4.4}$$

and are called *Hamilton's equations.*

Now let us write the partial differential equation obtained when we replace p_i by $\partial W/\partial q_i$, $i = 1, 2, 3$, in (2.4.3):

$$H\left(q_1, q_2, q_3, \frac{\partial W}{\partial q_1}, \frac{\partial W}{\partial q_2}, \frac{\partial W}{\partial q_3}\right) = h. \tag{2.4.5}$$

This equation is known as the *Hamilton-Jacobi equation.* The function W in (2.4.5) needs to be determined. More precisely, we need to find a solution

$$W(h, \alpha_2, \alpha_3, q_1, q_2, q_3)$$

of the Hamilton-Jacobi equation, such that W depends on three constants: h, and two other arbitrary constants, α_2, α_3. If we succeed in finding such a function, then the general integral of the system (2.4.4) of canonical equations of motion can be written in the form

$$\left.\begin{array}{lll} \dfrac{\partial W}{\partial h} = t + \beta_1, & \dfrac{\partial W}{\partial \alpha_2} = \beta_2, & \dfrac{\partial W}{\partial \alpha_3} = \beta_3, \\ \dfrac{\partial W}{\partial q_1} = p_1, & \dfrac{\partial W}{\partial q_2} = p_2, & \dfrac{\partial W}{\partial q_3} = p_3, \end{array}\right\} \tag{2.4.6}$$

where β_1, β_2, β_3 are new arbitrary constants. Formulas (2.4.6) give the connection of the coordinates q_1, q_2, q_3 and momenta p_1, p_2, p_3 with the time t and the six arbitrary constants h, α_2, α_3, β_1, β_2, β_3. Hence, these formula solve the dynamical problem completely. (To bring the investigation to its end one needs to derive from the system (2.4.6) the explicit dependences of the coordinates and momenta on time and the arbitrary constants, which usually is not an easy task.)

5. Integration

Thus, following the recipe of the Hamilton-Jacobi method, let us write the generalized momenta for our problem:

$$\left.\begin{aligned} p_\lambda &= \frac{\partial T}{\partial \dot{\lambda}} = c^2\,\frac{\lambda^2 + \mu^2}{1 + \lambda^2}\,\dot{\lambda}, \\ p_\mu &= \frac{\partial T}{\partial \dot{\mu}} = c^2\,\frac{\lambda^2 + \mu^2}{1 - \mu^2}\,\dot{\mu}, \\ p_w &= \frac{\partial T}{\partial \dot{w}} = c^2(1 + \lambda^2)(1 - \mu^2)\dot{w}. \end{aligned}\right\} \tag{2.5.1}$$

Using these relations to express velocities in terms of momenta and substituting the resulting expressions in the formula (2.3.5) for T, we obtain the transformed kinetic energy in the form

$$\widetilde{T} = \frac{1}{2c^2}\left(\frac{1+\lambda^2}{\lambda^2+\mu^2}\,p_\lambda^2 + \frac{1-\mu^2}{\lambda^2+\mu^2}\,p_\mu^2 + \frac{1}{(1+\lambda^2)(1-\mu^2)}\,p_w^2\right) \tag{2.5.2}$$

and the Hamiltonian in the form

$$H = \widetilde{T} - U, \tag{2.5.3}$$

where $U(\lambda,\mu)$ is given by formula (2.3.4). Now we can write the Hamilton-Jacobi equation, which after multiplication by $c^2(\lambda^2+\mu^2)$ reads

$$(1+\lambda^2)\left(\frac{\partial W}{\partial \lambda}\right)^2 + (1-\mu^2)\left(\frac{\partial W}{\partial \mu}\right)^2 + \left(\frac{1}{1-\mu^2} - \frac{1}{1+\lambda^2}\right)\left(\frac{\partial W}{\partial w}\right)^2 =$$

$$= 2fMc\lambda + 2hc^2(\lambda^2+\mu^2). \tag{2.5.4}$$

We shall seek the solution of this equation in the form

$$W = W_1(\lambda) + W_2(\mu) + W_3(w) \tag{2.5.5}$$

and set

$$\frac{dW_3}{dw} = \alpha_3 = \text{const} \quad (W_3 = \alpha_3 w). \tag{2.5.6}$$

Then the Hamilton-Jacobi equation (2.5.4) will be satisfied if

$$(1+\lambda^2)\left(\frac{dW_1}{d\lambda}\right)^2 = \frac{1}{1+\lambda^2}\,\alpha_3^2 + 2fMc\lambda + 2hc^2\lambda^2 + 2\alpha_2, \tag{2.5.7}$$

$$(1-\mu^2)\left(\frac{dW_2}{d\mu}\right)^2 = -\frac{1}{1-\mu^2}\,\alpha_3^2 + 2hc^2\mu^2 - 2\alpha_2, \tag{2.5.8}$$

where α_2 is a new constant. From these relations one can readily obtain $W_1(\lambda)$ and $W_2(\mu)$ by quadratures; this yields W in the form

$$W = \int \frac{\sqrt{L(\lambda)}}{1+\lambda^2}\,d\lambda + \int \frac{\sqrt{M(\mu)}}{1-\mu^2}\,d\mu + \alpha_3 w, \tag{2.5.9}$$

where

$$\left.\begin{aligned} L(\lambda) &= \alpha_3^2 + 2(1+\lambda^2)(hc^2\lambda^2 + fMc\lambda + \alpha_2), \\ M(\mu) &= -\alpha_3^2 + 2(1-\mu^2)(hc^2\mu^2 - \alpha_2). \end{aligned}\right\} \tag{2.5.10}$$

We can now complete the integration of the problem by quadratures, returning to formulas (2.4.6). Unfortunately, the equations obtained in this way cannot be inverted directly (i.e., solved with respect to the coordinates). For this reason we are forced to deviate somewhat from the traditional Hamilton-Jacobi method. Namely, we shall use simultaneously the relations (2.4.6) and (2.5.1) for the momenta. This yields

$$\left.\begin{aligned} p_\lambda &= \frac{\partial W}{\partial \lambda} = \frac{\sqrt{L(\lambda)}}{1+\lambda^2} = c^2\,\frac{\lambda^2+\mu^2}{1+\lambda^2}\,\dot{\lambda}, \\ p_\mu &= \frac{\partial W}{\partial \mu} = \frac{\sqrt{M(\mu)}}{1-\mu^2} = c^2\,\frac{\lambda^2+\mu^2}{1-\mu^2}\,\dot{\mu}, \\ p_w &= \frac{\partial W}{\partial w} = \alpha_3 = c^2(1+\lambda^2)(1-\mu^2)\dot{w}. \end{aligned}\right\} \tag{2.5.11}$$

Now let us introduce a new independent variable τ, related to t via

$$dt = c^2(\lambda^2+\mu^2)d\tau. \tag{2.5.12}$$

Then equations (2.5.11) become

$$\left.\begin{aligned} \frac{d\lambda}{d\tau} &= \sqrt{L(\lambda)}, \\ \frac{d\mu}{d\tau} &= \sqrt{M(\mu)}, \\ \frac{dw}{d\tau} &= \alpha_3\,\frac{\lambda^2+\mu^2}{(1+\lambda^2)(1-\mu^2)}. \end{aligned}\right\} \tag{2.5.13}$$

The first two equations immediately yield the quadratures

$$\int \frac{d\lambda}{\sqrt{L(\lambda)}} = \tau + \beta_\lambda, \quad \int \frac{d\mu}{\sqrt{M(\mu)}} = \tau + \beta_\mu, \tag{2.5.14}$$

where β_λ and β_μ are new arbitrary constants. If we succeed in inverting these quadratures, i.e., in obtaining explicit expressions for $\lambda(\tau)$ and $\mu(\tau)$, the right-hand side of the third equation in (2.5.13) becomes an explicit function of τ, and then

$$w = \alpha_3 \int \frac{\lambda^2+\mu^2}{(1+\lambda^2)(1-\mu^2)}\,d\tau + \beta_w, \tag{2.5.15}$$

which together with $\lambda(\tau)$ and $\mu(\tau)$ gives a parametric (the parameter being τ) equation of the trajectory (β_w is the last of the needed arbitrary constants). Finally, from (2.5.12) we obtain the explicit dependence of t on the same parameter τ:

$$t = c^2 \int (\lambda^2+\mu^2)d\tau, \tag{2.5.16}$$

which completes the solution of our problem.

6. The region of motion of a satellite

Thus, continuing our study, the first task at hand is to invert the quadratures (2.5.14), which unfortunately is far from easy. Since the polynomials under the square roots are of degree 4, the inversion can be accomplished by means of *Jacobi elliptic functions*, but, depending on the values of the roots of the polynomials, each time in a different way. Difficulties also arise when one attempts to calculate the roots of polynomials of degree 4 in terms of the coefficients of the polynomial, although for a qualitative study of trajectories such a calculation can be avoided.

It follows from the definition of the coordinate system used that the variables range in the domain

$$\left.\begin{aligned} -\infty < w < \infty,\\ 0 \leq \lambda < \infty,\\ -1 \leq \mu \leq 1, \end{aligned}\right\} \tag{2.6.1}$$

and hence when we analyze the motion we need to pay attention only to the values of w, λ, μ that fall in the domain (2.6.1) where the motion is defined. With this in mind, let us investigate qualitatively a typical "satellite" trajectory.

Let the value of the energy constant satisfy $h < 0$. Then $M(\mu) \to \infty$ as $\mu \to \pm\infty$ and $M(\mu) < 0$ for $\mu = \pm 1$ (this is readily seen from the expression (2.5.10) of the polynomial $M(\mu)$). Since for actual motions it necessarily holds that $M(\mu) > 0$, as well as $L(\mu) > 0$ (otherwise we would get complex values for coordinates and velocities), it follows that in the interval $-1 < \mu < 1$ of definition of μ there must be two real roots of $M(\mu)$, μ_1 and μ_2. The polynomial $M(\mu)$ is biquadratic, and consequently to each positive root corresponds a negative root of the same modulus. We conclude that the graph of $M(\mu)$ has the shape shown in Figure 2.3.

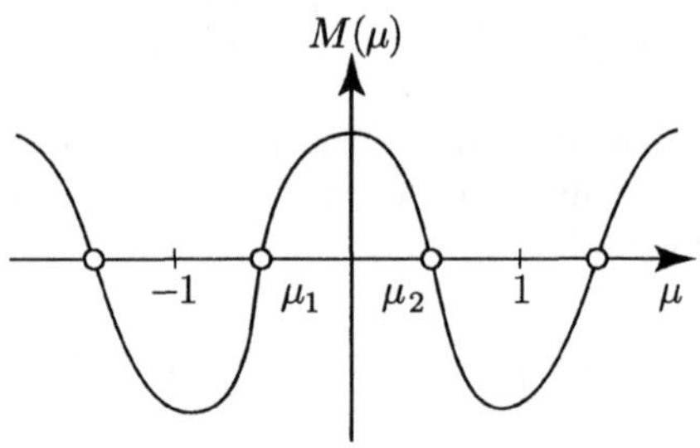

FIGURE 2.3. Graph of the polynomial $M(\mu)$

In an actual motion $-1 \leq \mu_1 \leq \mu \leq \mu_2 \leq 1$, $\mu_1 = -\mu_2$, and the values $\mu = \mu_1$ and $\mu = \mu_2$ correspond to one and the same hyperboloid; a negative [resp., positive] value of μ corresponds to a motion in the region $z < 0$ (Southern hemisphere) [resp., $z > 0$ (Northern hemisphere)]. Therefore, in the motion under consideration

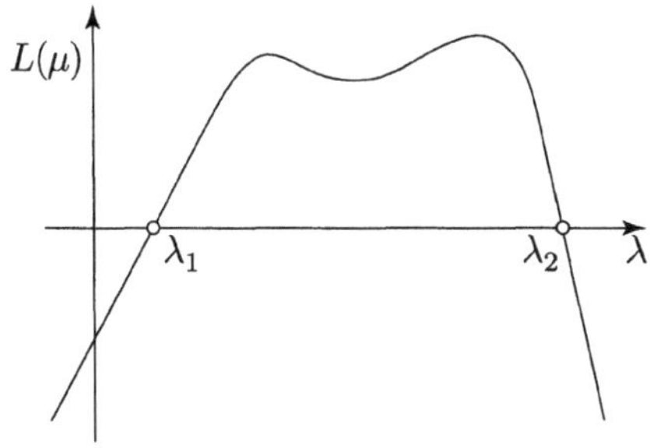

FIGURE 2.4. Graph of the polynomial $L(\lambda)$ (case of two real roots)

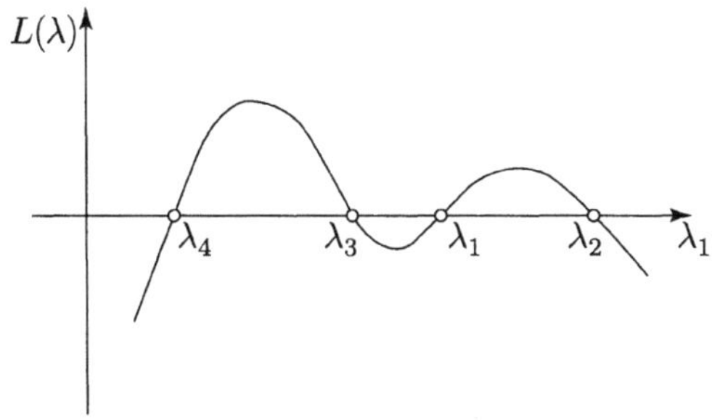

FIGURE 2.5. Graph of the polynomial $L(\lambda)$ (case of four real roots)

$\mu^2 \leq \mu_1^2 = \mu_2^2$, which corresponds to motion "outside" the hyperboloid $\mu^2 = \mu_1^2 = \mu_2^2$.

Now let us examine the polynomial $L(\lambda)$ (see (2.5.10)), which is written in more detail as

$$L(\lambda) = 2hc^2\lambda^4 + 2fMc\lambda^3 + (2hc^2 + 2\alpha_2)\lambda^2 + 2fMc\lambda + \alpha_3^2 + 2\alpha_2.$$

Under the same assumption that $h < 0$, it holds that $L \to -\infty$ as $\lambda \to \pm\infty$. Further, from the analysis of the polynomial $M(\mu)$ it is clear that $M(0) = -\alpha_3^2 - 2\alpha_2 > 0$, whence $\alpha_3^2 + 2\alpha_2 < 0$, and the more so $\alpha_2 < 0$. But then in $L(\lambda)$ all the coefficients of even powers of λ (including the free term) are negative, while all the coefficients of odd powers are positive. Consequently, $L(\lambda) < 0$ for $\lambda \leq 0$. All real roots lie on the half-line $\lambda > 0$. There are either two or four such (real) roots. By Viète's theorem, the product of the four roots is equal to the free term divided by the coefficient of the highest order term:

$$\lambda_1\lambda_1\lambda_3\lambda_4 = \frac{\alpha_3^2 + 2\alpha_2}{2hc^2} > 0,$$

and so there are either two or four real positive roots. If there are only two real roots, say, λ_1 and λ_2, the graph of $L(\lambda)$ has the shape shown in Figure 2.4. Hence, for an actual motion $L(\lambda) > 0$, which is possible only for $\lambda_1 \le \lambda \le \lambda_2$, i.e., the motion is confined between the ellipsoids $\lambda = \lambda_1$ and $\lambda = \lambda_2$. If there are four real roots (Figure 2.5.), then nonetheless during the whole motion λ remains between two roots, say, between λ_1 and λ_2.

Let us sum up.

Under the only condition that $h < 0$, *the motion always takes place outside the hyperboloid* $\mu_2^2 = \mu_*^2$ $(-\mu_* \le \mu \le \mu_*)$ *and inside the layer bounded by the two ellipsoids* $\lambda = \lambda_1$ *and* $\lambda = \lambda_2$ $(0 < \lambda_1 \le \lambda \le \lambda_2)$, i.e., the motion takes place inside the ellipsoidal annulus shown in Figure 2.6.

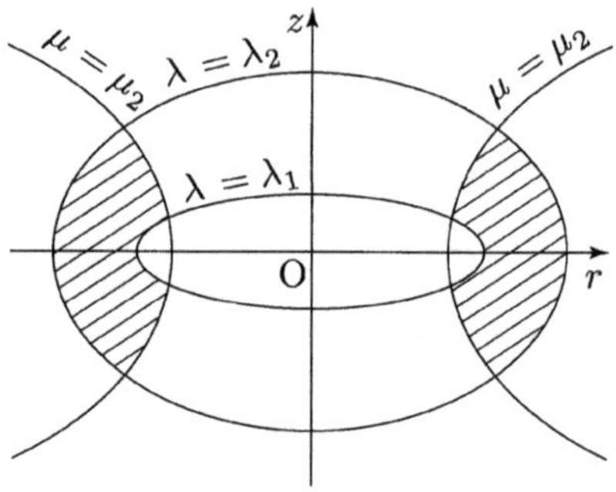

FIGURE 2.6. Region of motion of the satellite

This result can be better grasped if one resorts to the approximate (evolution) portrait of the motion obtained earlier. Recall that in the orbital plane the osculating ellipse executes a slow precession, retaining its shape, so that the distance from the satellite to the center of the Earth is never smaller than r_π or larger than r_α. But the orbital plane itself executes a slow precession around the Earth's axis, so that the annulus $r = r_\pi$ sweeps a band of angular width $2i$ on the sphere of radius $r = r_\pi$ (where i is the inclination angle of the orbital plane to the equatorial plane). Similarly, the annulus $r = r_\alpha$ sweeps a band of the same angular width on the sphere of radius $r = r_\alpha$. One concludes that the evolution motion takes place in the spheroidal annulus shown in Figure 2.7.

This annulus is very similar to the ellipsoidal annulus described above, which represents the exact region where motion is defined, but nonetheless is different from it. This difference is extremely small. Indeed, the eccentricity of the, say, inner ellipsoid is $\varepsilon_1 = \sqrt{1/(1+\lambda_1^2)}$. But from (2.3.2) one derives the relation $r_e = c^2(1+\lambda^2)$, where r_e is the distance from the satellite to the center of the Earth at the moment when it crosses the equatorial plane. Therefore,

$$\varepsilon_1^2 \sim \frac{c^2}{r_e^2} = \left(\frac{R^2}{r_e^2}\right)|I_2| < |I_2| \sim 10^{-3},$$

and thus the ellipsoidal annulus is very close to a spheroidal one.

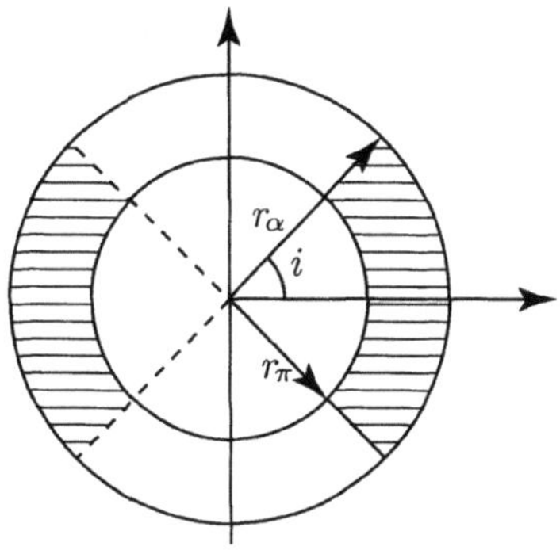

FIGURE 2.7. Region of evolution motion of the satellite

A careful analysis of the possible cases of bounded motion ($h < 0$) reveals not only satellite trajectories, which lie entirely above the Earth's surface, but also ballistic trajectories, which lie only partially above the Earth's surface (for such trajectories the inner boundary ellipsoid $\lambda = \lambda_1$ lies under the Earth's surface). It is also possible that a trajectory lies *entirely* on a single ellipsoid (and not in the space between two ellipsoids); this corresponds to the existence of multiple (nonsimple) roots for the polynomial $L(\lambda)$ (for example, when $\lambda_1 = \lambda_2$).

Referring the reader for details to the original source [2.1], we confine ourselves to discussing a particular case. It will help us understand how the analysis of integrable problems can be pursued to "almost" the final stage – explicit formulas which, for example, in the problem at hand, give the dependences $\lambda(\tau)$ and $\mu(\tau)$. Whenever such explicit formulas are available, they allow one to carry out a detailed analysis of the properties of trajectories and push this analysis to the realm of numerical results, the latter being what we mean by the "final stage!"

To these ends we will need some information on Jacobi elliptic functions.

7. Jacobi's elliptic functions

An integral of the type $\int dx/\sqrt{G(x)}$, where $G(x)$ is a polynomial of degree 3 or 4, can be reduced to an integral of the following form:

$$F(k, \varphi) = \int_0^\varphi \frac{d\varphi}{\sqrt{1 - k^2 \sin^2 \varphi}} = \int_0^z \frac{dz}{\sqrt{(1 - z^2)(1 - k^2 z^2)}} = \overline{F}(k, z). \qquad (2.7.1)$$

This integral is called the *Legendre normal elliptic integral of the first kind*. Here $0 \le k \le 1$; the number k is called the *modulus of the elliptic integral*. The definite integral

$$K(k) = F(k, \pi/2) \qquad (2.7.2)$$

is called the *complete integral of the first kind*. $K(k)$ is a monotonically increasing function of its argument, and $K(0) = \pi/2$, $K(1) = \infty$. The function $K(k)$ is tabulated; for large values of k it is convenient to use the expansion

$$K(k) = \frac{\pi}{2}\left\{1 + \left(\frac{1}{2}\right)^2 k^2 + \left(\frac{1\cdot 3}{2\cdot 4}\right)^2 k^4 + \cdots + \left(\frac{(2n-1)!!}{2n!!}\right)^2 k^{2n} + \cdots\right\}. \tag{2.7.3}$$

Here, as usual, the double factorial denotes the product of all odd [resp., even] integers from 1 to $2n-1$ [resp., $2n$] inclusively. Let us denote

$$u = F(k, \varphi). \tag{2.7.4}$$

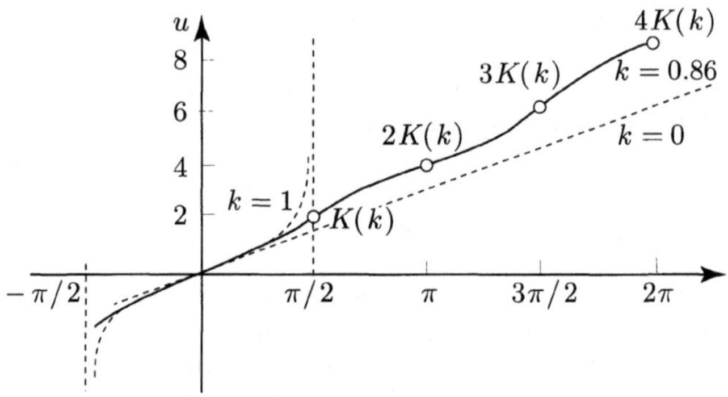

FIGURE 2.8. Graph of the function $u(\varphi)$

The dependence of u on φ for a few values of k is shown in Figure 2.8. The problem of the inversion of the elliptic integral $F(k, \varphi)$ is by definition the problem of finding the inverse dependence of φ on u instead of the known dependence of u on φ. The quantity φ, given as a function of u and k via relation (2.7.4), is called *amplitude* and is denoted by

$$\varphi = \operatorname{am} u. \tag{2.7.5}$$

The *elliptic sine function* ($\operatorname{sn} u$) is defined to be the sine of the amplitude, i.e.,

$$\operatorname{sn} u = \sin(\operatorname{am} u). \tag{2.7.6}$$

Similarly one defines the *elliptic cosine function*

$$\operatorname{cn} u = \cos(\operatorname{am} u). \tag{2.7.7}$$

One also introduces the function

$$\operatorname{dn} u = \sqrt{1 - k^2 \operatorname{sn} u}, \tag{2.7.8}$$

called the *delta amplitude*. The functions $\operatorname{sn} u$, $\operatorname{cn} u$, and $\operatorname{dn} u$ are called the *Jacobi elliptic functions*.

A number of problems of mechanics, starting with that of the motion of a pendulum, are solved in terms of these functions. The problems of the dynamics of space flight considered in this and the next essay also lead to Jacobi elliptic functions. The properties of the elliptic sine and cosine functions resemble in part those of the trigonometric sine and cosine functions. However, we should emphasize that an elliptic functions is actually a functions of *two* arguments, u and k. Usually k is omitted from notations, but the functions do depend on it! The functions $\operatorname{sn} u$ and $\operatorname{cn} u$ are periodic with period $K(k)$, while $\operatorname{dn} u$ is periodic with period $2K(k)$; also, $\operatorname{sn} u$ and $\operatorname{cn} u$ take values in the segment $[-1, 1]$, while $\operatorname{dn} u$ takes values in the segment $[\sqrt{1-k^2}, 1]$.

The following table shows a few values of the elliptic functions.

	$u = 0$	$u = K$	$u = 2K$	$u = 3K$	$u = 4K$
$\operatorname{sn} u$	0	1	0	-1	0
$\operatorname{cn} u$	1	0	-1	0	1
$\operatorname{dn} u$	1	$\sqrt{1-k^2}$	1	$\sqrt{1-k^2}$	1

Figure 2.9 shows the graphs of the functions $\operatorname{sn} u$, $\operatorname{cn} u$, and $\operatorname{dn} u$ for some values of k. The quantity $K(k)$ plays the same role for elliptic functions as the quantity $\pi/2$ for trigonometric ones. However, as we already mentioned, K may take any value in the interval $\pi/2 \le L < \infty$, depending on the value of the modulus k. For $k = 0$ the elliptic functions degenerate into trigonometric functions of period 2π:

$$k = 0: \quad \operatorname{sn} u = \sin u, \quad \operatorname{cn} u = \cos u, \quad \operatorname{dn} u = 1.$$

For $k = 1$ the elliptic functions are expressible in terms of nonperiodic functions, specifically, in terms of hyperbolic functions:

$$k = 1, \quad \operatorname{sn} u = \tanh u, \quad \operatorname{cn} u = (\cosh u)^{-1}, \quad \operatorname{dn} u = (\cosh u)^{-1}.$$

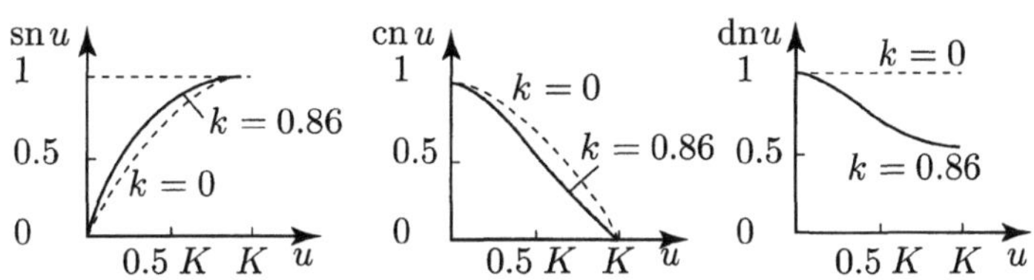

FIGURE 2.9. Graphs of the functions $\operatorname{sn} u$, $\operatorname{cn} u$, $\operatorname{dn} u$

It is useful to list the following properties of the Jacobi elliptic functions:

$$\left.\begin{array}{c} \operatorname{sn}(-u) = -\operatorname{sn} u \qquad \operatorname{cn}(-u) = \operatorname{cn} u \\ \operatorname{dn}(-u) = \operatorname{dn} u, \\ \operatorname{sn}(2K-u) = \operatorname{sn} u, \qquad \operatorname{cn}(2K-u) = -\operatorname{cn} u, \\ \operatorname{dn}(2K-u) = \operatorname{dn} u, \qquad \operatorname{sn}^2 u + \operatorname{cn}^2 u = 1, \\ \operatorname{dn}^2 u + k^2 \operatorname{sn}^2 u = 1, \qquad \operatorname{dn}^2 u - k^2 \operatorname{cn}^2 u = 1 - k^2. \end{array}\right\} \tag{2.7.9}$$

Also, the following differentiation rules hold:

$$\left.\begin{array}{l} \dfrac{d}{du}\operatorname{sn} u = \operatorname{cn} u \operatorname{dn} u, \\ \dfrac{d}{du}\operatorname{cn} u = -\operatorname{sn} u \operatorname{dn} u, \\ \dfrac{d}{du}\operatorname{dn} u = -k^2 \operatorname{sn} u \operatorname{cn} u. \end{array}\right\} \tag{2.7.10}$$

We are now ready to study an interesting particular satellite motion.

8. The motion of a polar Earth satellite

Consider formulas (2.5.13) and (2.5.10). If $\alpha_3 = 0$, then $dw/d\tau = 0$, $w \equiv w_0$, i.e., the motion takes place in the fixed meridian plane. Moreover,

$$L(\lambda) = -2|h|c^2(1+\lambda^2)\left[\lambda^2 - \frac{fM}{|h|c}\lambda + \frac{1}{c^2}\left|\frac{\alpha_2}{h}\right|\right], \tag{2.8.1}$$

and

$$M(\mu) = -2|h|c^2(1-\mu^2)\left[\mu^2 - \frac{1}{c^2}\left|\frac{\alpha_2}{h}\right|\right]. \tag{2.8.2}$$

From (2.5.14) one derives the relation

$$\tau + \beta_\mu = \int_0^\mu \frac{d\mu}{\sqrt{M(\mu)}} = \int_0^\mu \frac{d\mu}{\sqrt{2|\alpha_2|(1-\mu^2)\left[1 - c^2\left|\dfrac{h}{\alpha_2}\right|\mu^2\right]}}.$$

Denoting $u = \sqrt{2|\alpha_2|}(\tau + \beta_\mu)$ and $c^2|h/\alpha_2| = k_\mu^2$, we get

$$u = \int_0^\mu \frac{d\mu}{\sqrt{(1-\mu^2)(1-k_\mu^2\mu^2)}}.$$

Examining the polynomial $M(\mu)$ in its general form one concludes (Figure 2.3) that among its four roots two are smaller than 1 in modulus, while the other two are strictly larger than 1. Consequently, $\frac{1}{c^2}\left|\frac{\alpha_2}{h}\right| > 1$ and $k_\mu < 1$. Moreover, k_μ^2 is of the same order as the Earth's oblateness: $k_\mu^2 \sim I_2^2$. Indeed,

$$|h| \sim V^2, \qquad |\alpha_2| \sim \left(\frac{d\mu}{d\tau}\right)^2 \sim \left\{\frac{d\left(\dfrac{R}{c}\right)}{\dfrac{dt}{R^2}}\right\}^2,$$

whence

$$c^2\left|\frac{h}{\alpha_2}\right| \sim \frac{c^4}{R^4} = I_2^2.$$

But if $k < 1$ the last integral is already an elliptic integral of the first kind in Legendre normal form, which allows us right away to write

$$\mu = \operatorname{sn}\left(\sqrt{2|\alpha_2|}(\tau + \beta_\mu)\right) \tag{2.8.3}$$

The period of the coordinate μ with respect to τ is

$$T_\mu = \frac{4K(k_\mu)}{\sqrt{2|\alpha_2|}}.$$

Now consider the coordinate $\lambda(\tau)$. The real roots of the polynomial $L(\lambda)$ given by (2.8.1) are

$$\lambda_1 = \frac{fM}{2|h|c} - \sqrt{\left(\frac{fM}{2|h|c}\right)^2 - \frac{1}{c^2}\left|\frac{\alpha_2}{h}\right|},$$

$$\lambda_2 = \frac{fM}{2|h|c} + \sqrt{\left(\frac{fM}{2|h|c}\right)^2 - \frac{1}{c^2}\left|\frac{\alpha_2}{h}\right|},$$

and the integral in question can be written in the form

$$\int_{\lambda_1}^{\lambda} \frac{d\lambda}{\sqrt{(\lambda^2+1)(\lambda_2-\lambda)(\lambda-\lambda_1)}} = \sqrt{2c^2|h|}\,(\tau + \beta_\lambda). \tag{2.8.4}$$

This integral can also be reduced to normal form, but it takes more effort. Omitting the calculations, we give the final result:

$$\lambda = \frac{A + B\operatorname{cn}\sigma(\tau+\beta_\lambda)}{C + D\operatorname{dn}\sigma(\tau+\beta_\lambda)} \tag{2.8.5}$$

where

$$A = \lambda_2 - \lambda_1\sqrt{\frac{\lambda_2^2+1}{\lambda_1^2+1}}, \qquad B = \lambda_2 + \lambda_1\sqrt{\frac{\lambda_2^2+1}{\lambda_1^2+1}},$$

$$C = 1 - \sqrt{\frac{\lambda_2^2+1}{\lambda_1^2+1}}, \qquad D = 1 + \sqrt{\frac{\lambda_2^2+1}{\lambda_1^2+1}},$$

$$\sigma = \sqrt{2c^2|h|\sqrt{(\lambda_1^2+1)(\lambda_2^2+1)}}.$$

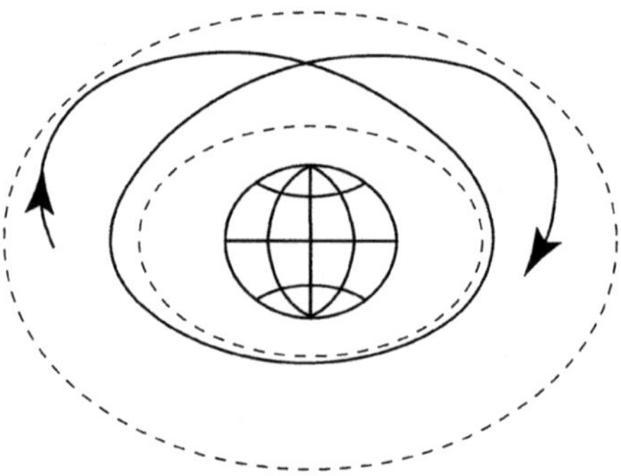

FIGURE 2.10. Trajectory of a polar satellite

Hence, the period of the function $\lambda(\tau)$ with respect to τ equals

$$T_\lambda = \frac{4K(k_\lambda)}{\sigma};$$

moreover,

$$k_\lambda^2 = \frac{1}{4}\,\frac{(\lambda_2-\lambda_2)^2 - \left[\sqrt{\lambda_2^2+1} - \sqrt{\lambda_1^2+1}\right]^2}{\sqrt{(\lambda_2^2+1)(\lambda_1^2+1)}}.$$

Since $T_\lambda \neq T_\mu$, the motion is not periodic: when λ reaches again the initial value, μ is slightly "shifted." Accordingly, the trajectory has a "rosétte" shape (Figure 2.10). The difference $T_\lambda - T_\mu$ characterizes the rate of the motion of the "semi-major axis of the osculating ellipse" of the orbit.

Let us point out that there exist almost circular ("quasi-circular") orbits $\lambda = \text{const}$, corresponding to the equality $\lambda_1 = \lambda_2$.

All that was said above is very likely sufficient for a first acquaintance with our problem; for details the interested reader is referred to the original sources, a complete list of which can be found in V. G. Demin's book [2.3].

Third Essay

Yet Another Reincarnation of an Old Problem

A man of culture gnaws a bone only slightly, and then throws it under the table.

Fafik[1]

(from *Thoughts of great and average men, and of Fafik the dog*)

1. And what sort of a problem is that?

This essay is devoted to the following problem. Suppose that a particle moves in the Newtonian gravitational field of an attracting center, but, in addition, the particle is subject to a "perturbing acceleration" of constant magnitude and direction (Figure 3.1). This problem was considered already by Lagrange, then in the XIXth century by the French mathematicians Charlier and Saint-Germain, and also by I. V. Meshcherskiĭ. They have shown that the problem is integrable by quadratures, but did not carry out the integration to the end, since the form of the quadratures that they obtained was not convenient for inversion. After that the above problem was thrown under the table and apparently saw no further development.

This integrable, but nearly-forgotten problem was resurrected to a new life in the 1960s. "Low-thrust" (ion, plasma, etc.) engines are being designed and developed for the space flights of the not-so-distant future; the task of such engines is to impart a spaceship a small reactive acceleration for a very long time interval (for more details see the following essays). One can imagine the situation in which an engine generates an acceleration of constant magnitude and direction. Then the motion of the spaceship can be described in the framework of the aforementioned classical problem of Charlier and Saint-Germain.

An important application of this problem is the study of the influence of the *radiation pressure* (pressure of light) on the orbit of a satellite. As is know, solar radiation generates a small, yet noticeable pressure on the illuminated parts of a body; as a matter of fact, there are satellites for which the force of radiation pressure is the main perturbing force. For example, American satellites in the Echo family are huge inflated balloons (up to 20–40 meters in diameter). Due to the small mass and large surface of such satellites, the perturbations of their motion caused by the radiation pressure are considerable. Over a (relatively) short time interval,

[1]Fafik the terrier is a contributor of the Polish journal "Przekrój."

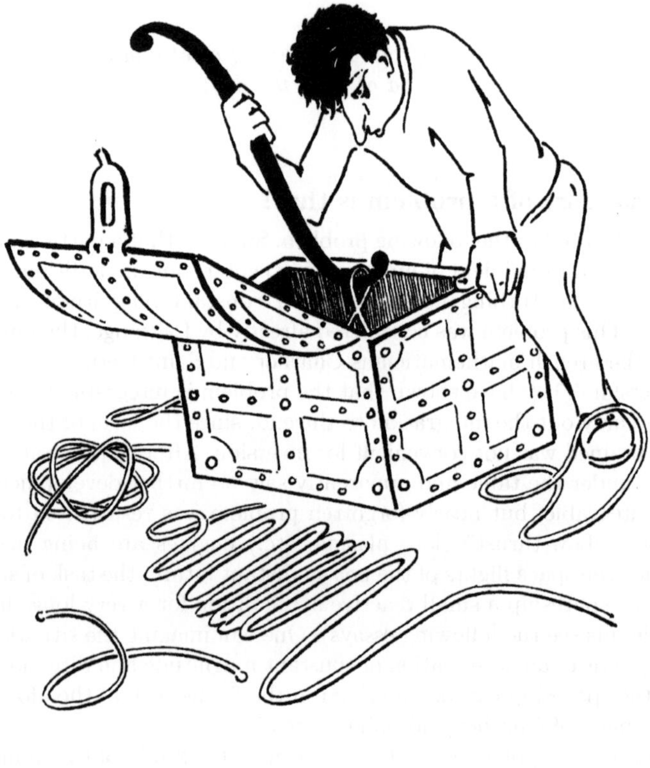

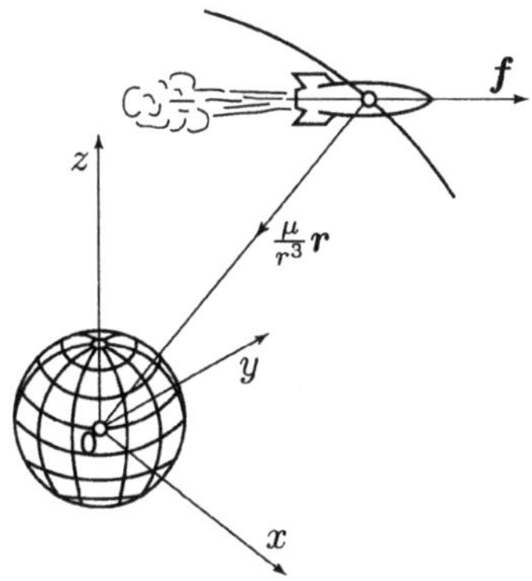

FIGURE 3.1. Diagram of acting forces

during which the Sun does not shift considerably in its annual motion on the celestial sphere, one can assume that the acceleration produced by the radiation pressure is constant in magnitude and direction (pointing "from the Sun to the body"). Then we are led again to the model of motion described by our classical problem.

It is therefore not surprising that in recent years the problem of motion in a Newtonian field in the presence of an additional acceleration vector was rescued from the attic and cleaned of the dust of centuries [3.1]–[3.6], [3.9].[2] The exposition below follows mainly the author's papers [3.1], [3.2].

2. Briefly on the equations of motion and their integration

Let us attach to the Newtonian center of attraction O the Cartesian coordinate system $Oxyz$ (Figure 3.1) and introduce dimensionless variables by the relations

$$\{r\} = \frac{\{r_D\}}{r_0}, \quad \{v\} = \frac{\{V_D\}}{\sqrt{r_0 g_0}}, \quad t = \frac{t_D}{\sqrt{r_0/g_0}}, \quad \{f\} = \frac{\{f_D\}}{g_0},$$

where $g_0 = \mu/r_0^2$ and the remaining notations are as follows: $\{r\}$, $\{v\}$, $\{f\}$, and t are the dimensionless coordinates, components of the velocity, components of the reactive acceleration, and time, respectively; $\{r_D\}$, $\{V_D\}$, $\{f_D\}$, and t_D are the corresponding dimensional variables; r_0 is a fixed (for example, initial) distance to the center of attraction; and g_0 is the acceleration of gravity at distance r_0 from the center of attraction.

[2]Reference [3.9] is a joke of the author. The following stanza from Marhsak's *The trunk* is aluded to: *An idiot, walking on a swamp, Found, hanging in an aspen tree, a trunk. At which he said: Oh! Lucky me, Now the richest of all I will be ...*, and so on. In the end, only a piglet tail is found in the trunk.

Let us direct the x-axis of the fixed coordinate system along the constant reactive acceleration vector $\mathbf{f}$. Then the equations of motion take on the form

$$\ddot{x} = -\frac{x}{r^3} + f, \quad \ddot{y} = -\frac{y}{r^3}, \quad \ddot{z} = -\frac{z}{r^3}, \quad r = \sqrt{x^2 + y^2 + z^2}\,. \tag{3.2.1}$$

If the term f were absent in the first equation, we would simply have the equations of Keplerian motion. As it will be become evident soon, the presence of the parameter f results in an abundant diversity of motions.

Equations (3.2.1) possess the following first integrals:

the *energy integral*,

$$\frac{1}{2}\left(\dot{x}^2 + \dot{y}^2 + \dot{z}^2\right) - \frac{1}{r} - fx = h; \tag{3.2.2}$$

the *area integral* (relative to the x-axis, i.e., the direction of the vector $\mathbf{f}$),

$$z\dot{y} - y\dot{z} = -k_0; \tag{3.2.3}$$

a third integral, which generalizes the so-called "*Laplace integral*" (or "*Runge-Lenz integral*") of Keplerian motion,

$$\dot{x}r\dot{r} - \frac{x}{r} - \frac{3}{2}fx^2 - \frac{1}{2}fr^2 - 2hx = c. \tag{3.2.4}$$

This integral will be useful later in our exposition; here we merely wish to emphasize that the existence of sufficiently many first integrals of the equations of motion (in the present case – three) is a condition for the integrability of these equations. Hence, when one sets out to integrate some system of equations it is useful to find, whenever possible, all its independent (i.e., that are not consequences of one another) first integrals. The existence of the three first integrals (3.2.2)–(3.2.4) allows us to reduce the problem studied here to quadratures.

Of course, as in the problem of two fixed centers, the integration could be carried out by the Hamilton-Jacobi method, by selecting an appropriate curvilinear coordinate system. However, since we do have the three first integrals (3.2.2)–(3.2.4) at our disposal, we may attempt to carry out the integration directly, by transforming these integrals to the requisite form. Namely, let us introduce the new variables

$$u = r - x, \quad v = r + x, \tag{3.2.5}$$

so that

$$r = \frac{1}{2}(u + v), \quad x = \frac{1}{2}(u - v), \tag{3.2.6}$$

and also a third variable φ, defined via the relations

$$\cos\varphi = \frac{y}{r_1}, \quad \sin\varphi = \frac{z}{r_1}, \quad r_1 = \sqrt{y^2 + z^2}\,. \tag{3.2.7}$$

Now in the first integrals (3.2.2)–(3.2.4) replace x, y, z by the new variables u, v, φ given by (3.2.5)–(3.2.7) and then replace t by a new parameter τ which increases monotonically with t according to the rule

$$d\tau = \frac{dt}{r}. \tag{3.2.8}$$

Then the integrals (3.2.2)–(3.2.4) can be readily solved with respect to the derivatives $du/d\tau$, $dv/d\tau$, $d\varphi/d\tau$, and a subsequent quadrature completes the integration. Omitting the "kitchen work" needed to carry out the integration, we will give here only the final results.

It turns out that the motion obeys the following quadratures:

$$\int \frac{dv}{\sqrt{V(v)}} = \tau + c_1, \quad \int \frac{du}{\sqrt{U(u)}} = \tau + c_2, \tag{3.2.9}$$

where

$$\left.\begin{aligned} V(v) &= fv^3 + 2hv^2 + 2(1+c)v - k_0^2, \\ U(u) &= -fu^3 + 2hu^2 + 2(1-c)u - k_0^2. \end{aligned}\right\} \tag{3.2.10}$$

As we already know, integrals of the type (3.2.9) can be inverted by means of Jacobi elliptic functions. This inversion yields the parametric equations of motion in a rotating plane that contains the radius vector $\mathbf{r}$ and the x-axis. The angle φ of rotation of this plane around the x-axis is given, according to (3.2.3) and (3.2.7), by the formula

$$\varphi - \varphi_0 = k_0 \int_0^\tau \frac{r\,d\tau}{r^2 - x^2}, \tag{3.2.11}$$

where, of course, r and x are considered as already known functions (as a result of (3.2.6) and the inversion of the integrals (3.2.9)). From relation (3.2.8) we obtain

$$t - t_0 = \int_0^\tau r\,d\tau, \tag{3.2.12}$$

and so the problem is completely reduced to quadratures.

Note that the system of curvilinear coordinates u, v, φ used above in the integration is paraboloidal: the coordinate planes $u =$const, $v =$const are paraboloids of revolution with a common axis – the x-axis of the Cartesian coordinate system. In the plane $\varphi =$const the coordinate grid is given by the family of parabolas $u =$const, $v =$const, as shown in Figure 3.2.

One could have guessed beforehand that the paraboloidal coordinate system is the most natural for the problem at hand. Indeed, sufficiently far from the center of attraction $f \gg 1/r^2$ (since $\mathbf{f}$ is constant and $1/r^2$ can be arbitrarily small). Hence, in the far region the motion will be close to the motion due only to the action of a constant acceleration vector $\mathbf{f}$ (the Newtonian acceleration $1/r^2$ will be small). But it is known that the motion in a homogeneous force field (i.e., in a field such that $\mathbf{f} =$const) takes place along parabolic trajectories (and the axis of these trajectories is directed along the vector $\mathbf{f}$).

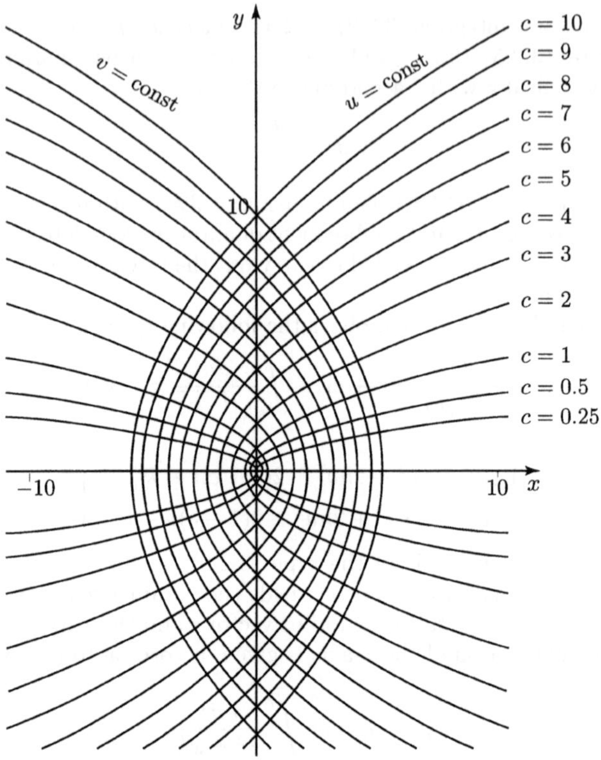

FIGURE 3.2. Paraboloidal coordinate system

3. Plane motion

In what follows we will confine ourselves to the case of plane motion (in the (x, y)-plane). For this case $k_0 = 0$ and $\varphi = \varphi_0$. The polynomials (3.2.10) can be written as

$$\left.\begin{aligned} V(v) &= \ \ fv(v-v_2)(v-v_3), \\ U(u) &= -fu(u-u_2)(u-u_3), \end{aligned}\right\} \tag{3.3.1}$$

and their roots are given by

$$v_1 = 0, \quad v_2 = \frac{-h + \sqrt{h^2 - 2f(1+c)}}{f}, \qquad v_3 = \frac{-h - \sqrt{h^2 - 2f(1+c)}}{f},$$

$$u_1 = 0, \quad u_2 = \frac{h + \sqrt{h^2 + 2f(1-c)}}{f}, \qquad u_3 = \frac{h - \sqrt{h^2 + 2f(1-c)}}{f}.$$

Therefore, the plane problem has the unquestionable virtue that all the roots of the polynomials (3.3.1) (and consequently the moduli of the elliptic functions as well as other parameters describing the motion) are readily expressible in explicit form through the initial conditions (i.e., through the values of the constants h and c). For that reason the classification of motions is conveniently carried out in the (h, c)-plane for a fixed value $f \neq 0$. This is done without special effort, since – let me repeat – the explicit dependences between the coordinates u, v, and τ and the constants h, c are known: $u(\tau, h, c)$, $v(\tau, h, c)$. However, these dependences, described by Jacobi elliptic functions, take different form in different cases (and indeed there are numerous such cases). We shall not give these dependences here. Rather, we shall describe the conclusions of the analysis, referring the reader for details to [3.1] and [3.2]. For the characteristics of trajectories only the fact that the trajectories are given by combinations of such dependences $u(\tau)$ and $v(\tau)$, schematically shown in Figure 3.3, is of importance. The variety of types displayed by these dependences is explained by the fact that for different values of h and c the roots of the polynomials (3.3.1) have different signs and different locations in the complex plane (i.e., are real or complex). Let us add that if $v(\tau) \to \infty$ when $\tau \to \tau^*$, then also the real-valued time $\tau \to \infty$ when $\tau \to \tau^*$, so that v can attain an infinite value only over an infinite time interval.

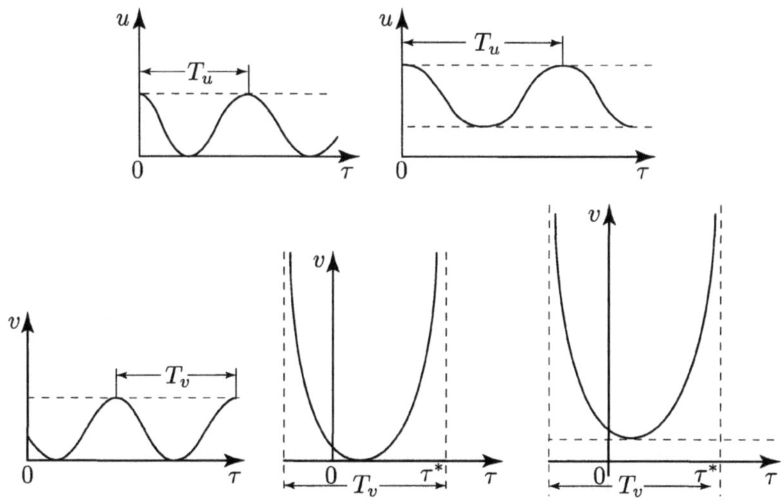

FIGURE 3.3. Possible types of dependences $u(\tau)$ and $v(\tau)$

There are two [resp., three] qualitatively distinct possible types of dependence $u(\tau)$ [resp., $v(\tau)$] (Figure 3.3). The character of the trajectory is determined by the following two factors: (1) the precise combination of functions $u(\tau)$, $v(\tau)$ from among the six logically possible choices (Figure 3.3) that describes the trajectory in question, and (2) the value of the ratio T_u/T_v of the periods of the functions $u(\tau)$ and $v(\tau)$. Since explicit expressions are available for $u(\tau)$ and $v(\tau)$ (in terms of Jacobi functions), the questions (1) and (2) can always be answered.

4. Description of the trajectories of plane motion

The trajectories of our problem can be divided into four basic types:

I. Unbounded self-intersecting trajectories that do not encircle the center of attraction. The range of parameters for trajectories of type I is

$$c < -1.$$

II. Unbounded self-intersecting trajectories that encircle the center of attraction. The range of parameters for these trajectories is

$$-1 < c < 1, \quad \frac{h}{\sqrt{f}} > -\sqrt{2(1+c)}\,.$$

III. Unbounded trajectories with no self-intersection. The range of parameters for these trajectories is

$$c > 1, \quad \frac{h}{\sqrt{f}} > \sqrt{-2(1-c)}\,.$$

IV. Bounded trajectories. Such trajectories exists for the following values of the parameters:

$$-1 < c < 1, \quad \frac{h}{\sqrt{f}} < -\sqrt{2(1+c)}\,.$$

In this region, however, there can also exist trajectories of type I.

In the remaining part of the $(c, h/\sqrt{f})$-plane (specifically, for $c > 1$, $h/\sqrt{f} < \sqrt{-2(1-c)}$) there are no actual motions. The partition of the $(c, h/\sqrt{f})$-plane into the regions I, II, III, IV is shown in Figure 3.4, which also depicts schematically the various types of trajectories (i.e., types of trajectories characteristic for each given region of the $(c, h/\sqrt{f})$-plane). Let us examine the properties of the trajectories corresponding to values of $(c, h/\sqrt{f})$ inside the regions I–IV and on their boundaries. We will pay attention only to the most interesting trajectories.

Trajectories of type I are unbounded and when $t \to \infty$ they tend to some parabola $u = u^*$; such trajectories do not encircle the center of attraction and are sufficiently "smooth;" for large values of c they look like arcs of arbitrary width (including very narrow ones). It is very interesting that there exist "self-returning" trajectories, which lead to some point u_k, v_k with velocity zero; however, there is no equilibrium at the point u_k, v_k (the forces do not vanish at that point), and *the motion continues along the same trajectory, but in the opposite direction*! This is similar to walking for a long time on a path which suddenly dead-ends, so that one is forced to "about-face forward march!" along the same path. Two spaceships launched at different moments of time on such a trajectory will unavoidably collide! (This, of course, is a purely mathematical situation. In real life the captains of the spaceships should manage in time to change the magnitude or the direction

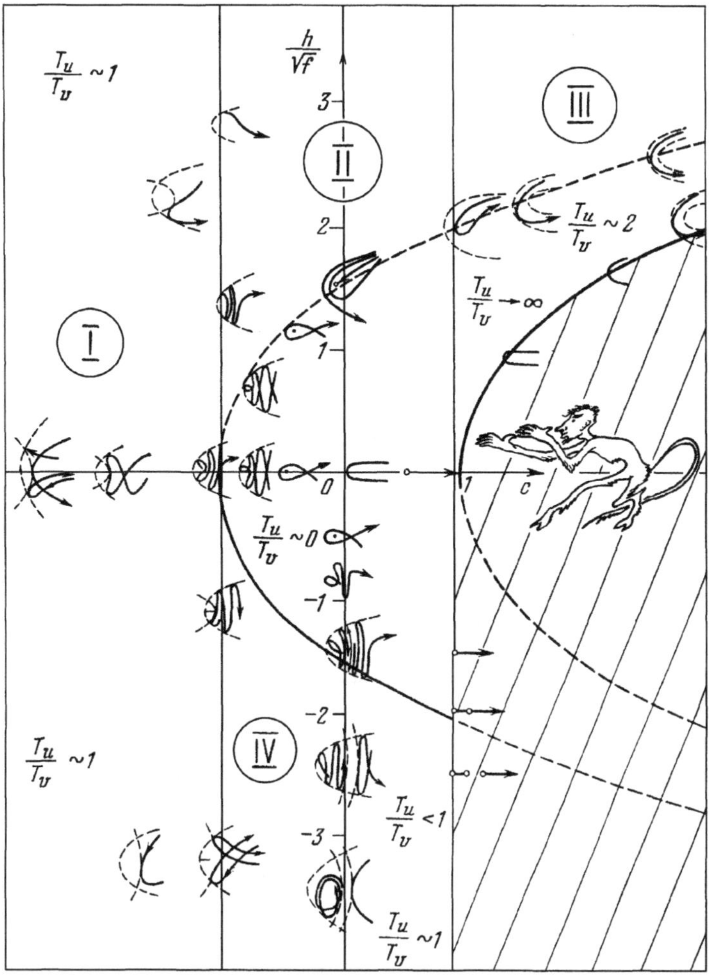

FIGURE 3.4. Partition of the $(c, h/\sqrt{f})$-plane into regions I, II, III, IV (including the shape of trajectories)

of the thrust **f** and safely pass clear of one another; but how to do this is already another problem). Figure 3.5 shows the trajectories just described. If the value of c is not very large, the trajectories may self-intersect (Figure 3.6) and even have a snake-like shape (Figure 3.7).

Trajectories of type II differ from the previous ones only by the fact that they may encircle the center of attraction ("self-returning" trajectories are therefore excluded). Figures 3.8–3.11 show some trajectories of this type.

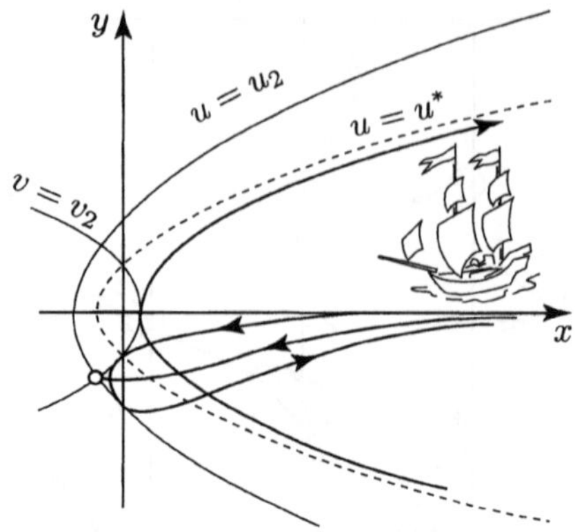

FIGURE 3.5. Trajectories of type I

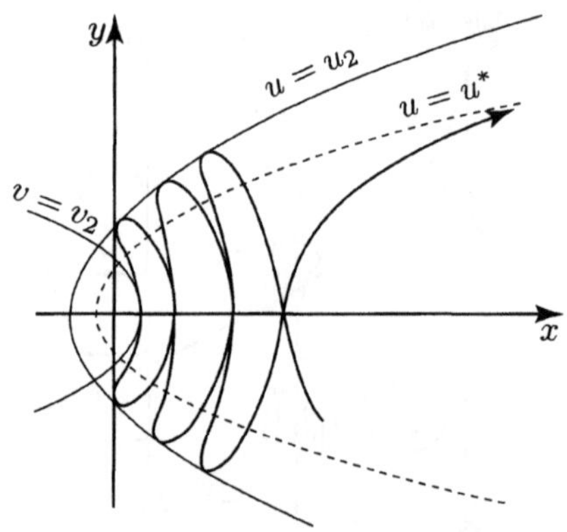

FIGURE 3.6. Trajectory of type I with self-intersections

Trajectories of type III are very "smooth," have no self-intersections (Figures 3.12, 3.13) and encircle the center of attraction. Among them there are *purely parabolic* trajectories, in particular, radial trajectories (on which the motion takes place along the direction of the vector **f**).

Let us return to the snake-like trajectory of type II (Figure 3.11). It has a noticeable "thickening" of the loop in its "middle," so to say, section. It is interesting to observe what happens with such a trajectory when we pass from region II to region IV (of bounded motions). On the boundary between these two regions our trajectory bifurcates into three new ones (Figure 3.14). First of all, at the place of thickening there arises a *periodic trajectory* $v = \text{const} = v_3$. The particle moves along a *bounded segment of this parabola*, namely, along the segment enclosed inside some parabola $u = u_2$. It oscillates back and forth along the curve $v = v_3$ between two points with coordinates $u = u_2$, $v = v_3$. To the

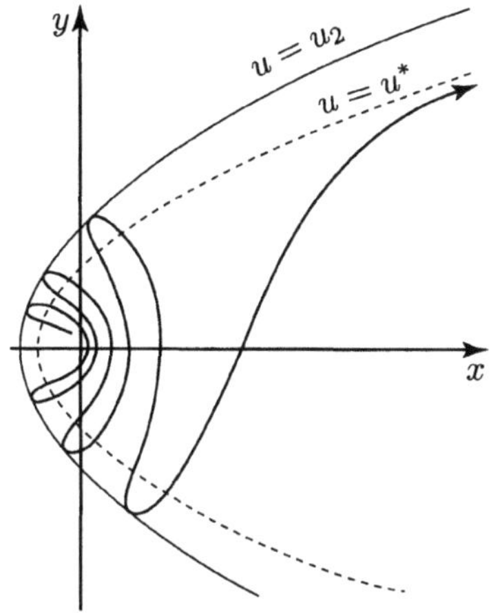

FIGURE 3.7. Snake-like trajectory of type I

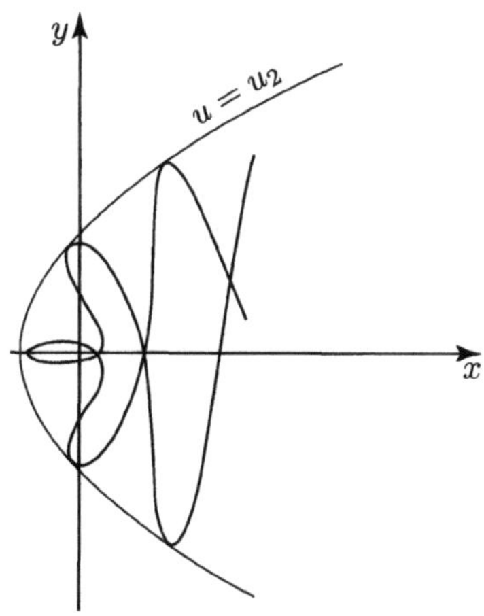

FIGURE 3.8. Trajectory of type II with self-intersections

left this trajectory is approached asymptotically as $t \to \infty$ by a bounded snake-like trajectory. As it approaches the parabola $v = v_3$, the loops of the snake-like trajectory get increasingly closer to one another. The global behavior of this snake-like trajectory is as follows: while oscillating, it moves away from the parabola $v = v_3$, makes one lap around the center of attraction, and then, as already mentioned, it approaches asymptotically and in an oscillatory manner an arc of the parabola $v = v_3$ as $t \to \infty$. Although our trajectory is indeed bounded, it does not admit a convenient description in terms of osculating elements: the perturbations

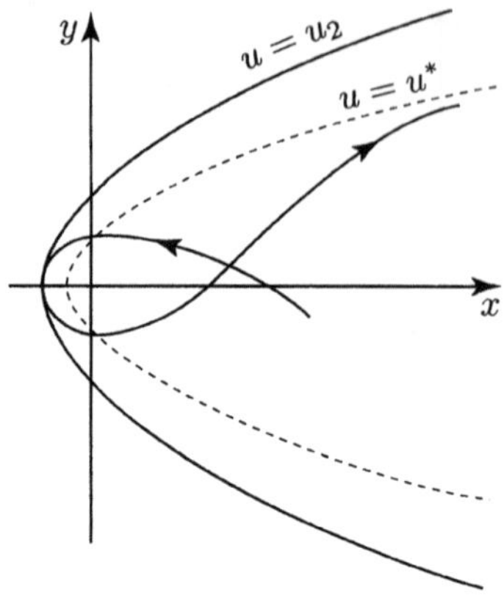

FIGURE 3.9. Loop-like trajectory of type II

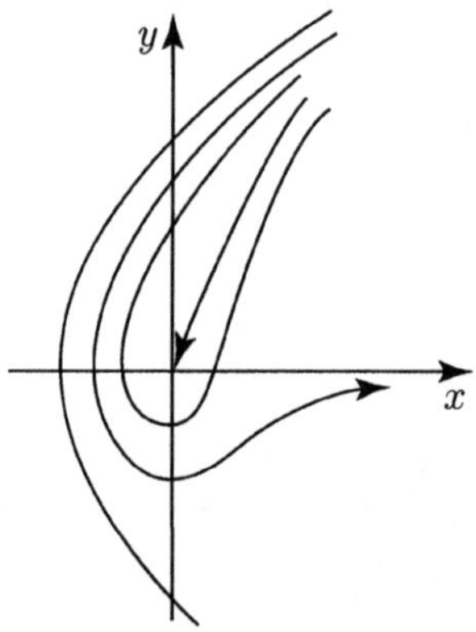

FIGURE 3.10. Trajectories of type II

caused by the supplementary acceleration **f** are rather large, and as we have seen the trajectory has nothing in common with an "osculating ellipse."

Finally, to the right of the arc $v = v_3$ there is one more trajectory, which moves away: it first oscillates like a snake and then, as $t \to \infty$, it tends to some parabola $u = u^*$ (and, together with the latter, goes to infinity; Figure 3.14).

Now let us consider the trajectories in the region of bounded motions (as we already mentioned, in that region there can also exist unbounded trajectories).

Trajectories of type IV. As we move away slightly from the boundary between regions II and IV inside region IV of the $(c, h/\sqrt{f})$-plane we encounter the trajectories shown in Figure 3.15. The bounded trajectory has a pronounced snake-like character due to the essential influence of the large supplementary acceleration **f**. This trajectory behaves as follows: outruning the center of attraction, the parti-

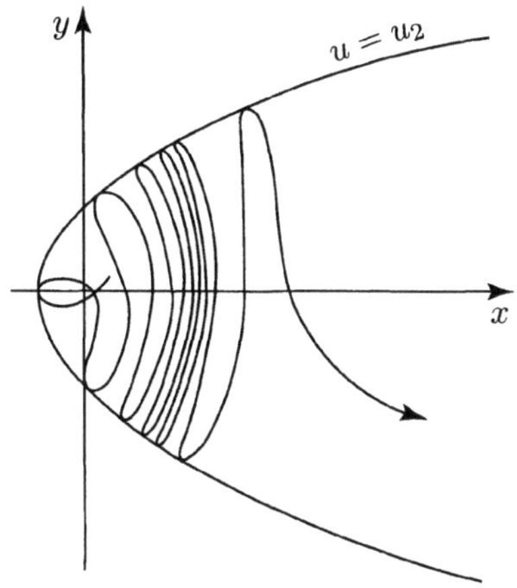

FIGURE 3.11. Snake-like trajectories of type II

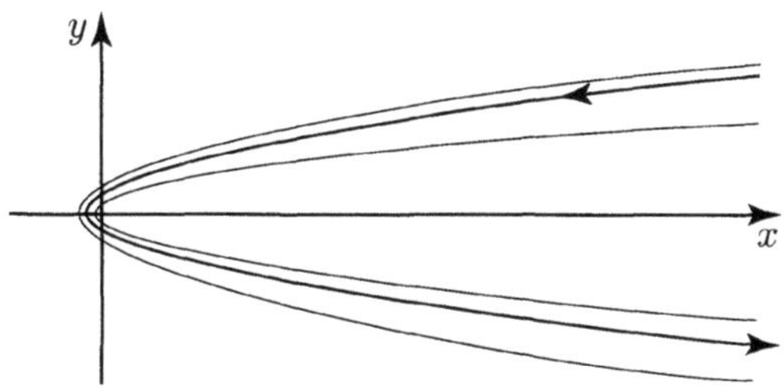

FIGURE 3.12. Trajectory of type III

cle oscillates between the branches of the parabola $u = u_2$, moving so that the value of v increases till it reaches $v = v_3$; after this boundary is reached, the particle retraces its motion "backward" (with respect to v), approaching the center of attraction along a snake-shaped trajectory, then outruns again the center of attraction, and so on.

In region IV, along with bounded trajectories there exist unbounded ones (see Figure 3.15). They are unbounded because they lie far from the center of attraction (as a consequence of which the influence of the supplementary acceleration **f** dominates that of the center of attraction).

As the value of the quantity $|h|/\sqrt{f}$ increases, such a bounded trajectory changes its shape substantially, becoming quasi-elliptic. This is due to the fact that the influence of **f** is small compared with that of the Newtonian force of attraction. A trajectory of this type can be described over a considerable portion of it in terms of slowly (in the mean) varying osculating elements. The shape of such a trajectory is shown in Figure 3.16.

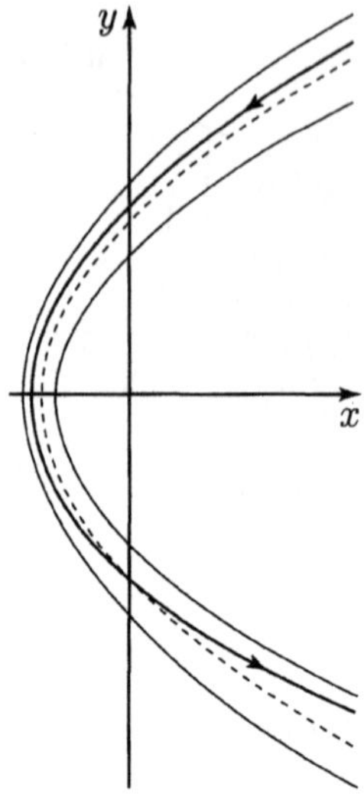

FIGURE 3.13. Trajectory of type III

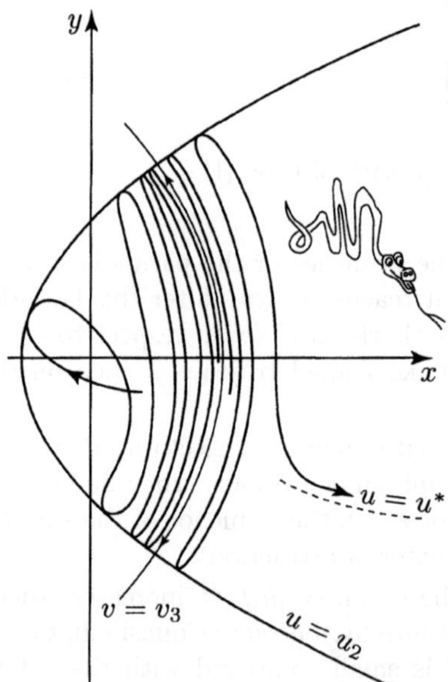

FIGURE 3.14. Trajectories of on the boundary of the domains II and IV

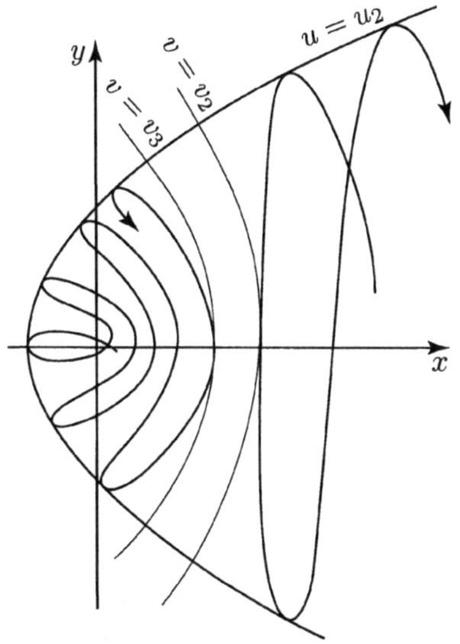

FIGURE 3.15. Bounded and unbounded trajectories in region IV

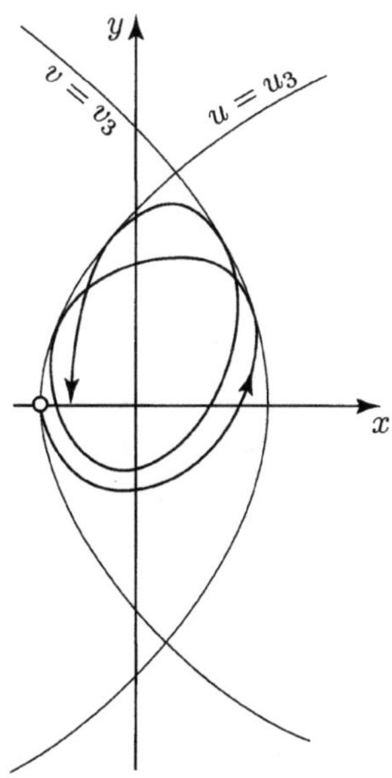

FIGURE 3.16. Quasi-elliptic trajectory in region IV

In the language of osculating elements one can say that the semi-major axis of the orbit remains practically constant in magnitude, while the direction of the line of apsides changes. Moreover, *as the axis of apsides moves away from the direction of* **f** *(from the x-axis) the eccentricity increases*: the trajectory evolves so that it becomes increasingly squeezed. The point of tangency of the trajectory to the boundary $v = v_3$ on Figure 3.16 moves along this boundary toward the point where the parabola $v = v_3$ intersects the boundary parabola $u = u_3$. If during its motion the particle reaches the critical point $v = v_3$, $u = u_3$, then its velocity at that point vanishes and subsequently *the particle will move along the trajectory that it has already traced, but in the opposite direction* (Figure 3.17).

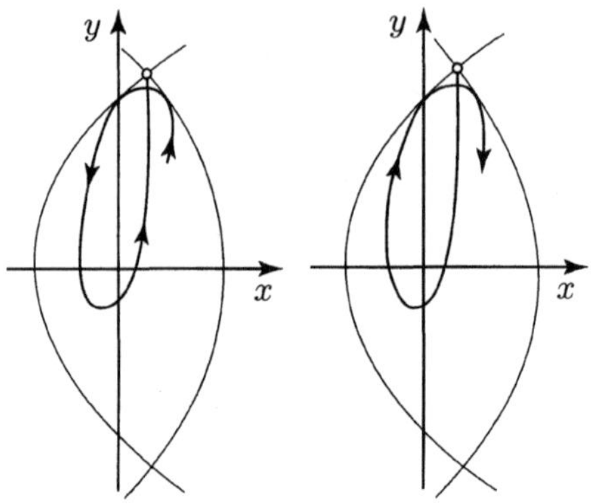

FIGURE 3.17. Switch of the direction of motion

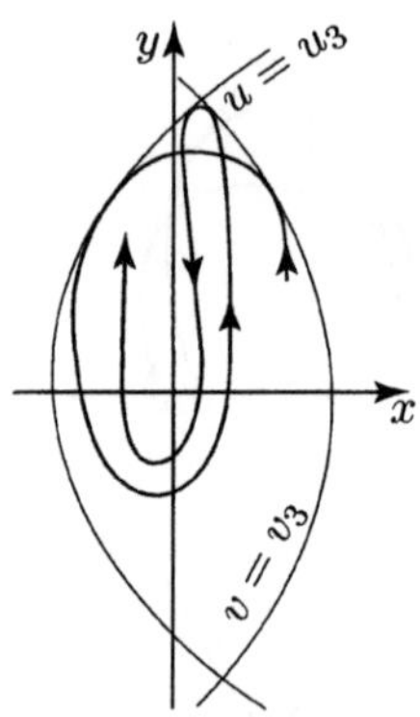

FIGURE 3.18. Switch of the direction of motion in the general case

In the general case the trajectory does not pass through the critical point (i.e., the probability of that event is small); nevertheless, in the vicinity of the critical point the direction of motion switches as shown in Figure 3.18, and the

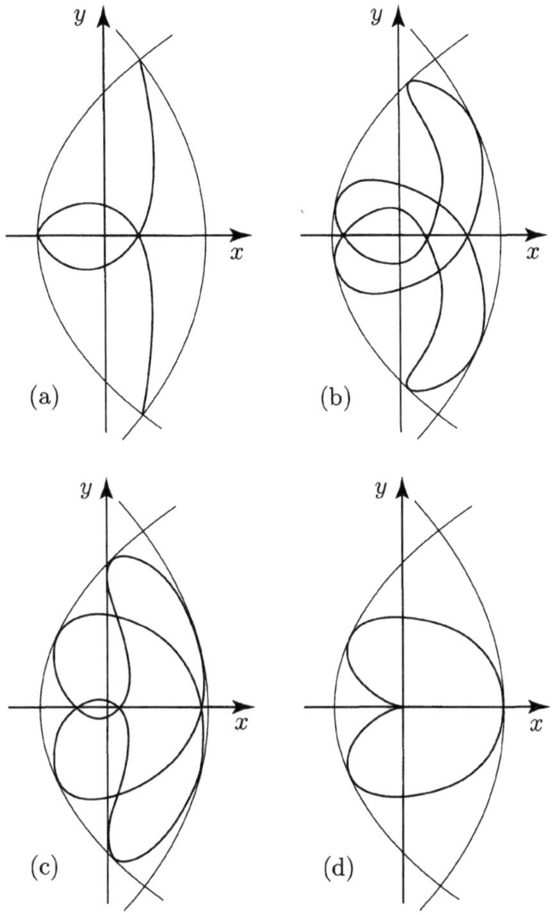

FIGURE 3.19. Periodic trajectories ($T_u/T_v = 1/2$)

trajectory evolves in the opposite order. *To the forward* [resp., *backward*] *motion of the particle there corresponds a forward* [resp., *backward*] *motion of the line of apsides of the osculating ellipse; when the line of apsides moves away from* [resp,. *toward*] *the direction of the vector* **f**, *the osculating eccentricity increases* [resp., *decreases*].

The switch of the direction of motion into the opposite one renders the description of the motion in terms of osculating elements more difficult, since at the switch time the osculating elements undergo a jump [3.6].

In practice the moving particle may not "survive" till the switch time, because due to the considerable "squeezing" (increase of eccentricity) that it experiences, the trajectory passes dangerously close to the surface of the central body (say, the Earth) and at some time later hits that surface [3.6]. For trajectories in three-dimensional space this danger is considerably less significant or does not exist at all, since a bounded motion always takes place outside the region bounded by two paraboloids of revolution, $u = u^*$, $v = v^*$ (the whole Earth can hide inside that region an thus become unreachable).

Finally, let us describe briefly some periodic trajectories. Since throughout region IV it holds that $T_u/T_v < 1$, this ratio can also be a rational number, which in the case of bounded trajectories means precisely that the trajectory in question is bounded. The shape of periodic trajectories depends not only on the ratio T_u/T_v, but also on the difference of the initial phases $u(\tau)$ and $v(\tau)$.

A few periodic trajectories are shown in Figures 3.19. Some of these trajectories are "nonphysical," since they pass through the center of attraction. Particularly interesting are the trajectories that pass through the point at which the velocity becomes zero. They are shaped as open loops, along which the particle alternately moves forward and backward (Figure 3.19(a)). Figure 3.19 displays trajectories for which $T_u/T_v = 1/2$. Trajectories of type 3.19(a) are obtained when the moment at which the function $v(\tau)$ attains its minimum and that at which the function $u(\tau)$ attains its maximum coincide (we shall call such trajectories "opposite-phase" trajectories). Figures 3.19(b) and 3.19(c) show trajectories with different phase shifts, while the trajectory shown in Figure 3.19(d) is "in phase:" the minimum of $v(\tau)$ is attained simultaneously with the minimum of $u(\tau)$. This last trajectory passes through the center of attraction.

Figure 3.20 shows trajectories in phase and in opposite phase with the ratio $T_u/T_v = 1/3$; finally, Figure 3.21 shows trajectories with $T_u/T_v = 2/3$: in phase (a) and opposite phase and in "1/4 phase" (b).

The material of this essay illustrates quite well the richness of information offered by exactly integrable problems. Snake-like trajectories that approach asymptotically an oscillatory motion along an arc of parabola; "turning" trajectories, on which the particle "suddenly" starts moving in the opposite direction; oscillatory motions along a bounded arc of a parabola; jumps of the osculating elements of a slowly evolving ellipse and switching of forward and backward motion; the possibility of the moving particle falling on the central body after a large number of revolutions around it; amazing "whiskered" periodic trajectories – the existence of all these effects would have been hard to predict with certainty by just "superposing" two such simple motions as the motion in a Newtonian central force field and the motion in a homogeneous force field. However, by integrating the problem in closed form we were able to reveal all these effects, and in all honesty we have to admit that a good half of them came as a surprise.

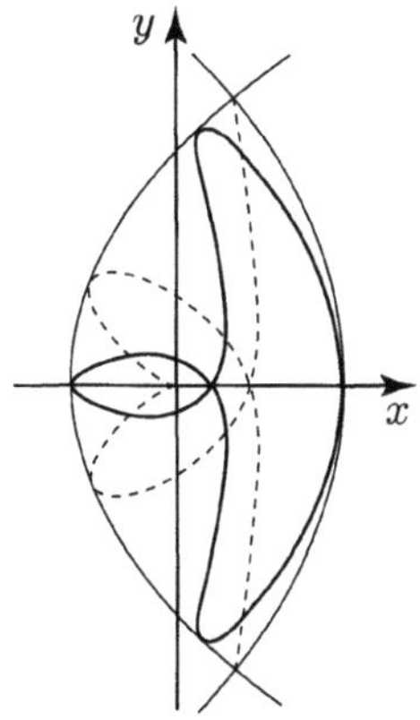

FIGURE 3.20. Periodic trajectories ($T_u/T_v = 1/3$)

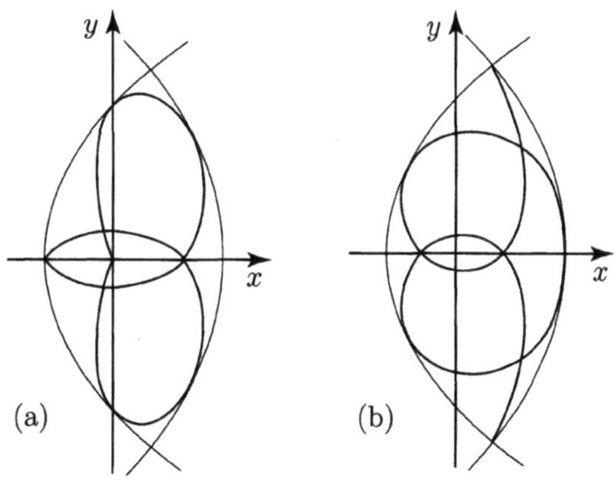

FIGURE 3.21. Periodic trajectories ($T_u/T_v = 2/3$)

5. A few words about the influence of radiation pressure on the motion of Earth satellites

If a satellite flies high enough above the Earth, and is large enough in size, but at the same time has a sufficiently small mass, then the pressure of the Sun's light rays (known as *radiation pressure*) has a very strong effect on its motion. Such conditions are satisfied, for example, in the case of the American Echo balloon-satellites, which are 30–40 meters wide inflated shells.

The satellites in the Echo family revolve around Earth several times in 24 hours, and several tens of times in a one–to–two week period. This time interval is long enough for the tendency to a short-period evolution of the orbit of such satellites to become evident. But over the same time interval the Sun does not shift too much on the celestial sphere, and consequently the direction of the radiation-pressure force acting on the satellite may be considered to be approximately fixed in space. Furthermore, the distance of the satellite to the Sun practically does not change, and so the magnitude of the radiation-pressure force is practically constant. Thus, the problem of studying of the motion of such a satellite reduces to the setting discussed in the present essay. There is however an essential correction to be introduced: *the satellite may periodically pass through the shadow of the Earth, where radiation pressure does not act.*

The circumstance pointed out above introduces new qualitative and quantitative specific features in the evolution of the motion of a satellite. These features were thoroughly examined by Yu. A. Chernikov [3.7], M. L. Lidov [3.8] and other researchers, and we will not repeat here their detailed analysis. Rather, we shall attempt, using the laws of motion that we already know, to understand the qualitative tendencies of the evolution of the motion of satellites that are subject to the influence of radiation pressure and pass periodically through the shadow of the Earth. To this end we shall use the information we possess about the evolution of bounded trajectories (Figure 3.16) and the energy integral (3.2.2), which we write here in the form

$$-\frac{1}{2a} - fx = h, \tag{3.5.1}$$

where a is the *osculating* semi-major axis of the orbit.

Since the magnitude f of the acceleration generated by radiation pressure is very small, bounded trajectories will have a quasi-elliptic shape.

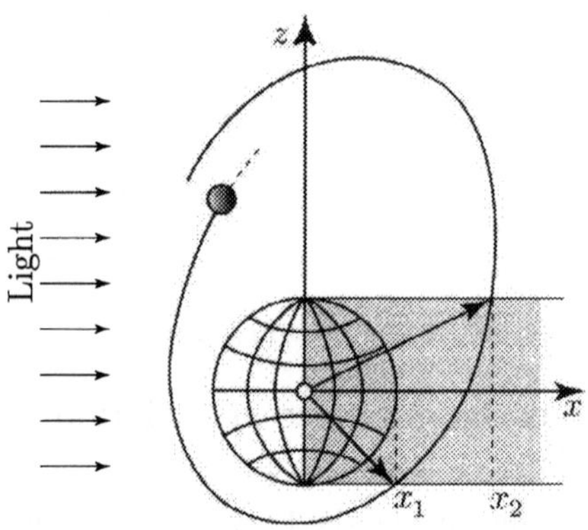

FIGURE 3.22. Passage through Earth's shadow

We will assume that the Earth's shadow is cylindrical and that the motion is planar, so that everywhere, except for the segment of orbit defined by $x_1 < x < x_2$ (Figure 3.22), the satellite is subject to the perturbing action of an acceleration $\mathbf{f}$ of constant magnitude and parallel to the x-axis. On the segment $[x_1, x_2]$ this acceleration is turned off: $|\mathbf{f}| \equiv 0$. Since for bounded motions $h < 0$, we can write

$$\frac{1}{2a} = |h| - fx \tag{3.5.2}$$

and work with the representation of the motion in the $(x,\ 1/2a)$-plane. Figure 3.23 shows the family of lines (3.5.2) for fixed f and different values of h. From our analysis of bounded trajectories we know that for constant $\mathbf{f}$ and fixed h the value of x cannot leave some interval $-x^* < x < x^*$. Therefore, in the $(x, 1/2a)$-plane the representing point will oscillate along a segment of a line (3.5.2) so that x remains in the segment $[-x^*, x^*]$, reaching its endpoints from time to time; the swing on each cycle will be of course different, but for our analysis this is not relevant.

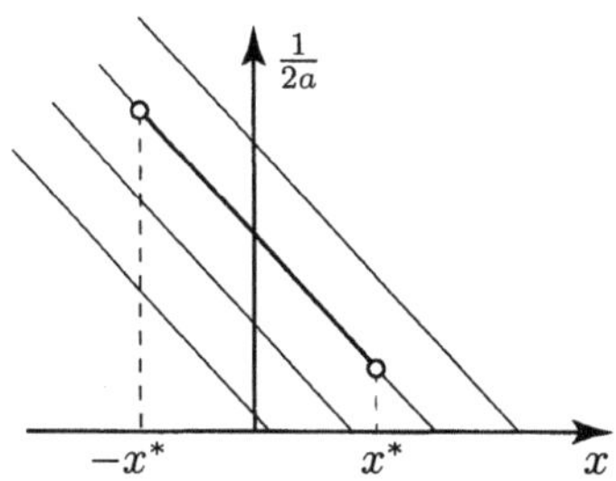

FIGURE 3.23. Representation of the motion in the $(x, 1/2a)$-plane

Now suppose that for a fixed h the motion starts at some point x_0, and at another point $x_1 > 0$ the acceleration $\mathbf{f}$ is "turned off," i.e., the satellite enters Earth's shadow. From that moment on the motion in the $(x, 1/2a)$-plane takes place along the line $1/(2a) = \text{const}$ (Figure 3.24), until the value $x = x_2$ is reached, when the satellite leaves the shadow.

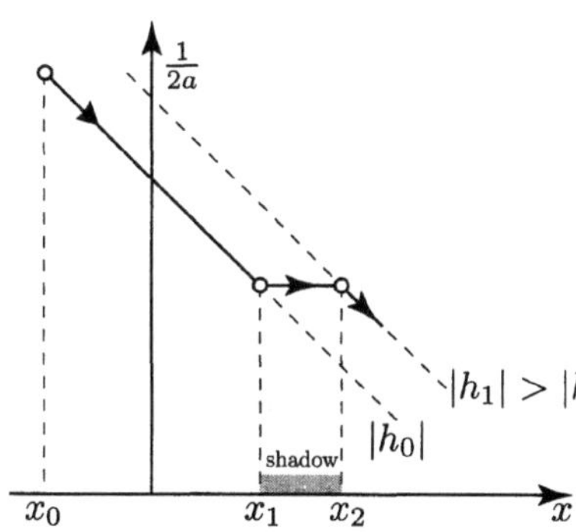

FIGURE 3.24. Interpretation of the influence of the shadow

Consider the situation depicted in Figure 3.22, in which *the point where the satellite enters the shadow is closer to the Earth then the point where it leaves the shadow*. Then after the satellite enters the shadow the coordinate x continues to grow $(x > x_1)$, and, as seen in the $(x, 1/2a)$-plane representation, *after leaving the shadow the motion of the satellite takes place at a higher energy level* $|h|$. This means that *on the average the value of the semi-major axis of the orbit decreased.* Here we are faced with a new qualitative feature, caused by the influence of the shadow. While in the absence of "deviations" of the acceleration the semi-major axis executes only short-period oscillations, entering and leaving the shadow results in systematic (long-period) variations of the semi-major axis. The possible chain of transitions to increasingly higher energy levels is displayed in Figure 3.25. Now let us observe that entering the shadow occurs closer to the Earth than leaving it only (in the situation shown in Figure 3.25) in forward motion; as we already know, forward motion is accompanied by a forward motion of the line of apsides, which in the situation at hand means that the line of apsides moves away from the direction of the acceleration vector $\mathbf{f}$. Our earlier analysis of the bounded motion revealed that such a motion of the line of apsides is accompanied by an increase in eccentricity. It follows that *the long-period increase of the eccentricity*, caused by the action of radiation pressure, *is accompanied by a long-period decrease in the magnitude of the semi-major axis of the orbit*, caused by the influence of Earth's shadow. Analyzing the backward motion of the satellite one discovers in exactly the same manner that a decrease of the eccentricity is accompanied by an increase of the semi-major axis of the orbit. We are thus led to the following general rule governing the motion of a satellite under the influence of radiation pressure and periodic passage through Earth's shadow:

The line of apsides of the orbit of an Earth satellite oscillates with a long period about the Sun–Earth direction. The motion of the semi-major axis away from [resp., *toward*] *this direction is accompanied by its decrease* [resp., *increase*] *and an increase* [resp., decrease] *in eccentricity.*

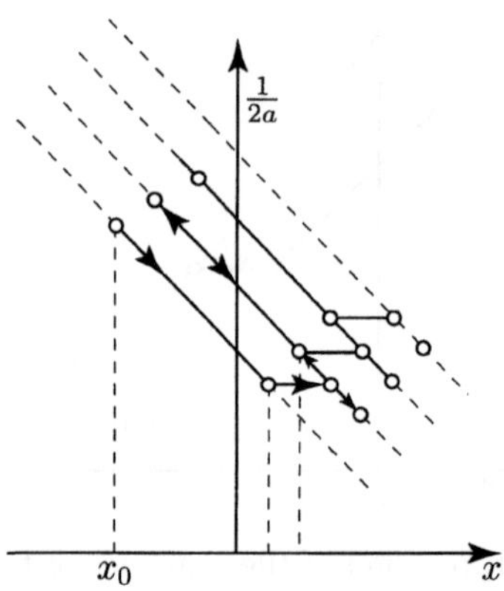

FIGURE 3.25. Interpretation of the evolution of the trajectory

The connection between eccentricity and the semi-major axis was discovered by J. Meeus by examining observation data of the motion of Echo satellites. Theoretically this effect was justified by Yu. A. Chernikov. Here we found a simple qualitative explanation of the effect.

Let us remark that, in principle, the influence of the shadow can be such that the semi-major axis will grow to very large dimensions, and possibly become infinite, and then the satellite will escape the Earth's gravity field and fly into the cosmic space! This situation is not excluded, at least theoretically. It is instructive to examine Figure 3.25 and follow on it the evolution in opposite order. Such an evolution leads to a transition to lower and lower energy levels $|h|$, which corresponds to an increase of the semi-major axis. Only a special investigation may reveal under what conditions this increase continues without bound. Over long time intervals one needs also to take into consideration the motion of the Sun.

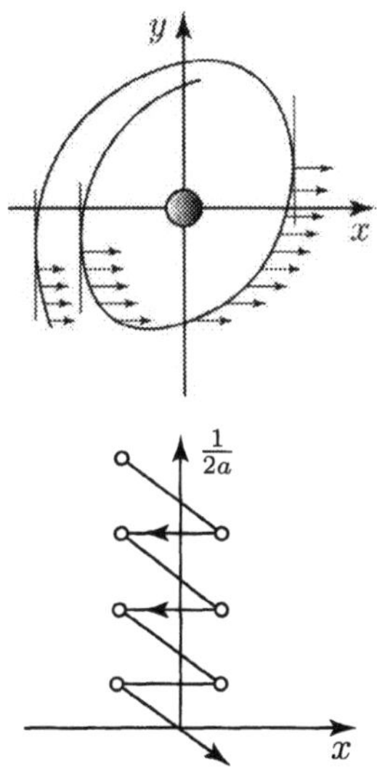

FIGURE 3.26. Satellite escape under periodic turning-on of acceleration

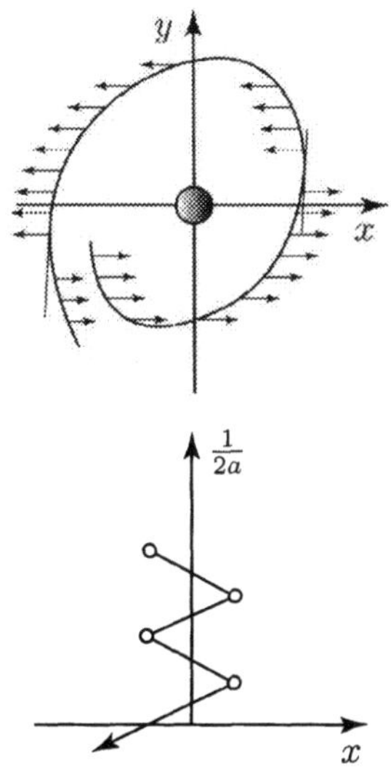

FIGURE 3.27. Satellite escape under periodic switching of acceleration

Referring the reader for details to the aforementioned works of M. L. Lidov [3.8], Yu. A. Chernikov [3.7], and others, we will conclude this essay by touching upon one other question. As we have already indicated, the acceleration **f** can be "organized" in artificial manner (using the reactive force of a ion or plasma engine). This means that we can arbitrarily turn it on and off. The foregoing arguments demonstrate sufficiently clearly that we can "design" a systematic process of turning on and off a small reactive acceleration (and even of changing its direction) so that the satellite will be spun into a spiral motion and hurled into space beyond the sphere of influence of the Earth's gravitational field. Figures 3.26 and 3.27 show such trajectories and the corresponding representation in the $(x, 1/2a)$-plane.

Fourth Essay

Motion of the Worlds

> The Old Gray donkey, Eeyore, stood by himself in a thistly corner of the forest, ... and thought about things. Sometimes he thought sadly to himself, "Why?" and sometimes he thought, "Wherefore?" and sometimes he thought, "Inasmuch as which?"
>
> A. A. Milne, *The World of Pooh*

1. One more time about the "Laplace Theorem" and other [Serious] "Things"

An artificial satellite flying around the Earth is subject, in addition to other forces, to the attraction of the Moon and of the Sun. As it turned out, the action of these perturbations on the orbit of a satellite is astonishing, and even, to a certain extent, unexpected ... But let's not jump ahead of the story; for two hundred years celestial mechanics has been busy with precisely the study of the motion of heavenly bodies under the action of forces of mutual attraction. The problem of the motion of an Earth satellite can be regarded as a particular case of the "n-body problem," about which one can draw information from the classics of celestial mechanics: they have plenty to say on this subject!

Classical celestial mechanics deals with the situation in which the mass of one of the bodies (the Sun!) is considerably larger than the masses of the other mutually-attracting bodies; hence, in a first approximation one can consider that all those bodies (planets) move around the Sun along Keplerian orbits; the question is then: what is the evolution of these orbits under the effect of the mutual attraction of the planets? And here is the question of all questions: will the present configuration of the Solar System remain unchanged forever? In other words, is the Solar System stable? Could the present motions of the planets change drastically? Could a heavenly body leave its own "circle?" Could heavenly bodies fall on one another or on the Sun? And could they escape from the native Solar System to infinity?

Here we have in mind stability (or instability) under the action of only the Newtonian forces of mutual attraction between planets. The planets are considered as particles with the masses equal to the planetary masses, as customary in the classical formulation of the "n-body problem."

Thus, let us consider a system of $n+1$ particles M_s of masses m_s ($s = 0, 1, 2, \dots, n$), which attract one another according to Newton's law. We shall assume that the mass m_0 is very large compared with the sum of the remaining masses. In the Solar System m_0 is the mass of the Sun; in the three-body system Earth–Moon–artificial satellite, m_0 is the mass of the Earth.

Henceforth, for the sake of simplicity, we will speak about the Solar System (unless otherwise stipulated).

The assumption made above has no effect on the accuracy of the equations of motion; it allow us, however, to distinguish in these equations the principal part and the perturbing part.

If one takes for the origin of the Cartesian coordinate system the point M_0, the equations of motion of the ith particle (with coordinates x_i, y_i, z_i) take the following form (see [4.1]):

$$\ddot{x}_i = \frac{\partial U_i}{\partial x_i}, \quad \ddot{y}_i = \frac{\partial U_i}{\partial y_i}, \quad \ddot{z}_i = \frac{\partial U_i}{\partial z_i}, \tag{4.1.1}$$

where

$$\left.\begin{aligned} U_i &= \frac{f(m_0+m_i)}{r_i} + U_i^*, \\ U_i^* &= f\sum_{j=1}^{n}{}' m_j\left(\frac{1}{\Delta_{ij}} - \frac{x_ix_j+y_iy_j+z_iz_j}{r_j^3}\right). \end{aligned}\right\} \tag{4.1.2}$$

Here the prime appended to the sum symbol indicates that the term with $j = i$ is omitted; $r_i = \sqrt{x_i^2+y_i^2+z_i^2}$ is the distance from the particle M_i to the central point M_0, and $\Delta_{ij} = \sqrt{(x_i-x_j)^2+(y_i-y_j)^2+(z_i-z_j)^2}$ is the distance between the particles M_i and M_j. If only the masses m_0 and m_i were present, then the motion of the particle M_i relative to M_0 would take place on a Keplerian orbit, corresponding to the force function $U_0 = f(m_0+m_i)/r_i$. However, the presence of the other particles introduces perturbations in the motion of particle M_i. These perturbations are described by the perturbing part U_i^* of the force function U_i. Keeping in mind the expression (4.1.2) of the perturbing function U_i^* for future use, let us turn now our attention to the following important transformations.

In order to reduce the equations governing the perturbed motion of planets to a canonical form that is convenient in investigations, we need to choose a specific coordinate system for the analysis of the motion of each planet M_s (see [4.1]). Namely, we will use an orthogonal coordinate system x'_s, y'_s, z'_s with origin at the center of mass G_{s-1} of the system of particles $M_0, M_1, \dots, M_s$ and fixed directions of the axes. Then the equations of motion contain the so-called "reduced masses"

$$m'_s = m_s\,\frac{m_0+m_1+\cdots+m_{s-1}}{m_0+m_1+\cdots+m_{s-1}+m_s}. \tag{4.1.3}$$

Clearly, under the assumption that the mass m_0 is dominant, i.e., $m_0 \gg \sum_{i=1}^{s} m_i$, we have $m'_s \approx m_s$. In the Solar System all the points G_{s-1} lie inside the

Sun, close to its center. The center of mass of the Moon-Earth system lies in the interior of the Earth; moreover, the ratio of the Moon's mass m_{M} to the Earth's mass m_{E} is $m_{\mathrm{M}}/m_{\mathrm{E}} \approx 1/81$.

Consider the total force function U of the system of particles $M_0, M_1, \ldots, M_s$,

$$U = f \sum_{s<j} \frac{m_s m_j}{\Delta_{sj}} \tag{4.1.4}$$

where Δ_{sj} are the mutual distances between particles. In the new coordinates x'_s, y'_s, z'_s the force function U given by (4.1.4) can be written in the form

$$U = f \sum_{j=1}^{n} \frac{m_0 m_j}{r'_j} + U', \tag{4.1.5}$$

where

$$r'_j = \sqrt{x_j'^2 + y_j'^2 + z_j'^2} \qquad (r'_1 = \Delta_{01}).$$

Of course, in expression (4.1.5) the form of the perturbing function U' is different from the perturbing function U_i^* given by (4.1.2) because of the new coordinate system used (and also because U_i^* refers to one particle, whereas U' refers to the entire system).

We shall not write explicitly the expression of U', since in the sequel we will need only several of its general properties. The force function U' is small of second order in the perturbing masses [4.1]. If one discards it in the expression (4.1.5), then the resulting equations of motion of particle M_s will not depend on the coordinates of any other particle M_j, $j \neq s$. Hence, M_s will move in a Keplerian orbit, which in the coordinate system x'_s, y'_s, z'_s is described by the elements

$$p_s,\ e_s,\ \omega_s,\ ,\Omega_s,\ i_s,\ \tau_s, \tag{4.1.6}$$

so that i_s is the inclination of the orbital plane to the plane x'_s, y'_s and so on, as discussed in the first essay.

However, due to the interaction between masses the perturbing function U' must be taken into account in our system. Considering the perturbed motion in the osculating elements (4.1.6) we discover that, due to the presence of U', these elements will deviate from their unperturbed, constant values. To analyze these deviations it is convenient to introduce the Delaunay elements, in much the same manner as we proceeded in the first essay:

$$\left.\begin{aligned} L_s &= \sqrt{\mu'_s a_s}, & l_s &= n_s(t - \tau_s), \\ G_s &= \sqrt{\mu'_s a_s (1 - e_s^2)}, & g_s &= \omega_s, \\ H_s &= \sqrt{\mu'_s a_s (1 - e_s^2)} \cos i_s, & h_s &= \Omega_s. \end{aligned}\right\} \tag{4.1.7}$$

Here $\mu'_s = f(m_0 + m'_s)$, where m'_s is given by formula (4.1.3); the mean motion equals $n_s = \sqrt{\mu'_s/a_s^3}$, and the semi-major axis equals

$$a_s = \frac{p_s}{1 - e_s^2}\,. \tag{4.1.8}$$

Let us introduce the Hamiltonian

$$F = \sum_{s=1}^{n} \frac{\mu_s'^{\,2}}{2L_s^2} + U', \tag{4.19}$$

where, of course, we assume that U' is expressed in terms of the Delaunay elements. Then the equations of the perturbed motion take on a canonical form:

$$\left.\begin{aligned} &\frac{dL_s}{dt} = \frac{\partial F}{\partial l_s}, \quad \frac{dG_s}{dt} = \frac{\partial F}{\partial g_s}, \quad \frac{dH_s}{dt} = \frac{\partial F}{\partial h_s}, \\ &\frac{dl_s}{dt} = -\frac{\partial F}{\partial L_s}, \quad \frac{dg_s}{dt} = -\frac{\partial F}{\partial G_s}, \quad \frac{dh_s}{dt} = -\frac{\partial F}{\partial H_s}, \\ &\qquad\qquad s = 1, 2, \ldots, n. \end{aligned}\right\} \tag{4.1.10}$$

These are the exact equations of motion, and they admit several exact first integrals. It is well known that the problem of the motion of n mutually attracting particles has 10 classical first integrals. Since the forces acting in the system can be regarded as internal forces, the total momentum of the system is conserved (which upon taking the components with respect to the axes of the coordinate system yields three first integrals); the center of mass of the system moves uniformly and rectilinearly (three more first integrals); the angular momentum of the system is also preserved (so that its components provide three additional first integrals); finally, conservation of energy yields the 10th first integral.

The integrals specifying the motion of the center of mass of our system are not very interesting, but from the remaining four first integrals one can definitely extract relevant information about the motion. In particular, it turns out that *for the motion to be bounded it is necessary* (*but not sufficient*!) *that the energy constant be negative* (which is precisely what we shall assume from now on). A simple proof of this assertion can be found, for example, in G. N. Duboshin's book [4.1] (which, incidentally, is widely used in the present essay as well).

From the three first integrals that describe the conservation of the angular momentum we will write in our variables (and use in our analysis) only one, namely

$$\sum_{s=1}^{n} m'_s \sqrt{\mu'_s p_s} \cos i_s = c. \tag{4.1.11}$$

Recall that (4.1.11) is an exact first integral of the equations of motion!

Further, noting that, as in the first essay, F is a periodic function of l_s, we expand F in an n-fold Fourier series:

$$F = \bar{F} + \\ + \sum_{k_1,\dots,k_n} \left(a_{k_1,\dots,k_n} \cos(k_1 l_1 + \cdots + k_n l_n) + b_{k_1,\dots,k_n} \sin(k_1 l_1 + \cdots + k_n l_n) \right), \tag{4.1.12}$$

where

$$\bar{F} = \frac{1}{(2\pi)^n} \int_0^{2\pi} \cdots \int_0^{2\pi} F \, dl_1 \cdots dl_n \tag{4.1.13}$$

and the coefficients $a_{k_1,\dots,k_n}$, $b_{k_1,\dots,k_n}$ depend (generally speaking) on all the elements, except l_s; also, for each term in the sum at least one k_i is different from zero. Assume that there is no set of (positive or negative) integers

$$k_1, \dots, k_n \tag{4.1.14}$$

for which the resonance condition $\sum_{s=1}^{n} k_s n'_s = 0$ is satisfied. In other words, for all sets of integers (4.1.14)

$$\sum_{s=1}^{n} k_s n'_s \neq 0. \tag{4.1.15}$$

Then the first-approximation equations (in the sense of asymptotic methods) are obtained by averaging the right-hand sides of equations (4.1.10) – or, equivalently, the Hamiltonian (4.1.12) – with respect to the variables l_s, independently over each of them. This means that we need to replace the Hamiltonian (4.1.9) by its mean value (4.1.13) and then substitute the latter in equations (4.1.10). Since $\bar{F}$ does not depend on the variables l_s (we averaged over them), we conclude from (4.1.10) that $dL_s/dt = 0$ (indeed, $\partial \bar{F}/\partial l_s = 0$). Then from the definition (4.1.7) of L_s it follows that the semi-major axes of all orbits are constant:

$$a_s = \text{const}. \tag{4.1.16}$$

We arrived at what is known as Laplace's theorem on "invariability" (in the sense of absence of secular perturbations) of the semi-major axes of the planets of the Solar System. Generally speaking, this conclusion holds on a bounded time interval, because only on such an interval does the first approximation of asymptotic methods "work." More precisely, we can assert that $|a_s - a_{s0}| \sim \varepsilon$ over a time $t \sim 1/\varepsilon$, where the small quantity ε has the order of the strength of the perturbations and a_{s0} are the initial values of the semi-major axes.

The attempt to derive exact estimates encounters considerable difficulties; but it is exactly on such estimates that the answer to the question of whether the motion of our system is stable on an infinite time interval depends. We will come back to this question later (after all, we wrote this essay precisely to deal with this question). For the moment let us see what conclusion can be derived [in Eeyore's words, "Inasmuch as which?"] under the assumption that (4.1.16) holds.

Denote

$$\sum_{s=1}^{n} m'_s \sqrt{\mu'_s a_s} = \varphi \tag{4.1.17}$$

and subtract from this expression the integral (4.1.11). Then recalling (4.1.8) we obtain

$$\sum_{s=1}^{n} m'_s \sqrt{\mu'_s a_s} \left(1 - \sqrt{1 - e_s^2} \cos i_s\right) = \varphi - c,$$

or

$$\sum_{s=1}^{n} m'_s \sqrt{\mu'_s a_s} \, \frac{\sin^2 i_s + e_s^2 \cos^2 i_s}{1 + \sqrt{1 - e_s^2} \, \cos i_s} = \varphi - c. \tag{4.1.18}$$

But if (4.1.16) holds, then $\varphi - c = c_0$ is a new constant, equal to the value of the right-hand side of relation (4.1.18) at the initial moment of time. Thanks to the positivity of each term of the sum in the left-hand side of (4.1.18), we have $c_0 > 0$, and each term of the sum is smaller than c_0. This yields the estimate

$$\frac{\sin^2 i_s + e_s^2 \cos^2 i_s}{1 + \sqrt{1 - e_s^2} \, \cos i_s} \leq \frac{c_0}{m'_s \sqrt{\mu'_s a_s}} . \tag{4.1.19}$$

The fact that a_s is not a constant, but differs only slightly from one ($|a_s - a_{s0}| \sim \varepsilon$) does not invalidate estimate (4.1.19) – the latter can be obtained even when one takes this circumstance into account.

If the initial values e_{s0}, i_{s0} are very small, then the constant c_0 will also be quite small, and then (4.1.19) implies that *the quantities e_s, i_s remain small over the entire time interval considered*. In other words, "*Laplace's theorem on the stability of the Solar System*" holds:

Assume that the orbits of the planets, which all revolve in the same direction, are nearly circular and only slightly inclined relative to one another, and that the mean motions of the planets are incommensurable. Then over a sufficiently large time interval the semi-major axes of the orbits will remain close to their initial values, and the orbits will remain nearly circular and only slightly inclined relative to one another.

This theorem is a splendid result of classical celestial mechanics. Indeed, the planets in the Solar System move along nearly-circular orbits and lie almost in the same plane. Hence, by "Laplace's stability theorem," the motion will retain this character for a sufficiently long duration. So, for the time being we don't have to worry that planets in our system will colliding!

Let us remark that if at the initial moment of time all orbits are circular and lie in the same plane, then $c_0 = 0$, and from (4.1.19) it follows that the orbits will remain circular and coplanar. In other words, a planar circular system undergoes no evolution.

But the words "Laplaces's theorem on the stability of the planetary system" were not put in quotation marks for no reason. What does "stability till after-tomorrow" mean?! (Or, for that matter, till the next century?) We are interested in stability in the classical sense, that is, stability "forever!" Is "Laplace's theorem" valid for an infinite time interval? This question was answered only not so long ago in studies of contemporary mathematicians, in the first place of the Russian mathematician V. I. Arnold. We will return to this subject a bit later. To conclude this section, let us describe the character of the perturbed motion of the planetary system. To the results formulated in "Laplace's theorem" one can add the following facts.

Let $\mathbf{e}_k$ be the Laplace vector of the kth planet ($\mathbf{e}_k$ is directed toward the perihelion of the orbit along its semi-major axis and its magnitude is proportional to the eccentricity of the orbit, namely, $|\mathbf{e}_k| = \sqrt[4]{m_k^2 a_k}\, e_k$). Then the analysis of the perturbed motion shows that $\mathbf{e}_k = \sum_{j=1}^{n} \mathbf{e}_{kj}$, where each vector $\mathbf{e}_{kj}$ rotates uniformly with frequency ν_j^ω.

The frequencies ν_j^ω are very small. Therefore, the perihelia of all orbits move slowly, but steadily (in "secular manner") in the respective orbital planes. This motion is determined by the set of n frequencies ν_j^ω, and is such that the Laplace vector of each orbit is a sum of n vectors which rotate uniformly (with frequencies ν_j^ω). Furthermore, the plane of the kth orbit is inclined relative to the basic coordinate plane at the angle i_k; the line of intersection of these two planes is called the *line of nodes*. The secular variation of these quantities is described by a vector $\mathbf{i}_k$, directed along the line of nodes; $|\mathbf{i}_k| = \sqrt[4]{m_k^2 a_k}\, i_k$. The vector $\mathbf{i}_k$, too, is a sum of vectors that rotate uniformly with slow frequencies ν_j^Ω, but now the number of vectors is $n - 1$.

Such is, approximately, the perturbed motion of a system of n planets around a massive central body (Sun). The planets move always along slightly elliptic and slightly inclined orbits, and the magnitude of their semi-major axes is preseved; the orbital planes and the orbits themselves, in their planes, execute a secular precession, the rates of these precessions being described by the superposition of uniform rotations described above. This kind of motion is termed *Lagrangian motion of planets*.

2. Wouldn't you like to see the Moon fall on Earth?

The most amazing effects in the mechanics of space flight are revealed when one investigates of the problem of three or more mutually attracting bodies.

As a rule, the investigations of the motions of planets and other bodies of the Solar System in classical celestial mechanics have been somewhat utilitarian, i.e., adequate for orbits that lie almost in one plane and are nearly circular. Such an approach was of course justified by the needs of astronomical practice. Laplace's theorem is a reflection of this approach. The majority of researchers in classical celestial mechanics did not manifest great interest in "exotic orbits," i.e., orbits that are strongly inclined and strongly elliptic. The launching of the first spacecraft sparkled a natural interest in studies that impose no constraints on the shape and position of orbits relative to one another.

Let us address, for example, the problem of the motion of an artificial Earth satellite when the perturbations caused by the Moon are taken into account. In other words, let us consider the three-body problem "Earth–Moon–satellite." Since the mass of the satellite is negligibly small compared with the masses of the Earth and of the Moon, it is natural to consider that the satellite exerts no attraction on these two massive bodies. Accordingly, the Earth and the Moon (regarded as particles) will move in unperturbed Keplerian orbits with respect to their common center of mass. The only question is then how the the orbit of the satellite around Earth is perturbed. This formulation of the problem is termed *the restricted three-body problem*. If the satellite does not pass too close to the Moon, then the perturbations of its motion are small, and the arguments of the preceding section of this essay apply. In particular, they lead to the conclusion that the semi-major axis of the satellite's orbit does not change in the mean:

$$a = \text{const.} \tag{4.2.1}$$

Now let us consider the area integral (4.1.11), which in the present case reduces to a sum of only two terms, involving only elements of the Moon's orbit and of the satellite's orbit. But the Moon moves along an unperturbed orbit, and so it elements p_{M} and $\cos i_{\mathrm{M}}$ remain constant. Hence, the integral (4.1.11) involves only satellite elements:

$$\sqrt{p}\cos i = \text{const.} \tag{4.2.2}$$

Squaring both sides of (4.2.2) and using (4.2.1) (and also (4.1.8)), we obtain

$$(1 - e^2)\cos^2 i = c, \tag{4.2.3}$$

where the constant c is determined by the initial values e_0, $\cos i_0$ of the elements of the satellite's orbit. Let us remark that here the angle i can be regarded as the inclination of the satellite's orbital plane to the Moon's orbital plane.

As the foregoing analysis makes clear, in the mean motion, in addition to the integral (4.2.3), there also exist the integral

$$\bar{F} = \text{const} = h, \tag{4.2.4}$$

where $\bar{F}$ is defined by formula (4.1.13). In the present case this integral can be written as

$$e^2\left(\frac{2}{5} - \sin^2\omega\,\sin^2 i\right) = h_0. \tag{4.2.5}$$

Eliminating here e^2 by means of the integral (4.2.3), we reduce (4.2.5) to the form

$$\sin^2\omega = \frac{\cos^2 i\,(2-5h_0)-2c}{(1-\cos^2 i)(\cos^2 i - c)}\,. \tag{4.2.6}$$

Using relation (4.2.6) it is not hard to exhibit the dependence of $\sin^2\omega$ on $\cos^2 i$ for fixed values of c and for various values of h_0. The possible qualitative portraits of this dependence are shown in Figures 4.1 (a) and 4.1 (b). The representing point in the $(\sin^2\omega, \cos^2 i)$-plane traverses the whole given integral curve completely, oscillating along it with a large period between the extreme possible values of the parameters. This clearly shows that the longitude of perihelion ω may vary monotonically, but may also oscillate within bounded limits (we wish to point out that there exists the stationary orbit $\omega =$const, $i =$const, and hence $e =$const). As for the angle i, its magnitude does always oscillate between its two extreme values.

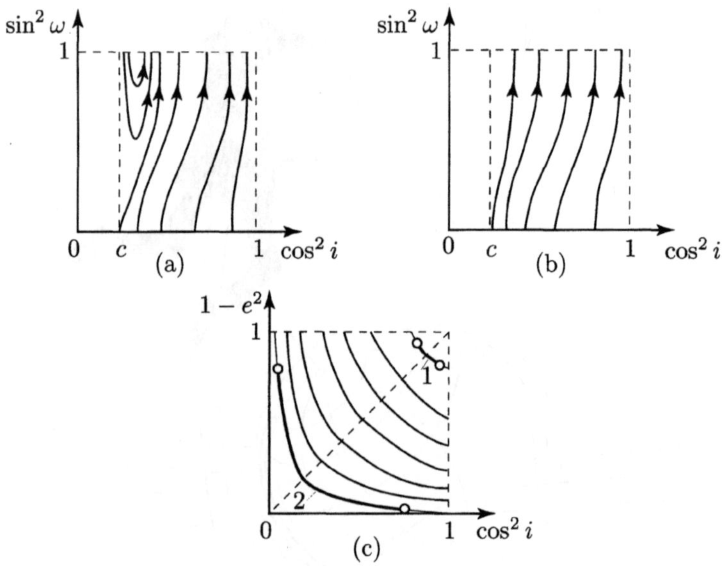

FIGURE 4.1. Integral curves of the evolution equations

Now let us consider the portrait of the integral curves given by (4.2.3) in the $(1-e^2, \cos^2 i)$-plane, for various values of the constant c (i.e., for various initial values e_0, i_0), shown in Figure 4.1 (c). The domain where the variables range is: $0 \le 1-e^2 \le 1$, $0 \le \cos^2 i \le 1$. The curves (4.2.3) in the variables $1-e^2$, $\cos^2 i$ are equilateral (rectangular) hyperbolas. During the evolution of the satellite's orbit its eccentricity e and inclination i change so that the representing point in the $(1-e^2, \cos^2 i)$-plane oscillates along the hyperbola corresponding to the value

$$c = c_0 = (1-e_0^2)\cos^2 i_0.$$

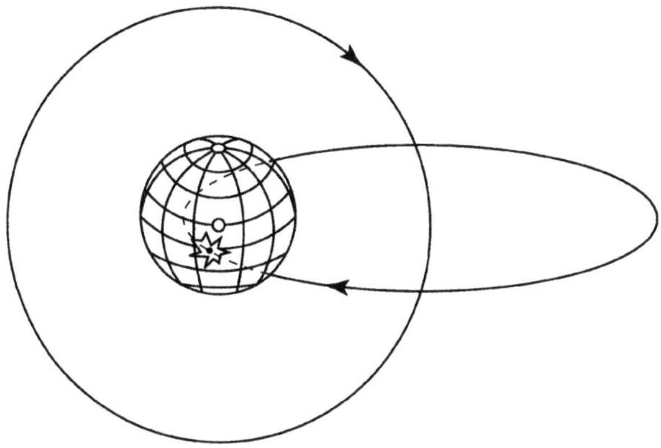

FIGURE 4.2. Orbits with different eccentricities

An analysis of the evolution equations and of the integral (4.2.6) shows that in a real motion the representing point traverses not the entire integral curve, but only a bounded arc on it; the limits within which the representing point oscillates in Figure 4.1 (c) depend on the value h of the integral (4.2.6); however, these limit can be sufficiently wide, so that a large arc of the integral curve (4.2.3) is traversed. In Figure 4.1 (c) the limits of the oscillations are schematically indicated by small circles and the segment of actual motion – by a thick line.

The right upper angle on Figure 4.1 (c) corresponds to orbits of small inclination ($i \approx 0$) and small eccentricity ($e \approx 0$). We see that if at the initial moment of time the orbit is nearly circular ($e_0 \approx 0$) and only slightly inclined to the orbital plane of the perturbing body ($i_0 \approx 0$), then the oscillations of the quantities e and i will be small; the satellite's orbit is stable and remains forever nearly-circular and of small inclination, in complete agreement with Laplace's theorem (curve 1 on Figure 4.1 (c)). In particular, an orbit that initially is planar and circular ($e_0 = 0$, $i_0 = 0$) remains forever planar and circular. Incidentally, as it follows from the integrals (4.2.3) and (4.2.5), for the planar case ($i \equiv 0$) of the problem at hand, any given value of the eccentricity remains constant ($e = e_0$). (Of course, this is no longer true in the planar n-body problem with $n > 3$.)

A completely different picture of the motion emerges for initial inclinations close to $\pi/2$. Then even an orbit that initially is circular approaches a highly oblate orbit (the eccentricity tends to a value close to 1), so that $1 - e^2 \to \epsilon \sim 0$. This situation is depicted by the curve 2 in Figure 4.1 (c). But if the semi-major axis remains constant, a grow in the eccentricity can only mean that the orbit becomes strongly squeezed. Figure 4.2 show orbits of an Earth satellite with the same semi-major axis, but different eccentricities. It is not hard to see that in the evolution process the orbit of the satellite may become so squeezed that it eventually crosses the Earth's surface. After several (or a few tens) revolutions the satellite may fall on the Earth simply due to the Moon's gravitational attraction!

Although in classical celestial mechanics and more recent studies (e.g., by N. D. Moiseev [4.2]) one can find material that would allow us to reach the above conclusion, it was only after the first artificial satellites were launched that the analysis of the evolution of arbitrary orbits began to attract wide interest.

A cycle of studies in this direction was conducted by M. L. Lidov, who, for example, determined the existence durations ("lifetimes") of satellites on strongly inclined orbits and carried out a detailed analysis of the perturbed motion. In particular we owe him the following convincing calculations that lead to truly catastrophic conclusions [4.3], [4.4].

Consider the Sun–Earth–Moon three-body system. As is known, the orbit of the Moon around the Earth lies almost exactly in the plane of the orbit on which the Earth revolves around the Sun (ecliptic plane). But what will happen if, without modifying the orbit of the Moon in any other way, we rotate it by 90° so that it becomes perpendicular to the Earth's orbital plane?

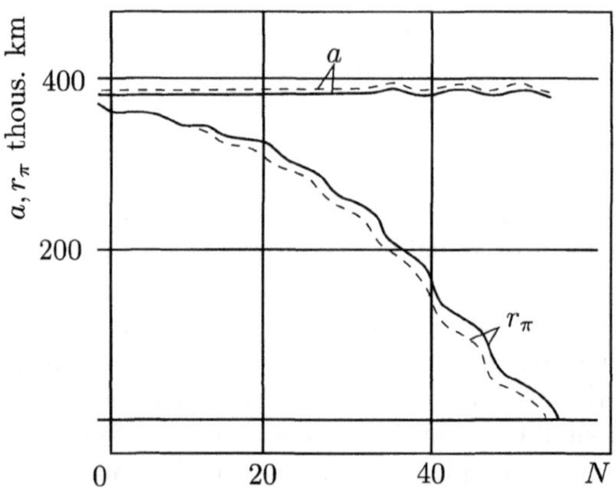

FIGURE 4.3. Evolution of elements of a "polar Moon"

The reader, of course, has already guessed the answer: the Moon will fall on the Earth! And as a matter of fact, very fast! Lidov carried out a precise computation (by numerical integration of the equations of motion on a fast computer). His results are displayed in Figure 4.3, which shows the values of the the semi-major axis a and of the distance r_π of the orbit's perihelion to the center of the Earth as functions of the number N of revolutions of the "Moon" around the Earth. We see that, indeed, the semi-major axis a remains practically constant, whereas r_π decreases swiftly and after only 55 revolutions becomes smaller than the radius of the Earth! In other words, on such an orbit the Moon would survive for only about four and a half years! (The representative estimate of the lifetime of such a Moon derived by Lidov from the averaged equations of motion is 52 revolutions. Comparing this estimate with the exact lifetime values one sees clearly that the averaged equations of motion yield not only valid qualitative results, but also correct quantitative estimates).

The discovery that it is possible for a satellite to fall on the central body under the action of gravitational forces only is undoubtedly one of the most interesting achievements of mechanics. Not many years ago wide circles of researchers still had poor intuition of such a possibility (or no intuition whatsoever). After the launch of the Soviet satellite Lunnik, which photographed the hidden face of the Moon, the author attended a public lecture by a respected scientist, who declared that Lunnik will fly in its orbit forever (because its perigee lies high above the Earth and consequently the atmosphere has no braking effect on the motion). This claim was made only a few days before Lunnik inevitably fell on Earth: its orbit was inclined about 90° to the ecliptic plane. Soon after that Lidov's works [4.3] and [4.4] were published, which contain his analysis of the evolution effects of mutually attracting bodies and calculations that revealed the phenomenon of "fall of a body on another body." These studies have attracted considerable attention.[1]

In the light of what has been said above it becomes clear that the fact that the majority of planets and satellites in the Solar System lie practically in the same plane is by no means an accident. Any body with a highly inclined orbit would have fallen on the Sun a long time ago.

True, there are satellites of Uranus whose orbits are highly inclined to the ecliptic plane. Lidov [4.3] has shown that the perturbations due to the noncentral character of the gravitational field of Uranus (i.e., due to its oblateness) compensate in a stable manner for the perturbations caused by the attraction of other bodies. Therefore, it is the strong oblateness of Uranus that saved its satellites from death. Actually, from the very fact that satellites of Uranus have stable orbits that are perpendicular to the ecliptic plane one could derive an estimate for the oblateness of this planet. For other planets the ratio of the perturbations due to oblateness to those due to external bodies is not large enough to prevent nonecliptic satellites from falling on the central body.

The reader, convinced by now that a Moon with an "edgewise placed" orbit may indeed fall on Earth, is probably worrying about the fate of our real Moon. Apparently, there is no reason to worry – the Moon will quite peacefully revolve forever. This roughly follows from "Laplace's theorem" and follows rigorously from the aforementioned investigations of V. I. Arnold. When we make this assertion we have in mind only forces of mutual attraction between "point masses;" however, in nature other forces act as well. "Tidal forces," due to the finite (rather than infinitesimal) extent of planets and the fact that they are nor rigid, are capable of slowing down their run in orbits and, consequently, modify these orbits. The fate of the Solar System depends to a large extent on these forces. If tidal forces are taken into account, then the conclusion that the orbit of the Moon is stable does no longer look so indisputable. For a detailed and captivating account of this

[1] The reader is encouraged to read A. Yu. Ishlinskiĭ's detailed survey on the achievements of mechanics [4.15].

question the reader is referred to V. G. Demin's book [4.5]. However, for artificial Earth satellites tidal forces are of lesser importance, and we will not dwell further upon them.

3. The region of weakly-perturbed motion

The entire foregoing analysis and the conclusions concerning the evolution of motion are of course valid only for weak perturbation of Keplerian motion. If in its motion an Earth satellite passes too close to the Moon, then the perturbations due to the Moon's attraction become so large that the orbit is sharply "pulled down" and the evolution (i.e., averaged) equations are no longer applicable. Therefore, the previous conclusions are true only under the condition that the satellite does not enter some neighborhood of the perturbing body. This condition is satisfied, for example, by any "inner" orbit (the satellite revolves around the Earth sufficiently "deep" inside the orbit of the Moon) or any stable "outer" orbit (the orbit of the Moon lies inside the orbit of the satellite); unstable outer orbits may also avoid the close vicinity of the Moon (but they may also fail to do so). However, for outer orbits the perturbations due to the attraction of the Sun play the dominant role. The existence of stable outer orbits speaks in favor of the possibility of existence of stable planetary systems around double stars that was mentioned in the first essay.

Let us define more precisely what we mean by the"neighborhood of the perturbing body" that the satellite must avoid in order for the evolution equations to remain valid.

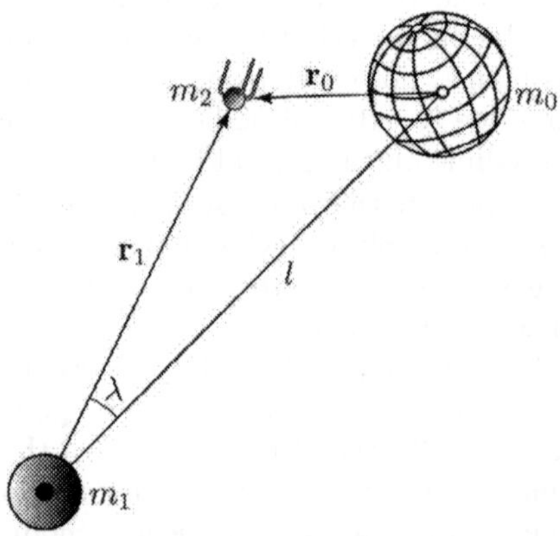

FIGURE 4.4. System of three particles m_0, m_1, m_2

Consider a system of particles of masses m_0 (Earth), m_1 (Moon) and m_2 (satellite) (Figure 4.4). Then to the force function (4.1.2) corresponds the background acceleration $\mathbf{a}_0$ of the particle m_2 due to the Earth's gravitational attraction, and the perturbing acceleration $\mathbf{f}_0$ due to the Moon's attraction:

$$\mathbf{a}_0 = -f\frac{m_0 + m_2}{r_0^3}\mathbf{r}_0, \qquad \mathbf{f}_0 = fm_1\left(\frac{\mathbf{r}_1 - \mathbf{r}_0}{l^3} - \frac{\mathbf{r}_1}{r_1^3}\right). \tag{4.3.1}$$

Here $\mathbf{r}_0$ [resp., $\mathbf{r}_1$] is the geocentric [resp., selenocentric] radius vector of the satellite and l is the distance between Earth and Moon. But the same motion of the satellite can be considered relative to the Moon, under the action of a background acceleration $\mathbf{a}_1$ and perturbing acceleration $\mathbf{f}_1$:

$$\mathbf{a}_1 = -f\frac{m_1 + m_2}{r_1^3}\mathbf{r}_1, \qquad \mathbf{f}_1 = fm_0\left(\frac{\mathbf{r}_0 - \mathbf{r}_1}{l^3} - \frac{\mathbf{r}_0}{r_0^3}\right). \tag{4.3.2}$$

If the quantity

$$\varepsilon_0 = \frac{|\mathbf{f}_0|}{|\mathbf{a}_0|} \tag{4.3.3}$$

is sufficiently small, then the motion is close to an unperturbed Keplerian motion around the Earth. Conversely, if the quantity

$$\varepsilon_1 = \frac{|\mathbf{f}_1|}{|\mathbf{a}_1|} \tag{4.3.4}$$

is sufficiently small, then the motion is close to an unperturbed Keplerian motion in the proximity of the Moon. The ratio

$$\varepsilon = \frac{\varepsilon_0}{\varepsilon_1} \tag{4.3.5}$$

will be taken as a measure of how close a motion is to Keplerian motion around the Earth. When ε_0 decreases, ε_1 increases and the motion becomes increasingly closer to a motion around the Earth and increasingly farther from a motion around the Moon. The indicated variation of ε_0 and ε_1 results in a decrease of ε.

Now let us consider the surface $\varepsilon = \text{const}$. This surface divides the space into regions that are "permitted" and "forbidden" for the application of the evolution equations; specifically, the "forbidden" region must cover the vicinity of the Moon, and in the "permitted" region the evolution equations describe the motion with an error of order ε over a time interval of order $1/\varepsilon$ (according to the general principles of asymptotic methods). We will call such a surface an *ε-surface.*

Squaring both sides of (4.3.5) and performing some simple transformations we obtain (provided also that we neglect the mass m_2, which is very small compared with m_0 and m_1) the equation

$$m_1^4 r_0^4\left(\frac{1}{l^4} + \frac{1}{r_1^4} - \frac{2}{l^2 r_1^2}\cos\lambda\right) = \varepsilon^2 m_0^4 r_1^4\left(\frac{1}{l^4} + \frac{1}{r_0^4} + \frac{2(lr_1\cos\lambda - l^2)}{l^3 r_0^3}\right), \tag{4.3.6}$$

where λ denotes the angle between the directions Moon–Earth and Moon–satellite (Figure 4.4) and

$$r_0^2 = l^2 + r_1^2 - 2r_1 l \cos\lambda. \tag{4.3.7}$$

Equations (4.3.9) and (4.3.7) give the ε-surface $r_1(\lambda)$

The complicated formula (4.3.6) is readily simplified. Namely, it is natural to expect that the ε-surface will bound near the Moon a region of small size (compared with the Moon–Earth distance), and so we can assume that r_1/l is small. Let us bring the terms in (4.3.6) to the common denominator and expand in powers of r_1/l. Retaining in this expansion only the terms of lowest order in r_1/l we obtain

$$\frac{1}{\varepsilon^2}\left(\frac{m_1}{m_0}\right)^4 = \left(\frac{r_1}{l}\right)^{10}(1 + 3\cos^2\lambda). \tag{4.3.8}$$

This surface of revolution is very close to a sphere. Indeed, when $\cos\lambda = 0$ we have $r = r_1^{\max}$, while for $\cos^2\lambda = 1$ we have $r = r_1^{\min}$; moreover, the ratio of the maximal dimension of the surface, $r_1^{\max}$, to its minimal dimension, $r_1^{\min}$, equals $\sqrt[5]{2} \approx 1.15$. Hence, for the ε-surface one can take the sphere

$$r_1 = l\left(\frac{m_1}{m_0}\right)^{2/5}\frac{1}{\sqrt[5]{\varepsilon}}. \tag{4.3.9}$$

Thus, if during its evolution the trajectory of our satellite does not enter the neighborhood of the Moon bounded by the sphere (4.3.9), then the evolution (averaged) equations will give a solution that differs from the true solution by a quantity of order ε, this being valid for a long time interval (generally speaking, of order $1/\varepsilon$).[2]

For $\varepsilon = 1$ equation (4.3.9) gives a sphere that is widely used in the mechanics of space flight under the name of the *sphere of influence* or *sphere of activity* [4.6]. If the satellite falls into the sphere of influence of the Moon, then it must already be considered a satellite of the Moon (for as long as it does not leave its sphere of influence). Outside this sphere the satellite can truly be regarded as a satellite of the Earth.

We speak here about the Moon–Earth system for the sake of definiteness, but one can consider other systems as well, for instance, the Earth–Sun system. Then formula (4.3.9) defines the ε-sphere (and the sphere of influence for $\varepsilon = 1$) of the Earth relative to the Sun. The table below gives the magnitudes of the radius r_1 of the ε-sphere and of the sphere of influence of the Moon (relative to the Earth) and of the Earth (relative to the Sun).

	$\varepsilon = 1$	$\varepsilon = 0.1$	$\varepsilon = 0.01$	$\varepsilon = 0.001$
Moon	66,280 km	105,050 km	166,360 km	263,790 km
Earth	924,820 km	583,490 km	368,460 km	232,370 km

[2]We will see that this time interval can also be infinite for finite ε.

In this table the radii of the ε-sphere are calculated for Earth satellites perturbed by the Moon (first row) or by the Sun (second row). Hence, when ε decreases the radii in the first row increase (the forbidden region around the Moon increases), whereas the radii in the second row decrease (the permitted region around the Earth increases). Naturally, to calculate the second row we must replace ε by ε^{-1} in formula (4.3.9). It is interesting to note that for $\varepsilon = 0.1$ there exists a permitted region surrounding both the Moon and the Earth, in which there can exist orbits of Earth satellites that are equally "weakly" ($\varepsilon \sim 0.1$) perturbed by the Moon and by the Sun. For $\varepsilon = 0.01$ there are no weakly-perturbed "outer" orbits that encircle both the Moon and the Earth (Figure 4.5). The values of the parameters used in calculating the preceding table are as follows:

	m_1/m_2	l
Moon (relative to the Earth)	81.35	384,300 km
Earth (relative to the Sun)	332,400	149,598,500 km

Let us remark that in the mechanics of space flight people also use the so-called *gravity spheres*, whose definition is based on other considerations [4.7].

The concept of sphere of influence will also be needed in our analysis of lunar fly-by orbits.

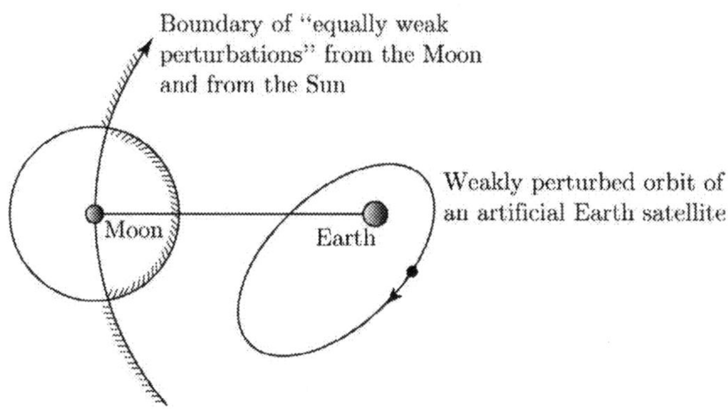

FIGURE 4.5. The region of weakly-perturbed motion

4. Stability of the Solar System

All the preceding conclusions concerning the stability of planetary orbits and their evolution were reached based on approximate, averaged, evolution equations and hence they cannot provide a rigorous answer to the problem of stability over an infinite time interval.

Recall that when we pose the question of the "stability of the Solar System" we have in mind the mathematical problem of stability of trajectories in the problem of n particles that attract each other according to Newton's law; the mass of one of the particles (the Sun) is considerably larger than the masses of all the others. No other forces besides the Newtonian force of attraction are considered.

In this formulation the problem was the subject of investigations of a number of prominent mathematicians for two centuries, but their efforts did not achieve the goal.

However, by 1961 the Russian mathematician V. I. Arnold was already reporting his fundamental results, which represented a giant step towards the solution of the stability problem. Arnold's results, based on a number of ideas of A. N. Kolmogorov, are of a general nature, dealing with the behavior of solutions of canonical (Hamiltonian) systems. The conclusions about the behavior of the Solar System follow as a particular case.

To explain this, let us consider a canonical system with the Hamiltonian

$$H(p,q) = H_0(p) + \varepsilon H_1(p,q,\varepsilon),$$

where p, q are n-dimensional vectors, ε is small, and H_1 is periodic in q with period 2π. If $\varepsilon = 0$, then the motion is described by the equations

$$\dot{p} = 0, \quad \dot{q} = \omega(p),$$

where $\omega = \partial H_0/\partial q = (\omega_1, \dots, \omega_n)$, and is represented on the n-dimensional torus $p =$const by a "winding" $q(t)$ of the torus.

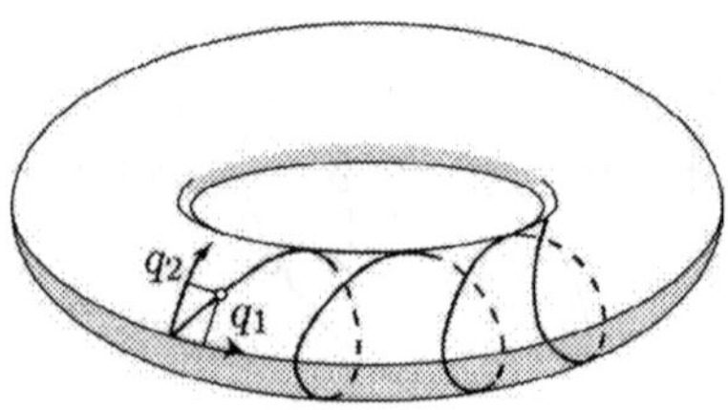

FIGURE 4.6. Torus carrying a conditionally-periodic unperturbed motion

For the two-dimensional case the situation is shown in Figure 4.6. On the two-dimensional torus ("bagel surface") we use the longitude q_1 and latitude q_2 as coordinates. The simultaneous uniform change of q_1 with rate ω_1 and of q_2 with rate ω_2 gives a trajectory that (as one can show) winds densely[3] around the torus provided that ω_1 and ω_2 are incommensurable. This kind of motion is

[3]This means that sooner or later the moving point will find itself inside any arbitrarily small domain on the torus.

called *conditionally periodic* or *quasiperiodic*. If, however, $\omega_1/\omega_2 = m/n$, where m and n are integers, then after the time $t = 2\pi m/\omega_1 = 2\pi n/\omega_2$ the trajectory closes itself. In that case the motion is periodic. Similarly, in the n-dimensional case, if $\omega_1 k_1 + \cdots + \omega_n k_n = 0$, where k_i are integers and at least two of them are different from zero, the trajectory is periodic, whereas if $\omega_1 k_1 + \cdots + \omega_n k_n \neq 0$ (for all choices of integers $k_1, \ldots k_n$ such that $k_1^2 + \cdots + k_n^2 \neq 0$) the motion is conditionally periodic and the trajectory $q_i(t)$ $(i = 1, \ldots, n)$ fills densely the n-dimensional torus p_i =const.

Now let us "turn on" the "perturbing part" εH_1 of the Hamiltonian H. Assume that ε is sufficiently small. Can we assert that the motion remains close to the original conditionally-periodic or periodic motion? In other words, are the original tori carrying the unperturbed motion preserved or destroyed?

If the original tori are not destroyed, but only "slightly deformed," then the perturbed motion remains forever close to the unperturbed one. In the perturbed motion $(\varepsilon \neq 0)$ $p_i \neq$const ("deformed torus"), but the p_i are close to the values p_i^0 =const ("nondeformed torus") corresponding to the unperturbed motion ($\varepsilon =$ 0.). To convince ourselves that the trajectory of the perturbed motion deviates only slightly from that of the unperturbed motion, it would suffice to construct (that's all!!!), in a neighborhood of the unperturbed motion, series in ε that converge for small ε and describe the perturbed motion. This "that's all" proved to be an enormous challenge for mathematicians for a duration of two hundred years, and only Arnold managed to cook this tough "bagel" (the torus carrying the perturbed motions).

Following an idea of Kolmogorov, Arnold constructed successive approximations of the type used in "Newton's method" (the analogue of the method of tangents for calculating roots of algebraic equations). The kth approximation of this method gives an error of order ε^{2^k}, which ensures the rapid convergence the method as $k \to \infty$. In the practical construction of the proof Arnold had to circumvent a number of serious difficulties. Let us mention several of them.

1. Resonances. If in the perturbed motion $\sum k_i\omega_i = 0$, then the perturbed trajectory does not necessarily lie on a torus: the original torus is destroyed. Even if one would succeed in constructing convergent series that describe the perturbed motion, this could be done not in the entire phase space, but only in certain "layers" of it. More precisely, it turns out that the construction is possible in the vicinity of perturbed tori carrying conditionally-periodic (but not periodic!) motions.

A second difficulty due to the presence of resonances is that the series describing the perturbed motion contain divisors (denominators) of the form $\sum k_i\omega_i$, which are small when the frequencies are close to a resonance. The presence of the small divisors prevents the series from converging. These difficulties were overcome by a theorem of Kolmogorov, the complete proof of which was given by Arnold [4.8]. The Newton-type convergence (like ε^{2^k}) "defeats" the harmful influence of the small divisors.

However, the application of the results of the analysis of Hamiltonian systems to the problem of stability of planetary orbits faces further obstacles, which were also circumvented by Arnold in the proof of his development of Kolmogorov's theorem. Let us mention these obstacles.

2. Proper degeneracy. This term is used to designate the fact that the number of frequencies in the perturbed motion can be larger than that in the unperturbed motion. Recall that the "Lagrangian motion of planets" involves more frequencies than the Keplerian motion: in addition to the fast "frequencies" of the motion of planets in their Keplerian orbits there are also the slow "frequencies" of precession of the planets' perihelia and of the ascending nodes of their orbits.

A path circumventing this obstacle is to take as unperturbed motion the "new" motion (for instance, instead of the Keplerian motion, consider the more accurate Lagrangian motion).

3. Limit degeneracy. This term refers to the following situation: among the n-dimensional tori $p =$const of the unperturbed motion there can be tori of dimension $k < n$, for instance, an equilibrium position ($k = 0$) or a single-frequency periodic motion ($k = 1$).

After a voluminous and difficult work, Arnold arrived in a rigorous and consistent manner to a theorem (see [4.9]) whose essence can be formulated as follows:

The perturbed motion described by a Hamiltonian system is for most initial conditions conditionally-periodic and remains forever close to an appropriately chosen "unperturbed" conditionally periodic motion.

Here the essential stipulation "for *most* initial conditions" means that *in the perturbed motion all tori of the unperturbed motion are preserved, except, possibly, for tori with resonant frequencies together with a small neighborhood of such tori.* The tori with resonant frequencies are, generally speaking, destroyed in the perturbed motion. The overall phase portrait of the perturbed motion can be represented in the form of conditionally-periodic motions on the preserved (slightly deformed) tori, with the gaps between these tori being filled by the remains of the destroyed tori (Figure 4.7).

Very little is known about the behavior of the trajectories in these "gaps."[4] If the problem under consideration has two degrees of freedom, then the energy integral $H =$const gives a three-dimensional manifold, which is divided by the tori carrying conditionally-periodic motions (Figure 4.7), and the trajectories in the gaps between these tori have nowhere to go (they cannot cross the tori carrying conditionally-periodic motions because motions depend in a one-to-one manner on the initial conditions). And since the width of a gap is small together with ε, *in a system with two degrees of freedom one can assert that even the resonant motions are stable.* In systems with a higher number of degrees of freedom the dimension of the tori carrying conditionally-periodic motions is smaller by at least 2 than

[4]Translator's note. This statement, of course, was valid at the time this book was written. See our note in the Additional Comments for this essay.

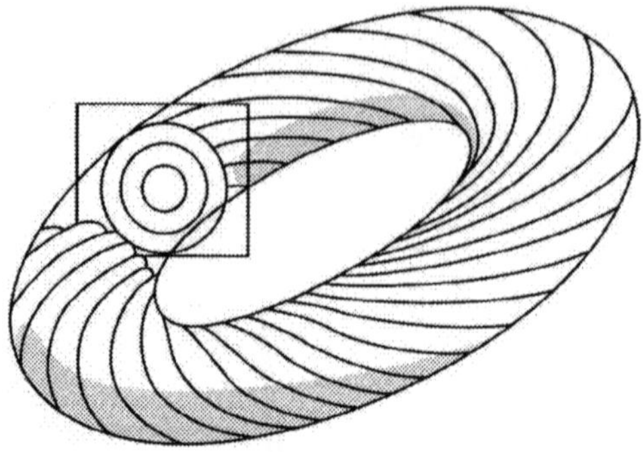

FIGURE 4.7. Tori carrying conditionally-periodic perturbed motions

the dimension of the phase space. These tori no longer divide the phase space and resemble a pine forest with trees growing arbitrarily close to one another, in which a squirrel, descending to the ground, may run between three pine trees for as long as it wishes (stability), but it may also ran away as far as it wishes (instability). Another squirrel may meanwhile run on the trunk of one of the pine trees in a conditionally-periodic motion, which remains stable even in a strong wind.[5]

The following strong result is an application of Arnold's theorem. Let the masses of the planets be sufficiently small compared with the mass of the central heavenly body (the Sun). If the eccentricities and inclinations of the orbits are sufficiently small, then *for most initial conditions the true motions of the planets are conditionally-periodic and remain forever close to corresponding suitably chosen Lagrangian motions. Moreover, the eccentricities and inclinations remain forever small, and the semi-major axes remain forever close to their initial values.* In other words, "Laplace's theorem" is valid for an infinite time interval.

Since earlier we have obtained "Laplace's theorem" by the averaging method, we conclude that there are cases (and very important ones!) in which the portrait of the motion furnished by the averaging method is close to the true portrait over the whole infinite time interval.

Let us state once more that the expression "most initial conditions" used above *excludes resonances between the frequencies of the motions.*

Arnold's results can hardly be overestimated. They were achieved by combining profound ideas with exhausting efforts to implement them. Just the formulation of individual lemmas and theorems takes up full pages of his work. Arnold himself uses rather strong statements in connections with this aspect of his work, for example: "... We will use the antiquated formal apparatus of dynamics" ([4.9], page 94).[6]

[5]See also N. N. Nekhoroshev's paper *The behavior of Hamiltonian systems that are close to integrable ones*, Funktional. Anal. i Prilozheniya **5** (1971), no. 4, 82–83. [MR **45** #3881]

[6]Eeyore, the old grey donkey, stood by the side of the stream, and looked at himself in the water. – "Pathetic," he said. "That's what it is. Pathetic." A. A. Milne, *The World of Pooh.*

To solve problems that have a two-century history is obviously not an easy task. Substantial contributions to the subject discussed here were made by prominent mathematicians of this century, among whom we mention H. Poincaré, G. D. Birkhoff, A. N. Kolmogorov, C. L. Siegel, and J. Moser. The Kolmogorov–Arnold–Moser theorem was decisive in hammering out the solution to the problem of the stability of motions of a system of gravitating bodies.

There remains of course the problem of the applicability of these results to concrete cases, for instance, to our concrete Solar System.

5. Is the Solar System resonant?

The difficulties with convergence of the series that describe perturbed motions in celestial mechanics arose largely because among the frequencies there are some that are nearly commensurable. Thus, for example, if the mean motion (frequency) ω_{Jup} of Jupiter's revolution around the Sun is taken to be 1, then the mean motion (frequency) ω_{Sat} of Saturn's revolution will be 0.4027, and so

$$2\omega_{\text{Jup}} - 5\omega_{\text{Sat}} = -0.0135. \tag{4.5.1}$$

Thus, the motions of Jupiter and of Saturn are nearly commensurable (resonance $\omega_{\text{Jup}}/\omega_{\text{Sat}} = 5/2$).

A. M. Molchanov proposed the following astonishing conjecture. In the evolution process – as he correctly argues – we should take into account small dissipative forces (i.e., forces that dissipate energy). In the Solar System these can be tidal forces, drag forces due to interplanetary dust, and other, perhaps not yet known to us dissipative forces.

We know that the perturbations due to the interaction between planets are very small. The dissipative forces are orders of magnitude smaller than even those small perturbations. But, acting over billions of years, the dissipative forces systematically change the orbits and steer the motion of planets to some stationary orbits – orbits that will remain practically unchanged for billions of years ahead.

To this point we have said nothing original: "all this is already known for a long time." The original part is that, according to Molchanov's conjecture, *these stationary orbits must be resonant*! In his words, "oscillating systems that have reached evolutionary maturity are unavoidably resonant, and their structure is given by a set of integers." In particular, Molchanov conjectured that *the Solar System is fully resonant*: the frequencies (mean motions) of revolution of the planets differ only slightly from frequencies for which Molchanov found a *full system of resonances* for the 9 known large planets. In addition, he discovered that similar resonances hold for some satellites of planets. Tables displaying his results were published in the international journal *Icarus*, devoted to problems concerning our Solar System [4.10], from which we borrowed Table 4.1.[7]

[7]The last rows of the table were constructed by employing the formal requirement that the corresponding determinant be equal to 1.

	Planet	ω_i^{O}	ω_i^{T}	$\Delta\omega/\omega$	n_1	n_2	n_3	n_4	n_5	n_6	n_7	n_8	n_9
1	Mercury	49.22	49.20	0.0004	1	−1	−2	−1	0	0	0	0	0
2	Venus	19.29	19.26	0.0015	0	1	0	−3	0	−1	0	0	0
3	Earth	11.862	11.828	0.0031	0	0	1	−2	1	−1	1	0	0
4	Mars	6.306	6.287	0.0031	0	0	0	1	−6	0	−2	0	0
5	Jupiter	1.000	1.000	0.0000	0	0	0	0	2	−5	0	0	0
6	Saturn	0.4027	0.400	0.0068	0	0	0	0	1	0	−7	0	0
7	Uranus	0.14119	0.14286	−0.0118	0	0	0	0	0	0	1	−2	0
8	Neptune	0.07197	0.07143	0.0075	0	0	0	0	0	0	1	0	−3
9	Pluto	0.04750	0.04762	−0.0025	0	0	0	0	0	1	0	−5	1

TABLE 4.1. Resonance relations in the Solar System

Table 4.1. gives integers n_i – positive, negative, and zero – such that

$$\sum_{i=1}^{9} n_i\omega_i = 0,$$

where ω_i are the frequencies (mean motions) of revolution of the large planets of the Solar System. Here one takes for ω_i some "theoretical values" ω_i^{T} of the frequencies, which satisfy exactly the resonance relations $\sum_i n_i\omega_i$, but side-by-side with these the table gives values ω_i^{O} that are obtained from actual astronomical observations, as well as the ratios

$$\frac{\Delta\omega}{\omega} = \frac{\omega_i^{\mathrm{O}} - \omega_i^{\mathrm{T}}}{\omega_i^{\mathrm{O}}},$$

which measure the deviations of the true frequencies from the resonant ones. One sees that these deviations are indeed small!

Similar tables are provided for the satellite systems of Jupiter, Saturn, and Uranus (Table 4.2).

These tables are quite convincing. In the worst case the deviations of the true frequencies from the resonant ones are no larger than 1.5 %. As a rule, resonance takes place for planets (or satellites of planets) that are closer to one another.

Since we are forced to give the frequencies only approximately as finite decimal fractions (i.e., as rational numbers), we can, generally speaking, always find integers n_i sufficiently large in the modulus, for which the resonance relations for frequencies are satisfied. However, Molchanov's tables contain *small* $|n_i|$, and not large ones, which also speaks in favor of his conjecture. For a given ν, let us denote by N the number of integers n_i given in Molchanov's tables for which $|n_i| = \nu$. The graph in Figure 4.8 shows the dependence of N on ν. We see that the values

Jupiter's satellites

	Satellite	ω_i^{O}	ω_i^{T}	$\Delta\omega/\omega$	n_1	n_2	n_3	n_4
1	Io	4.044	4.000	0.0110	1	−2	0	0
2	Europa	2.015	2.000	0.0075	0	1	−2	0
3	Ganymede	1.000	1.000	0.0000	0	0	−3	7
4	Callisto	0.4288	0.4285	0.0008	0	0	−1	2

Saturn's satellites

	Satellite	ω_i^{O}	ω_i^{T}	$\Delta\omega/\omega$	n_1	n_2	n_3	n_4	n_5	n_6	n_7	n_8
1	Mimas	16.918	16.800	0.0070	−1	0	2	0	0	0	0	0
2	Enceladus	11.639	11.600	0.0035	0	−1	0	2	0	0	0	0
3	Tethys	8.448	8.400	0.0057	0	0	−1	0	2	1	0	2
4	Dione	5.826	5.800	0.0045	0	0	0	−1	2	−1	0	−1
5	Rhea	3.530	3.500	0.0086	0	0	0	0	−1	2	2	0
6	Titan	1.000	1.000	0.0000	0	0	0	0	0	−3	4	0
7	Hyperion	0.7494	0.7500	0.0008	0	0	0	0	0	−1	0	5
8	Iapetus	0.2010	0.2000	0.0050	0	0	0	0	0	0	−1	4

Uranus' satellites

	Satellite	ω_i^{O}	ω_i^{T}	$\Delta\omega/\omega$	n_1	n_2	n_3	n_4	n_5
1	Miranda	6.529	6.545	−0.0025	−1	1	1	1	0
2	Ariel	3.454	3.454	0.0000	0	−1	1	2	−1
3	Umbriel	2.100	2.091	0.0043	0	0	−2	1	5
4	Titania	1.000	1.000	0.0000	0	0	1	−4	3
5	Oberon	0.6466	0.6364	0.0160	0	0	1	−2	0

TABLE 4.2. Resonance Relations in satellite systems of planets

$|n_i|$ are concentrated in the region where their values are small. It is therefore doubtful that the resonance relations discovered are accidental.

In connection with this, it may turn out that an important role in settling the question of the stability of the Solar System is played by the investigation of stability of precisely the resonant motions, which are discarded from the considerations in Arnold's theory (needless to say, this does not diminish the importance of that outstanding theory). On the other hand, Molchanov's conjecture generates more questions than answers. Is the system of "small" resonance numbers found by Molchanov unique, or one can choose another, no worse system? The magnitude of the deviation from zero of the resonance relations between the *observed* frequencies (rather than the chosen ones) may become significant compared with the smallest frequencies in the Solar System. For example, compare the mismatch 0.0135 in the resonance (4.5.1) with the frequencies of revolution of Saturn and Pluto, 0.07197 and 0.04750.

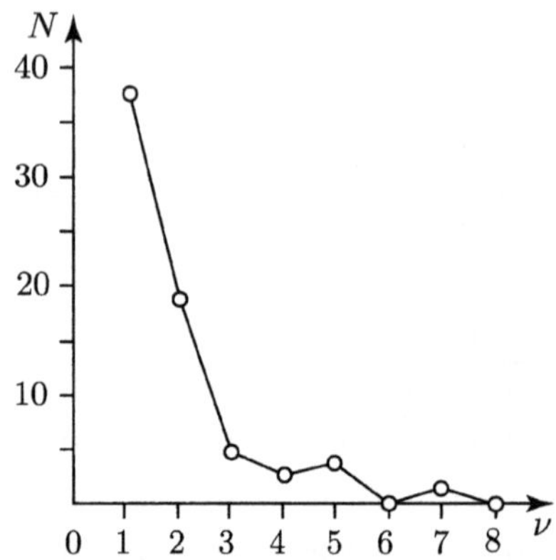

FIGURE 4.8. Distribution of the resonance numbers of the Solar System

Why did the Solar System, in its evolution, "sail" exactly toward such resonances and not others? Finally, what is the actual mechanism by which a system "sails" into a resonance regime? Remember that Molchanov's *conjecture* is for the moment just that, because it has not been established with mathematical rigour that under certain conditions (which precisely?) an oscillatory system necessarily enters a resonance regime (not to mention a regime of *full resonance*).[8]

It is well known from observations that asteroids "avoid" moving in certain resonant (with Jupiter) orbits; the gaps in Saturn's rings also have a resonant structure. The Moskovite mathematician A. D. Bruno has shown (see [4.11]) that in Hamiltonian systems the probability of an unstable periodic regime is the higher the smaller the *order of the resonance*.[9] Figure 4.9, which we borrowed from [4.12], presents the distribution of the number of asteroids according to the mean diurnal

[8]Rich and interesting information on similar phenomena in nature and technology and on the theory of such phenomena can be found in I. I. Blekhman's book *Synchronization of Dynamical Systems*, "Nauka," Moscow, 1971 (in Russian); see also I. I. Blekhman, *Synchronization in Science and Technology*, Amer. Society of Mechanical Engineers, 1988.

[9]Denote the mean diurnal motion of the asteroid and of Jupiter by n and n_{Jup}, respectively. Then the relative mean angular velocity of the asteroid is $n-n_{\text{Jup}}$. If the ratio $n_{\text{Jup}}/(n-n_{\text{Jup}}) = p/q$, where $q > 0$ and p, q are coprime integers, then, following Bruno's definition [4.11], the number q is called the *order* of the resonance.

motion n, with the value of n shown on the horizontal axis (in seconds of arc per day) and the number of asteroids in $5''$-intervals of mean diurnal motion shown on the vertical axis. At the top of the graph one sees the ratio $n/n_{\rm Jun}$, where $n_{\rm Jup} = 300''$ is Jupiter's mean diurnal motion. The order of the resonance is equal to the difference between the numerator and the denominator of these ratios. Bruno remarked that the character of the observed gaps (slits) in the distribution of the asteroids is qualitatively explained by his theory. The gap of order 3 ($n/n_{\rm Jup} = 5/2$) is complete, the possible periodic motion is always unstable – and there are no such motions. The higher-order gaps are not complete, and with the increase of the order they become less notable, since the probability of instability decreases: depending on the initial conditions, the motion may be stable as well as – with decreasing probability – unstable.

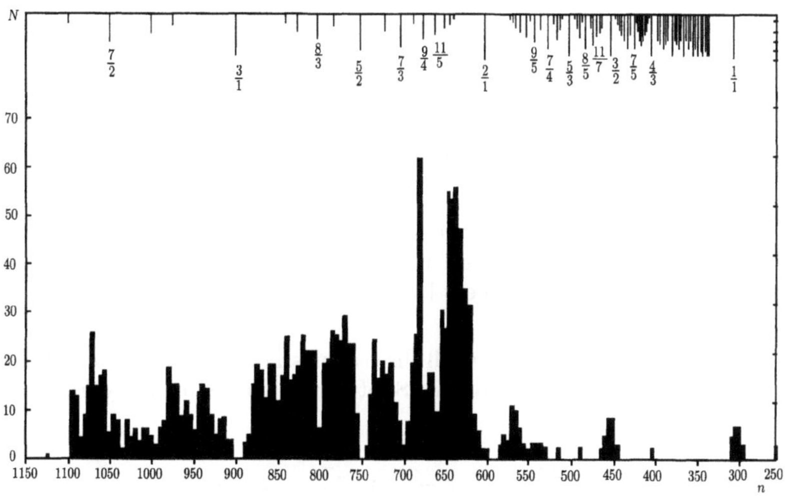

FIGURE 4.9. Distribution of the number of asteroids

How to reconcile the conjecture that the Solar System is resonant with the refusal of the asteroids to move along resonant orbits? Apparently, the answer lies in the fact that resonant motions are "singular trajectories" of the system and, by analogy with the "singular points" of differential equations, they can be stable or unstable. Consequently, our evolving system moves away from certain resonances (unstable ones) and sails toward other resonances (stable ones).

The special role played by resonances in the Solar System manifests itself also in the rotation of planets around their own axes. We all know that the Moon always shows the same side to the Earth – this is a beautiful example of $1:1$ resonance between the period of orbital revolution and the period of rotation around an axis (this is known also as *spin–orbit resonance*). We shall return later, in another essay, to the explanation of this effect. Relatively recently, after radar data on Mercury and Venus were processed, new facts about the rotation of these planets were revealed [4.13]. It turns out that the ratio of Mercury's period of axial rotation, T_{rot}, to its period of revolution, T_{rev}, equals 2/3! And already truly spectacular is the resonance of the rotation of Venus! Each time Earth and Venus get maximally close to one another, Venus is turned to the Earth with the same side, that is, the rotation of Venus is even in resonance not with its orbital motion, but with the motion of Venus relative to the rotating Sun–Earth axis!

According to recent radar data, the period of rotation of Venus is $T_\omega^{\mathrm{V}} = 243.24 \pm 0.1$ days, which within the limits of accuracy of observations coincides with the number of 243.16 days on which Venus, in each inferior conjuction, must be turned toward the Earth with the same side. The conjunction period is $\tau = 583.92$ days. Therefore, over this duration the Earth traverses along its orbit an arc

$$\alpha_{\mathrm{E}} \approx 2\pi + 2\pi \cdot 0.6 = \omega_{\mathrm{E}}\tau,$$

where ω_{E} is the mean diurnal motion of the Earth.

Similarly, in its orbital motion Venus traverses the arc

$$\alpha_{\mathrm{V}} = 2\pi \cdot 2 + 2\pi \cdot 0.6 = \omega_{\mathrm{V}}\tau;$$

rotating around its axis in *counter-wise* or *retrogade* manner (i.e., in the opposite direction with respect to its orbital revolution) with angular velocity ϖ, Venus spins around its axis over the time τ by an angle equal to $\beta = -2\pi \cdot 3 + 2\pi \cdot 0.6 = -\varpi\tau$.

This means that, after time τ, all three angles ω_{E}, ω_{V} and β acquire the same value $\approx 2\pi \cdot 0.6$ (up to multiples of 2π). Subtracting the last equality from the two preceding ones, we find that

$$5 \cdot \frac{2\pi}{\tau} = \omega_{\mathrm{V}} + \varpi, \quad 4 \cdot \frac{2\pi}{\tau} = \omega_{\mathrm{E}} + \varpi.$$

This means that over the time τ Venus performs 5 complete revolutions relative to its own orbital coordinate system (relative to the Sun–Venus axis) and 4 complete revolutions relative to the Sun–Earth axis. Since $2\pi/\tau = \omega_{\mathrm{V}} - \omega_{\mathrm{E}}$, the last two resonance relations lead to one and the same relation

$$\varpi = 4\omega_{\mathrm{V}} - 5\omega_{\mathrm{E}}.$$

Data on resonances in rotations of planets are collected in Table 4.3.

All of the above testifies about the special and nonaccidental role of resonances in the Motion of the Worlds.

It is said that in the process of earning general acceptance any good idea goes through three stages:

"This is not possible!"

"This is possible, but it is not yet proved."

"This was already known for a long time."

Molchanov's idea is presently somewhere halfway between the first and second stages.[10] His conjecture is surrounded by controversy and passion. And although the conjecture has not been confirmed, the mere fact that it was formulated has stimulated a wave of investigations in this direction; the scientific value of Molchanov's conjecture is therefore undeniable.

Molchanov's idea has something in common with earlier ideas of N. G. Chetaev.[11] We owe to Chetaev the following highly interesting thought [4.14]: "Stability, a fundamentally general phenomenon, should probably manifest itself in one form or another in the fundamental laws of Nature."

Developing this thought further, Chetaev arrived, in particular, at a hypothesis on the quantization of stable orbits of dynamics. He argued that stability is enjoyed only by some exceptional trajectories, much in the same way in which, in quantum mechanics, only some exceptional orbits of electrons are stable.

In conclusion, let us try to comprehend from the point of view of the "resonance principle" some well-known properties of motion in a central force field.

If a particle of mass m moves in a central force field with a force function $U(r)$ that depends only on the distance r from the particle to the origin, then the motion is planar and is completely determined by the laws of conservation of energy and momentum. Let us introduce polar coordinates r, φ in the plane where the motion takes place. Then the energy and area integrals are written as

$$\frac{m}{2}(\dot{r}^2 + r^2\dot{\varphi}^2) - U(r) = h, \tag{4.5.2}$$

and

$$r^2\dot{\varphi} = c, \tag{4.5.3}$$

respectively. Eliminating $\dot{\varphi}$ from (4.5.2) by means of relation (4.5.3), we arrive at the problem of investigating the following one-dimensional motion:

$$\frac{m}{2}\dot{r}^2 - U^*(r) = h, \quad U^*(r) = U(r) - \frac{mc^2}{2r^2}\,. \tag{4.5.4}$$

The quantity $U^*(r)$ is called the *effective* or (*amended*) *force function*. Denoting

$$\Phi(r) = \frac{2}{m}(h + U^*),$$

[10]This assertion was made in the 1972 edition of the book. For the present point of view see the Additional Comments at the end of this essay.

[11]Nikolai Gur′evich Chetaev (1902–1959), Corresponding Member of the Academy of Sciences of USSR, a prominent researcher in mechanics and mathematics, is the author of fundamental studies and ideas in the field of stability theory and analytical mechanics.

1. Moon	
Period of revolution around the Earth, in mean days	$T_0 = 27.32$
Period of axial rotation, in mean days	$T_\omega = 27.32$
Direction of rotation	direct*
Resonance	$\frac{T_\omega}{T_0} - 1 = 0$
2. Mercury	
Period of revolution around the Sun, in mean days	$T_0 = 87.97$
Period of axial rotation, in mean days	$T_\omega = 59 \pm 3$
Direction of rotation	direct
Resonance	$\frac{3T_\omega}{2T_0} - 1 \approx \pm 0.017$
3. Venus	
Period of revolution of Venus around the Sun, in mean days	$T_0^{\mathrm{V}} = 224.7$
Period of revolution of Earth around the Sun, in mean days	$T_0^{\mathrm{E}} = 365.24$
Period between the closest positions of Earth and Venus, in mean days	$\tau = 583.92$
Period of axial rotation of Venus, in mean days	$T_\omega^{\mathrm{V}} = 243.24 \pm 0.1$
Direction of rotation	opposite
Resonance	$\left(\frac{4}{T_0^{\mathrm{V}}} - \frac{5}{T_0^{\mathrm{E}}} - \frac{1}{T_\omega^{\mathrm{V}}}\right) T_\omega^{\mathrm{V}} \approx 0.001$

*Direct [resp., opposite, or retrograde] rotation means rotation in the direction of the orbital motion [resp., in the opposite direction].

TABLE 4.3. Resonances in the rotation of planets and their satellites we obtain from (4.5.4) the integrable equation

$$\dot{r} = \pm\sqrt{\Phi(r)}\,. \tag{4.5.5}$$

In an actual motion $\Phi(r) \geq 0$; if r_1 and r_2 are two consecutive roots of the equation $\Phi(r) = 0$ and if $\Phi(r) > 0$ for $r_1 < r < r_2$, then an actual motion that starts in the interval $r_1 \leq r \leq r_2$ will never leave it. The quantity r will typically oscillate with some period T_r between its extreme values r_1 and r_2. Furthermore, this picture is generally speaking valid for arbitrary values of the energy constant h ranging in some completely determined interval $h' \leq h \leq h''$.

Next, eliminating dt from (4.5.3) by means of relation (4.5.5) we conclude that over the time T_r the radius vector of the orbit rotates by an angle of $2\Delta\varphi$, where

$$\Delta\varphi = \int_{r_1}^{r_2} \frac{c\,dr}{r^2\sqrt{\Phi(r)}}\,. \tag{4.5.6}$$

If $\Delta\varphi = \pi$, then the trajectory is closed and in fact closes after one complete rotation of the radius vector. In the general case the trajectory closes if $\Delta\varphi = k\pi/n$, where k and n are integers.

But this is by no means obligatory! For arbitrarily taken values of the energy constant h and area constant c, the trajectory of a particle in a central force field is generally speaking not closed, representing, for example, a rosette with infinitely many petals, as shown in Figure 1.12 (a) of the first essay.

It this therefore even more staggering that in the Newtonian force field with the force function

$$U = \frac{\mu m}{r}, \tag{4.5.7}$$

which actually does exist in the Universe, the trajectories – Keplerian ellipses – are always closed for any negative values of the energy constant h; the orbits close after one complete rotation of the radius vector. This property does by no means follow from the conservation laws; rather, it is a consequence of the specific form of the effective force function

$$U^* = \frac{\mu m}{r} - \frac{mc^2}{2r^2} \tag{4.5.8}$$

corresponding to Newton's law of gravitational attraction. Except for the Newtonian force field, similar properties are enjoyed only by the field

$$U = -kr^2, \quad k = \text{const},$$

in which a particles executes harmonic oscillations.

Thus, the Newtonian force field is a unique phenomenon, one of the wonders of Nature. It is amazing how Nature did select a special, unique way of gravitational attraction, which ensures that the orbits are closed.

But what does it *mean* that an orbit is closed? It means that the period T_r of oscillations of the magnitude of the radius vector and the period T_φ of rotation of the radius vector by an angle of 2π are commensurable! Keplerian orbits enjoy the commensurability $T_r : T_\varphi = 1 : 1$. In other words, "at the core" of Keplerian motion – and hence of Newton's law of universal attraction – lies a resonance!

The *resonance principle* states that "evolutionary-mature systems are unavoidably resonant." One is prompted to ask whether Newton's law of gravitational attraction is not itself a product of the evolution of the Universe by virtue of the "resonance principle?" We owe this daring thought also to Molchanov.

Additional comments for this translation

More than 20 years have passed since the publication of the second edition of these Essays (1977), years rich in new investigations and results in the theory of dynamical systems. A quite typical phenomenon in dynamical systems is that of "deterministic chaos" – trajectories whose behavior is practically chaotic, despite the fact that the initial conditions and the parameters of the system have a deterministic nature [4.16], [4.17]. Special limit motions, termed chaotic attractors ("strange attractors") were discovered. This demonstrated that limit regimes in dynamical systems are not necessarily resonant. Therefore, the "resonance principle" in Molchanov's general formulation ("evolutionary-mature systems are unavoidably resonant") is not valid.

This, of course, does not preclude the existence of limit resonances in dynamical systems, in particular, of numerous resonances in the Solar System.

Concerning the rotational motion of Venus mentioned in this essay see the 6th Essay and the additional comments to it.

Translator's notes

1. In recent years the behavior of orbits in the gaps between the perturbed invariant tori in the Kolmogorov–Arnold-Moser theory, including the phenomenon known as "Arnold diffusion," has become a subject of intensive research. Since it is hard to compile a reasonably complete list of relevant references, we mention here only the paper by P. Lochak *Arnold diffusion; a compendium of remarks and questions*, in: Hamiltonian Systems with Three or More Degrees of Freedom (S'Agars, 1995), 168–183, NATO Adv. Sci. Inst. Ser. C Math. Phys. Sci., 533, Kluwer Acad. Publ., Dordrecht, 1999, where a critical analysis of the status of the theory in 1995 is made.
2. The interested reader may also consult the recently published book by C. D. Murray and S. F. Dermott *Solar System Dynamics,* Cambridge Univ. Press, Cambridge, 1999.

Fifth Essay

The Restricted Three-body Problem, Flight to the Moon, and Galactic Evolution

> *Mephisto*: ... O Moon, for you – my kiss. (Spreading his dark red cape on the ground, he sits down on it, and flies away with a whir through the window, which opens widely with a noise).
>
> A. V. Lunacharskiĭ, *Faust and the City*

1. The Hill surfaces

The opening up of the Earth–Moon–Earth route, which took place right in front of our eyes, did actually start long before the first artificial satellite was launched – on paper, on the draft table, in numbers crunched by fast computers. Already in 1957 V. A. Egorov published a voluminous work [5.1] containing results of an analysis of trajectories of lunar flights. In this short essay we will pause only briefly to discuss Egorov's investigations, with the aim of demonstrating how much can be obtained with very elementary means. A detailed account of these investigations can be found in the monograph [5.2]. A popularized, but in-depth exposition of Egorov's works is also given in V. I. Levantovskiĭ's book *On a Rocket to the Moon* [5.3].

The problem of lunar flight can be considered in the setting of the *restricted three-body problem*. Let m_1 be the mass of the Earth and m_2 the mass of the Moon. A projectile of mass m_0 moves under the attraction of these masses; it is assumed that m_0 is very small compared with m_1 and m_2, so that the influence of the attraction of the projectile on the Earth and on the Moon can be neglected. Then the Earth and the Moon move in familiar Keplerian orbits around their common center of mass. We will assume that these orbits are circular (i.e., consider the *circular three-body problem*). We take as unit of time $T/(2\pi)$, where T is the period of revolution on these circular orbits, and take the distance between Earth and Moon ($a = 384400$ km) as the unit of length. Then, by Kepler's third law, in the measurement units adopted

$$f(m_1 + m_2) = a^3 \left(\frac{T}{2\pi}\right)^{-2} = 1,$$

where f is the constant of gravitation.

Our task is to determine the trajectories of the particle m_0 in this setting. The differential equations governing the motion of m_0 can be written in the form

$$\left.\begin{aligned} \ddot{x} &= \quad 2\dot{y} + \frac{\partial \Phi}{\partial x}, \\ \ddot{y} &= -2\dot{x} + \frac{\partial \Phi}{\partial y}, \\ \ddot{z} &= \quad \frac{\partial \Phi}{\partial z}, \end{aligned}\right\} \tag{5.1.1}$$

where

$$\Phi = \frac{1}{2}(x^2 + y^2) + \frac{fm_1}{r} + \frac{fm_2}{\rho}. \tag{5.1.2}$$

Equations (5.1.1) are the equations of motion of the nonattracting particle m_0 in the restricted three-body problem. Here the coordinates x, y, z define the position of the particle m_0 in a rotating frame whose x-axis passes at all times through the centers of the Earth and the Moon and whose y-axis lies in the Moon's orbital plane and passes through the Earth–Moon center of mass.[1] The z-axis completes the orthogonal frame. Also,

$$r = \sqrt{(x - x_1)^2 + y^2 + z^2}\,, \quad \rho = \sqrt{(x - x_2)^2 + y^2 + z^2}\,,$$

where x_1 and x_2 are the constant coordinates of the points m_1 and m_2, situated on the x-axis; r [resp., ρ] is the distance from the nonattracting particle (projectile) m_0 to the Earth m_1 [resp., the Moon m_2].

Unfortunately, equations (5.1.1), which play a very crucial role in celestial mechanics and the mechanics of space flight, are not integrable. Perhaps a more cautious statement would be that until now no one has succeeded in integrating them in closed form. At any rate, at the present time the investigation of flight trajectories of a rocket in the Earth–Moon system must be carried out either by using some approximation tools, or by means of the numerical integration of the equations of motion.

True, some conclusions can nevertheless be drawn by analyzing the equations (5.1.1). It is readily seen that these equations admit a first integral, known as the *Jacobi integral*

$$\frac{1}{2}V^2 = \Phi + h, \tag{5.1.3}$$

where $V = \sqrt{\dot{x}^2 + \dot{y}^2 + \dot{z}^2}$ is the velocity of the projectile in the rotating coordinate system, h is an arbitrary constant, and $\Phi(x, y, z)$ is given by formula (5.1.2). Since $V^2 \geq 0$, in any real motion

$$\Phi(x, y, z) \geq -h. \tag{5.1.4}$$

[1] It is precisely the rotation of the coordinate system that is responsible for the terms $2\dot{y}$ and $-2\dot{x}$ in (5.1.1) and $\frac{1}{2}(x^2 + y^2)$ in (5.1.2). If these terms were absent, then equations (5.1.1), (5.1.2) would reduce to the already familiar integrable equations of the problem of two fixed centers.

The surfaces

$$\Phi(x, y, z) = -h, \tag{5.1.5}$$

which bound the region of possible motions, are called *Hill surfaces*, in the honor of the first scientist who studied them. For the Earth–Moon system $m_1 : m_2 = 81.30$, and for this case the Hill surfaces are sketched in Figure 5.1.

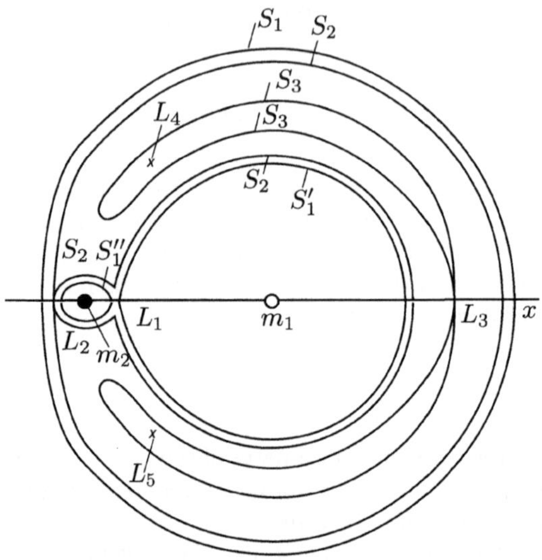

FIGURE 5.1. Section of the Hill surfaces by the Moon's orbital plane

This figure shows the section of the Hill surfaces (5.1.5) by the plane $z = 0$; several surfaces are shown, corresponding to different values of the energy constant of relative motion, h. As it turns out, for large negative values of h motion is possible only inside some noncontiguous surfaces S' and S'', which are close to spheres centered at the points m_1 and m_2, respectively, but also in the exterior of a certain surface S, which surrounds both S' and S''. The section of each of the surfaces S' and S'' is nearly circular. Although these surfaces are not shown in Figure 5.1, one can readily imagine how they look (by slightly shifting inward with respect to the surfaces S_1' and S_1'' and outward with respect to the surface S_1). Therefore, for such values of h the projectile m_0 can be either a satellite of the Earth, or a satellite of the Moon (and it can also move far away from the Earth–Moon system, without intersecting the boundary S).

However, no motions are possible in which the projectile first moves in the close vicinity of the Earth, and then, after some time, reaches the close vicinity of the Moon; in order for this to become possible it is necessary to increase the energy constant h. More precisely, for some value $h = h_1$ the surfaces S' and S'' become contiguous at a point L_1. Let us denote the corresponding surfaces by

S_1' and S_1'' (Figure 5.1). If we slightly increase the value of h further so that the difference $h - h_1$ is small, we discover that a narrow "wormhole" (neck) forms on the Hill surface around the point L_1. The region of possible motions of the projectile now includes a neighborhood of the Earth as well as a neighborhood of the Moon, connected by a narrow wormhole. Hence, trajectories which start near the Earth and penetrate the vicinity of the Moon through the wormhole near the point L_1 are not "forbidden." In other words, Earth–Moon flights become possible only for energy values $h > h_1$. Since the flight starts always very close to Earth (say, from the orbit of an Earth satellite located 200 km above the Earth's surface), increasing the energy h is tantamount to increasing the initial velocity V_0.

The initial position of the projectile on the orbit of an Earth satellite has practically no influence on the magnitude of the velocity V_0 that is required, say, to attain the energy value $h = h_1$ (we will denote this velocity by $V_0^{(1)}$; the corresponding numerical values will be given below).

Let us increase the initial velocity further to a value $V_0^{(2)}$ (i.e., increase the energy to a value $h_2 > h_1$), such that the surfaces S and S'' become contiguous at a point L_2 (on Figure 5.1 to this case there corresponds the single surface S_2). Clearly, increasing the energy further results in a second wormhole being formed around the point L_2. What this amounts to is that, if the energy is only slightly larger than h_2, then a trajectory of the projectile of the following kind is possible: it first moves in the vicinity of the Earth, then passes through the wormhole near the point L_1 to the vicinity of the Moon, and then passes through the wormhole near the point L_2 and leaves the Earth–Moon system. How many revolutions will the projectile make around the Earth before it reaches the vicinity of the Moon? Will the projectile revolve around the Moon, as its temporary satellite, before it flies away from the Earth–Moon system into the outer space? We do not know the answers to these questions. At this time we can only assert that for $h > h_2$ trajectories of this type are not forbidden; but without further investigations it is not clear whether they actually exist. This question will be me made somewhat clearer below. For the moment, an examination of Figure 5.1 reveals that for some $h = h_3$ the region where motion is forbidden has in each of the half-planes a "drop-shaped" component which shrinks to a point L_4 [resp., L_5] when h increases from h_3 to h_4.

When $h > h_3$ the projectile can escape to infinity near the point L_3 on the x-axis, while for $h > h_4$ (with $h_4 < 0$) it can escape to infinity along any direction in the (x, y)-plane. However, in the (x, y, z)-space the Hill surface, which bounds the motion, disappears when $h = 0$.

The points L_i $(i = 1, 2, 3, 4, 5)$ are called *libration points*. Table 5.1 shows the results of the computations of the distances r_i and ρ_i from the libration points to the Earth and to the Moon, respectively, the energies h_i corresponding to these points, and the velocities $V_0^{(i)}$ corresponding to these energies. In the computation of the values $V_0^{(i)}$ it was assumed that the initial point lies 200 km above the Earth's surface.

The distances from the points L_1 and L_2 to the Moon are equal to 58,000 km and 65,000 km, respectively, i.e., they both lie inside the Moon's sphere of influence (see the definition in the preceding essay), close to the boundary of that sphere (which has a radius of 66,000 km).

Libration point	r_i	ρ_i	h_i	$V_0^{(i)}$, in units $2\pi a/T$	$V_0^{(i)}$, in km/sec
L_1	0.8491539	0.1508461	−1.594067	10.60335	10.84890
L_2	1.1677237	0.1677237	−1.585991	10.60411	10.84968
L_3	0.9929263	1.9929263	−1.506062	10.61165	10.85738
L_4	1	1	−1.494001	10.61278	10.85854

TABLE 5.1.

Note how small ($\sim$ 1 m/sec) the difference between the velocities $V_0^{(1)}$ and $V_0^{(2)}$ (and also between $V_0^{(3)}$ and $V_0^{(4)}$) is. This means that a very small change in the initial velocity can result in a qualitative change of the trajectory: for $V_0 \approx V_0^{(1)}$ the trajectory is necessarily bounded, whereas for $V_0 \approx V_0^{(2)}$ it may become unbounded – the force of attraction of the Moon can eject the projectile (through the wormhole near L_2) from the Earth–Moon system.

2. Digression on libration points

The libration points L_i are singular points of the surface $\Phi(x, y, z) = h$, i.e., their coordinates are solutions of the system of equations

$$\frac{\partial \Phi}{\partial x} = 0, \quad \frac{\partial \Phi}{\partial y} = 0, \quad \frac{\partial \Phi}{\partial z} = 0. \tag{5.2.1}$$

Libration points have an important mechanical meaning. Indeed, comparing the equations of motion (5.1.1) and conditions (5.2.1) we see that the libration points are *relative equilibria*: if the initial velocity of the projectile (in the rotating coordinate system) vanishes, i.e., if $\dot{x} = \dot{y} = \dot{z} = 0$, and if the projectile is located in one of the libration points, then the projectile will remain in that libration point for the entire duration of its motion.

The question of the stability of these equilibrium positions was posed already by Lagrange. Will the projectile move forever in the vicinity of a libration point if it is slightly pushed away from that point and/or imparted a small velocity? It soon was discovered that the libration points L_1, L_2, L_3 are *unstable*.

The problem of the stability of the points L_4 and L_5 proved to be more difficult (these points are known as the *Lagrange triangular points* because they form triangular configurations together with the two attracting masses m_1 and m_2). It turned out that if one linearizes the equations of motion (5.1.1) near the triangular libration points, then a solution of the linearized equations will be bounded if and only if the ratio of the masses of the attracting points, $\varkappa = m_1/(m_1+m_2)$ (here one assumes that $m_1 < m_2$), satisfies the condition

$$\varkappa(1-\varkappa) < \frac{1}{27}\,. \tag{5.2.2}$$

If (5.2.2) does not hold, then the motion is unstable in the linear approximation. If (5.2.2) holds, then the motion is stable in the linear approximation; we know, however, that this may not be sufficient for stability.

The problem of the stability of Lagrange's triangular points remained unsettled for almost two hundred years. Progress in the investigation of this problem became possible only recently, after V. I. Arnold proved a theorem on the stability of equilibrium positions in Hamiltonian systems with two degrees of freedom. This allowed one to move forward in the study of the *planar* problem. Using Arnold's results, A. M. Leontovich showed [5.4] that for all values of the mass ratio $\varkappa$ satisfying condition (5.2.2), except, possibly, for a finite number of such values of $\varkappa$, the motion is stable. Subsequently A. Deprit and A. Deprit-Bartholomé [5.5] described the set of exceptional values in concrete terms. The solution of this problem was finalized by A. P. Markeev in 1969 [5.6]. Markeev's results alow us to state the following theorem.

The Lagrange triangular point are stable for all mass ratios that satisfy condition (5.2.2), *except for the two ratios*

$$\left.\begin{aligned} \varkappa &= \frac{15-\sqrt{213}}{30} = 0.0135160\ldots, \\ \varkappa &= \frac{45-\sqrt{1833}}{90} = 0.0242938\ldots, \end{aligned}\right\} \tag{5.2.3}$$

for which the Lagrange triangular points are unstable.

This theorem settles the two-century old problem of the stability of the libration points in the planar circular restricted three-body problem. The three-dimensional problem was also investigated in detail.

Let us point out that the stability of the libration points depends in essential manner on the influence of the so-called "gyroscopic" terms in the equations of motion. These terms are present because the motion is considered in a rotating (together with the Earth–Moon axis) coordinate system. The balance of the forces of attraction and the centrifugal forces is achieved precisely at the libration points.

In stability theory one proves that "gyroscopic" stability is destroyed by dissipative forces (for instance, the drag forces due to the presence of interplanetary matter). Hence, in the presence of dissipative forces a projectile that enters the vicinity of a libration point may remain near that point only temporarily – possibly for a very long time, but not forever.

It is therefore even more interesting to mention that in 1961 the Polish astronomer K. Kordylewski discovered accumulations of cosmic dust near the triangular libration points of the Earth–Moon system. These formations may possibly have an unstable character – part of the dust particles gradually leave the vicinity of the libration points under the influence of the perturbing forces, but is replaced by new dust particles arriving from the surrounding space.

Now let us return to the lunar flight problem.

3. Moon intercept trajectories and a method for their investigation

Thus, as we saw above, if in the vicinity of Earth one imparts to a projectile a speed of (in the rotating system of coordinates) $V_0^{(1)} = 10.84890$ km/s (see Table 5.1), then the projectile has the chance of sooner or later reaching the close vicinity of the Moon.

But does this happen sooner rather then later?

Of course, in practice one is interested in rapid lunar flights, and not flights that might drag for ... several centuries. Incidentally, numerical computations show that after it acquires the velocity $V_0^{(1)}$, the projectile will reach a distance of "merely" $\approx 260,000$ km from Earth (and 384,400 km from the Moon!), and then return back to the Earth; true, in the next revolution the projectile will reach a bit farther, but then return again to the Earth, and so on. Hundreds or thousands of revolutions around the Earth are apparently necessary for the projectile to get close to the "wormhole" near the libration point L_1 and fly through it to the Moon. Such a trip is both boring and lengthy. It would be far better to reach the Moon on the first lap in orbit. We will consider here the trajectories that intersect the Moon's sphere of influence during their first revolution (around the Earth). Such trajectories were termed by V. A. Egorov *encounter trajectories* ("*traektorii sblizheniya*"). We will use here the term *Moon intercept trajectories*.

Among the Moon intercept trajectories of particular interest are those in which the projectile does actually reach the Moon after its first lap around the Earth; according to Egorov's calculations, the minimal absolute velocity necessary for this to happen is $V_1 = 10.90525$ km/s. This value was obtained without taking into account the influence of the Moon's attraction, but the effect of this influence on the value of V_1 is small. Hence, V_1 should be regarded as a basic, characteristic velocity. It is instructive to compose a table of the characteristic initial velocities for flights in the Earth–Moon system (all computations were carried out for an initial altitude of 200 km above the Earth's surface; the values were recalculated in the nonrotating coordinate system):

$V_0^{(1)} = 10.86570$ km/s – in principle the Moon can be reached after many revolutions around Earth;

$V_0^{(2)} = 10.86640$ km/s – escape from the Earth–Moon system is possible (using the attraction of the Moon for additional acceleration);

$V_1 = 10.90525$ km/s – the minimal velocity necessary for reaching the Moon on the first lap of trajectory around the Earth;

$V_p = 10.99967$ km/s – the *parabolic* or *escape velocity*, which guarantees escape from the Earth–Moon system (regardless of the influence of the Moon's attraction).

We have already mentioned that the problem of calculating a lunar flight trajectory is very difficult, because even if one takes into account only the main factors – the attraction of the Earth and of the Moon – the problem cannot be solved analytically.

If under study were the motion of an Earth satellite with allowance for the attraction of the Moon, then, as we know from the preceding essay, we could use the fact that the perturbations due to the Moon are small and apply asymptotic methods of analysis.

In the present case this is not possible, because the influence of the Moon not only is not small, but is in fact dominant in the near-Moon leg of the flight.

However, we can take advantage of precisely this circumstance to construct an approximation algorithm for calculating lunar flight trajectories. We shall consider here only Moon intercept trajectories. Then the effect of Moon's gravity is small as long as the projectile does not enter the sphere of influence of the Moon; accordingly, the effect of the Moon's gravity will be fully neglected until that moment. Conversely, when the projectile moves inside the Moon's sphere of influence we will take into account the attraction of the Moon, but neglect completely the influence of Earth. In this manner the three-body problem reduces to two two-body problems, and as we already know, the two-body problem can be easily solved. Its solutions are Keplerian trajectories. The algorithm for calculating lunar approach trajectories is then as follows:

1) Use Kepler's laws to calculate the geocentric motion from Earth to the Moon's sphere of influence.
2) At the boundary of the Moon's sphere of influence recalculate the parameter of the geocentric motion into parameters of selenocentric motion.
3) Calculate the selenocentric Keplerian trajectory.
4) If needed, the calculation can be pursued further: after the projectile leaves the Moon's sphere of influence, one can calculate a new geocentric trajectory.

Such a method for calculating trajectories of celestial bodies was proposed already by Laplace (with applications to the motion of artificial celestial bodies in mind).

It turns out that this simple method of computation yields excellent estimates of the parameters of motion, which differ very little from the exact values. This was verified by comparing the results of the approximate calculations with those of the numerical integration of the equations of motion. The method described above has been successfully applied to the calculation of orbits of lunar spacecraft in works of V. A. Egorov [5.1], [5.2], of M. L. Lidov, D. E. Okhotsimskiĭ, and N. M. Teslenko [5.7], and of other researchers. Following Egorov's work, we shall discuss here in more detail an example of calculation of lunar flyby trajectories.

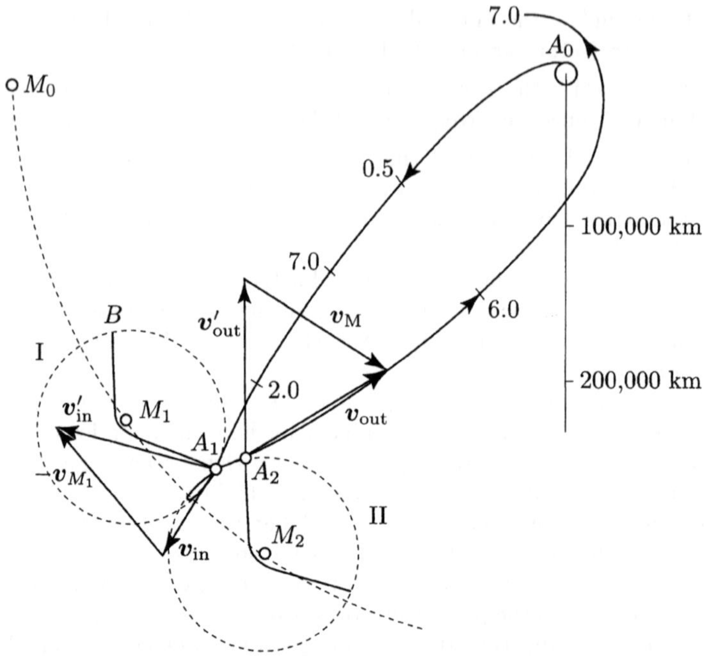

FIGURE 5.2. Approximate calculation of lunar flyby trajectories according to V. A. Egorov (flight durations are indicated in days)

Assume that the projectile starts at the point A_0 near the Earth (Figure 5.2) and moves along an arc of Keplerian ellipse in the direction of the Moon, which at the initial time is located at the point M_0. (Figure 5.2 shows a trajectory for a concrete case, with the flight time in days indicated along various trajectory segments). In the point A_1 the projectile intersects the Moon's sphere of influence (the dotted line in Figure 5.2). In the time interval from the beginning of the flight to this moment the Moon moved along its orbit from M_0 to the point M_1. The geocentric velocity of the projectile at the moment when its orbit intersects the Moon's sphere of influence is represented in Figure 5.2 by the vector $\mathbf{v}_{\text{in}}$. According to the agreement made above, at this moment of time we begin to regard the motion as selenocentric. The orbital velocity of the Moon is $\mathbf{v}_{M_1}$, and its magnitude is $|\mathbf{v}_{M_1}| = 10.2$ km/s. The velocity of the projectile *relative to the Moon*,

$$\mathbf{v}'_{\text{in}} = \mathbf{v}_{\text{in}} - \mathbf{v}_{M_1}, \tag{5.3.1}$$

is calculated by the usual formula: "the relative velocity equals the difference of the absolute velocity and the transport velocity." In our case the transport velocity is the velocity of the Moon, whence the vector formula (5.3.1). The vectors $\mathbf{v}'_{\text{in}}$ and $-\mathbf{v}_{M_1}$ are also shown in Figure 5.2.

Thus, the projectile enters the sphere of influence of the Moon with the velocity $\mathbf{v}'_{\text{in}}$.

V. A. Egorov discovered the following remarkable fact: *for any Moon intercept trajectory the entry velocity in the Moon's sphere of influence, $|\mathbf{v}'_{\text{in}}|$, calculated relative to the Moon, is always larger than the selenocentric escape (parabolic) velocity at the boundary of the sphere of influence.* This escape velocity, $V'_{\text{p}} = 0.383$ km/s, guarantees that the projectile will unavoidably leave the Moon's sphere of influence. Since $|\mathbf{v}'_{\text{in}}| > V'_{\text{p}}$, our projectile will *either fall on the Moon, or necessarily leave the Moon's sphere of influence, passing by and around the Moon on a hyperbolic trajectory.*

Figure 5.2 shows such a hyperbolic trajectory, which goes around the Moon inside its sphere of influence. The projectile flies on this trajectory from the point A_1 to the point B. But over the time spent by the projectile on the arc $\overset{\frown}{A_1B}$ the sphere of influence will itself move from position I to position II. When the projectile leaves the sphere of influence, intersecting it at the point B, the Moon occupies the position M_2, while the point B occupies position A_2. Consequently, relative to the Earth the trajectory A_1B, which *relative to the Moon* is hyperbolic, becomes the "strange" loop-like curve A_1A_2. Moving along this loop, the projectile leaves the Moon's sphere of influence with relative velocity $\mathbf{v}'_{\text{out}}$ (which is practically of the same magnitude as $\mathbf{v}'_{\text{in}}$, but, needless to say, has a different direction). The geocentric exit velocity $\mathbf{v}_{\text{out}}$ is calculated by the formula "the absolute velocity equals the sum of the relative velocity and the transport velocity", so that

$$\mathbf{v}_{\text{out}} = \mathbf{v}'_{\text{out}} + \mathbf{v}_{M_2}. \tag{5.3.2}$$

The vector $\mathbf{v}_{\text{out}}$ is shown in Figure 5.2. If its magnitude and direction are known, it is not difficult to calculate the Keplerian trajectory of return to the vicinity of Earth, the result being displayed in Figure 5.2 as well.

In this example we see how the method proposed by Egorov enables us to construct complicated trajectories of the three-body problem by using very modest means. As we already mentioned, the accuracy of this method is sufficiently high to make it usable in the preliminary phase of the design of Earth–Moon flight trajectories. In [5.1], [5.2] Egorov did calculate, analyze, and describe hundreds of Earth–Moon flight trajectories, obtained a complete classification of Moon intercept trajectories, and considered a number of other interesting problems of the dynamics of Lunar flight. A detailed and captivating exposition of this work can be found in the excellent popularization texts of V. I. Levantovskiĭ [5.3] and [5.8], after the reading of which Egorov's original book [5.2] should become more accessible. For a first acquaintance with these problems the reader is referred to the survey paper of L. I. Sedov [5.9] on the orbits of lunar rockets.

4. Galactic evolution

Now instead of a single individual particle m_0 whose motion is governed by equations (1.5.1), let us imagine a "cloud" made of a few hundreds or thousands of such particles. In other words, let us consider simultaneously several hundreds (or thousands) restricted three-body problems. Such a set of problems serves as a good mathematical model for the interaction and evolution of galaxies. Here, for example, the particle of mass m_1 plays the role of the core of the galaxy we are interested in; the cloud of nonattracting (but attracted) particles m_0 represents the galaxy itself (the stars, dust, or gas particles that initially revolve around the core m_1); finally, the mass m_2 models a "perturbing body," which may be a neighboring galaxy or a supermassive invisible body (or several such bodies). In this setting it is natural to consider the *hyperbolic* three-body problem: the mass m_2 flies in a hyperbolic orbit relative to the mass m_1, perturbing in its motion the initial motion of the cloud of particle-satellites of the mass m_1. The problem is then to determine the evolution of the structure of our cloud-galaxy under the described circumstances.

Astronomers are interested already for a long time in the remarkable features of the observed structure of galaxies. Many galaxies have a clear-cut spiral structure; photographs of some star systems show protracted "tails;" one also encounters pairs of close galaxies with "tails" pointing in opposite directions. Sometimes one can see "bridges" that appear to connect two galaxies to one another. Such pairs of galaxies are said to be *interacting*. The reader is referred to the *Atlas of Interacting Galaxies* compiled by B. A. Vorontsov-Vel′yaminov.[2]

After the discovery of interacting galaxies a number of hypotheses on the various forces of interaction were proposed to explain the observed structures. However, it is rather natural to analyze and explain first of all the gravitational effects. There is no doubt that galaxies close to one another interact gravitationally. A galaxy can also be perturbed gravitationally by material objects that are not optically or radio-astronomically visible: "dead" quasars undergoing gravitational collapse and/or old, extinct galaxies whose stars have exhausted their supply of nuclear energy. The existence of objects of this kind can be discovered precisely by observing the perturbations of the structure of a galaxy near which the objects in question have passed.

In 1971–72, at the Institute for Applied Mathematics of the Academy of Sciences of USSR, N. N. Kozlov, R. A. Syunyaev and T. M. Eneev have studied the influence of large massive objects on the evolution of galaxies. Their results are discussed in a number of papers [5.10]–[5.13]; our exposition follows their survey paper [5.10]. Kozlov, Syunyaev, and Eneev adopted the mathematical model of a

---

[2]Incidentally, Professor Vorontsov-Vel′yaminov is the author of the wonderful popularization book *The Universe* (*Essays about the Universe*), which has seen a large number of editions, e.g., Mir Publishers, Moscow, 1985. We are fortunate that this book exists. Many researchers have decided already back in their school years what their interests and future scientific fate will be thanks to this book – the present author is one of them.

galaxy described at the beginning of this section, and assumed that the particle-satellites form a disk-shaped "cloud" whose density decreases with the distance from the core. In the initial (unperturbed) state all the satellites move in the same direction along different circular orbits. The process of gravitational interaction was modeled on a computer and the results were displayed graphically on a monitor; a film of these images was made to demonstrate how the structure of galaxies changes under the influence of gravitational perturbations.

We should mention that the staggering volume of computations needed in order to model the evolution of galaxies can be handled only by very fast and powerful computers: in the problem described above one needed to integrate 1,000 systems of systems of six differential equations, in order to describe the motion of 1,000 particle-satellites. This explains why investigations of this kind were initiated only recently, despite the fact that astronomers have been studying the structure of galaxies for many decades. The monitor of a powerful computer is a particularly impressive investigation tool. Its screen allows us to see with our own eyes what once was considered to be forever inaccessible to the human eyes: the change of the shape of a galaxy with time, the whole process of galactic evolution. And not only once, but under many different conditions. The pictures given below show a few frames of a film illustrating the work carried out by T. M. Eneev and his collaborators. Unfortunately, these frames can give only a modest glimpse into the ravishing picture that unravels on the monitor in front of the investigator's eyes: a live, breathing portrait of an evolving galaxy, glimmering with the green brilliance of its hundreds of "stars."

Let us examine Figures 5.3 and 5.4. Each shows six frames for a given variant of a flight by the perturbing body. In reality each variant takes up a few thousands frames. The first frames show the direction of rotation of the galaxy and the direction of motion of the perturbing body in its orbit. Each frame is "dated," the time T being measured in billions of years. The value $T = 0.00$ corresponds to the moment when the flying body is the closest to the core of the galaxy.

The frames in Figure 5.3 demonstrate the flyby of a perturbing body in its projection on the plane of the unperturbed galaxy. A body of mass equal to the mass of the galaxy moves perpendicularly to the galactic plane with hyperbolic velocity (which in the present case is twice the parabolic velocity), at a distance of 40 kps[3] from the center of the galaxy. The galaxy has a mass equal to 10^{11} solar masses and a radius of 36 kps. The period of revolution of a particle in an orbit of radius equal to 36 kps is 2 billion years. The flyby of the massive body results, first of all, in the formation of two sharply defined spiral galactic arms. Furthermore, the perturbing action of the body on some of the particles turns out to be so strong that some particles are ejected from the galaxy's sphere of attraction, while others rapidly change their direction of motion and are pulled down toward the central region of the galaxy or even toward its center. The unperturbed plane of the

[3] 1 kiloparsec (kps) = 3,529 light-years = $30.84 \cdot 10^{15}$ km.

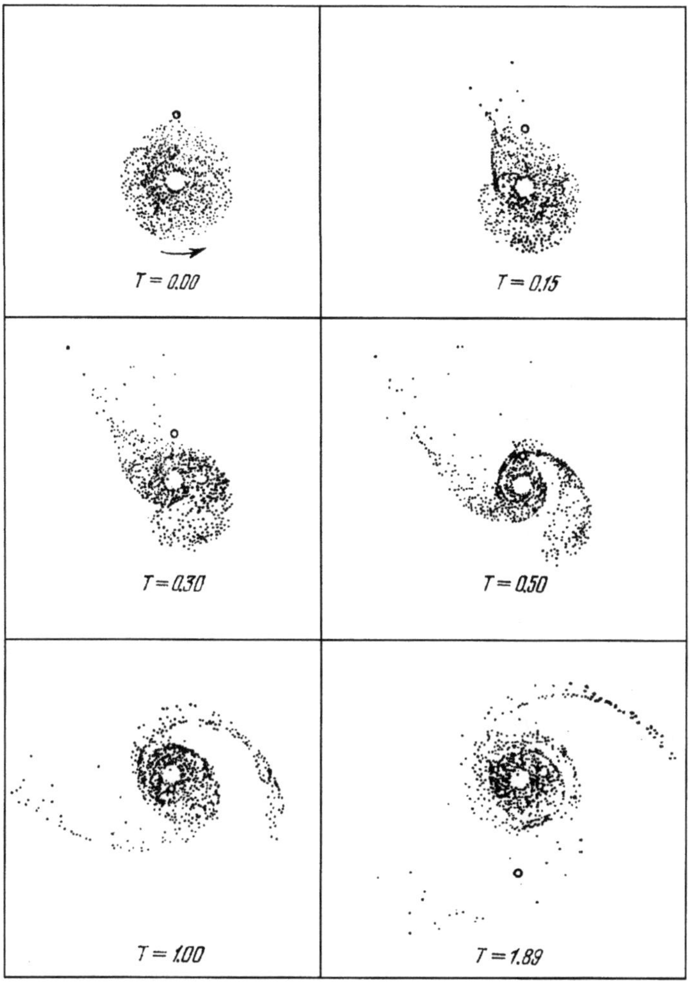

FIGURE 5.3. Evolution of a galaxy caused by a perturbing body (shown as a small circle) flying-by perpendicularly to the original galactic plane. The time T is measured in billions of years

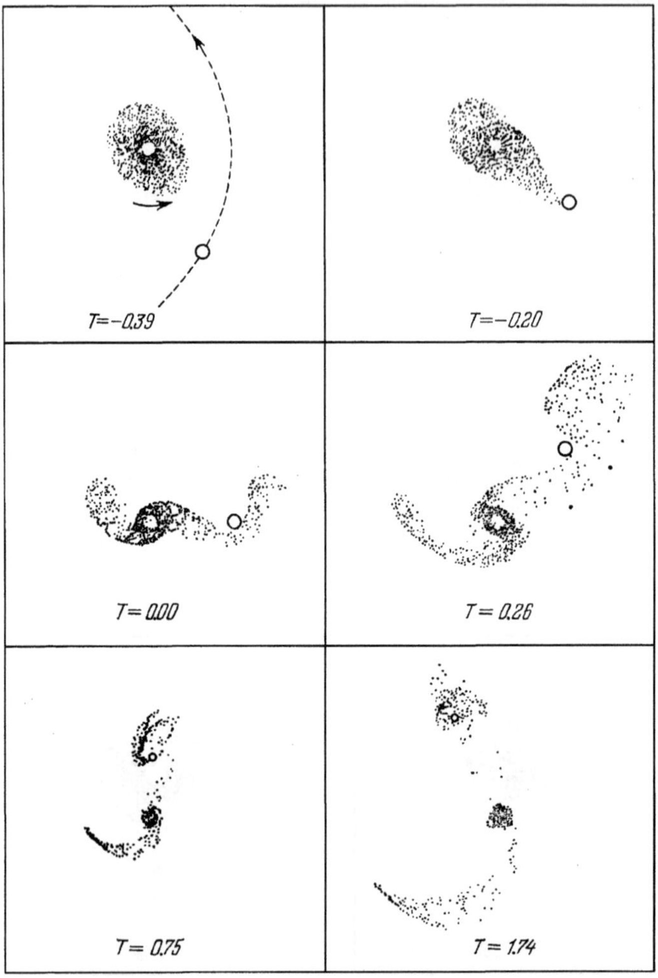

FIGURE 5.4. Evolution of a galaxy caused by a massive perturbing body (shown as a small circle) flying-by in the galactic plane (in the last two frames the scale is half of the one used in the first four). The time T is measured in billions of years

galactic disk also undergoes a distortion: the structure becomes three-dimensional (of course, the planar projection in Figure 5.3 does not show this).

Figure 5.4 shows several frames corresponding to the flyby of a perturbing body in the plane of the galactic disk, in the direction of its rotation. The mass of the perturbing body is assumed to be 4 times larger than the mass of the galaxy near which it flies. The body moves with parabolic velocity and approaches the galaxy from a distance equal to twice the radius of the galaxy (72 kps). As it turns out, the body "captures" and carries away with it 1/5 of the galactic matter. Therefore, a "dead" quasar or some other massive compact object that flies near a galaxy and captures part of its matter may create a new galaxy. The captured matter may form an isolated cluster of stars and gas, connected with the galaxy by an arch – a "bridge."

The authors of the paper [5.10] discussed here conclude the paper with the following cautious statement: "The mathematical modeling of the processes of gravitational interaction of galaxies helped one to discover effects which resemble actual structures observed in photographs of galaxies. At this time one cannot assert that these are precisely the mechanisms responsible for the formation of tails and intergalactic bridges. However, it is already clear that gravitational interaction has a strong influence on the dynamics of the evolution of star systems. Further investigations will allow one to establish the true scales of this influence and its genuine role in galactic evolution."

The importance of the work of Eneev and his collaborators undoubtedly goes beyond the limits set by this cautious self-evaluation. Here is what V. G. Demin says on this subject in the second edition of his book *The Fate of the Solar System* [5.14]:

"In astronomy reigns the tendency to explain everything that is unclear from the position of new physical laws and discoveries. Unfortunately, however, this is done at a stage where the evolution of systems of celestial bodies has not yet been investigated with some degree of accuracy in the framework of the schemes of mechanics!...

"Eneev and his collaborators have set out to elucidate whether the formation of spiral arms of galaxies can be explained solely by mechanical processes. In contrast to other astronomical investigations, no interactions of a different physical nature were assumed to be involved. In spite of this, their results surpassed all expectations... The portraits of the evolution of a galaxy that they produced demonstrate rather convincingly that the development of spiral arms can be explained by a tidal effect."

Additional comments for this translation

The classical restricted three-body problem, one of the fundamental problems of celestial mechanics, did attract the attention of many researchers and continues to do so. One of the recent monographs on this subject is A. D. Bruno's book [5.15]. There the reader will find an extensive bibliography and surveys of earlier investigations; but its principal value resides in the original methods used and the results of the analysis of the restricted three-body problem. The main role is played by the generating solutions of the problem for $m_2 \to 0$ and their analysis (the Hénon diagram for symmetric periodic orbits [5.16] and its interpretation in terms of other parameters, etc.). In particular, Bruno [5.17] succeeded in constructing periodic trajectories such that each of them passes near the surface of the Earth as well as near that of the Moon. Particularly astonishing are the stable periodic solutions which pass near both Jupiter and the Earth that Bruno found in the Sun–Jupiter–spacecraft problem.[4]

[4]Translator's note. Other book on this subject are: Szebehely, V. *Theory of Orbits*: *The Restricted Problem of Three Bodies*, Academic Press, New York, 1967; Marchal, Ch., *The Three-Body Problem*, Studies in Astronautics, 4. Elsevier Science Publishers, B.V., Amsterdam, 1990; and Hénon, M., *Generating Families in the Restricted Three-Body Problem*, Lecture Notes in Physics, New Series: Monographs, 52. Springer-Verlag, Berlin, 1997

Sixth Essay

They are Waltzing in Orbits

How lovely these flares are that light up the dark
They climb their own peak and lean down to look
They are *dancing ladies* whose glances become eyes arm and hearts

Guillaume Apollinaire, *Wonder of war*
in: *Calligrammes*, translated by Anne Hyde Greet
Univ. of California Press, Berkeley, 1980

The Earth drifts along its orbit, majestically rotating around its axis. The large and small planets also spin around their axes. The alert and curious reader has undoubtedly observed tiny stars drifting on the night sky, alternately flashing brightly and dying out like candles: those are artificial satellites, which in their rotation show the observer different sides, alternately turning to the Sun a large shining side or one that reflects almost no light.

The dances of heavenly bodies are complex and diverse. Some swing steadily, other rotate slowly simultaneously around several of their axes, and other spin at a tremendous speed, like a puppy racing to catch its own tail.

The study of the laws governing the rotation of artificial satellites is the subject of a special branch of the mechanics of space flight – the dynamics of the motion of a satellite around its own center of mass [6.1].

This essays deals with some remarkable effects arising in the rotation and orientation of satellites.

1. Gravitational potential

For the problems with which we will be concerned here the motion of a satellite around its center of mass is essentially determined by the torques (moments) of gravitational forces. In their turn these torques are readily calculated in terms of the force function which describes the action of the Newtonian central force field on our satellite (taken with the opposite sign, the force function is called the *gravitational potential*). We know that if the satellite can be regarded as a particle of mass dm, then the force function dU is given by the formula

$$dU = \frac{\mu dm}{r} \tag{6.1.1}$$

where $\mu = fM$ (with f the constant of gravitation and M the mass of the Earth) and r is the distance from the "satellite" (the particle of mass dm) to the center of the Earth.

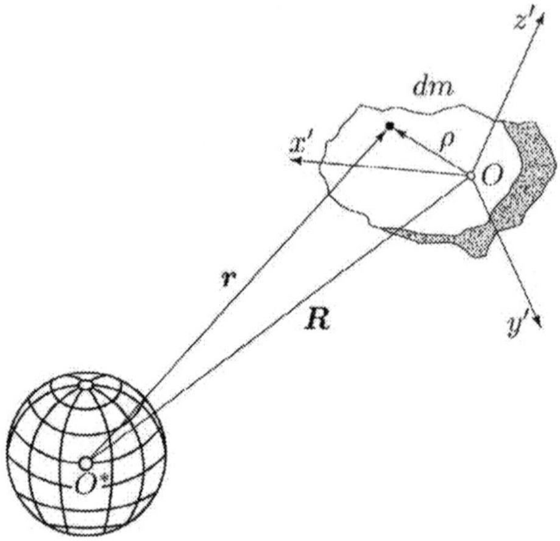

FIGURE 6.1. Regarding the calculation of the gravitational potential

However, in studying the motion of a satellite around it own center of mass an essential point is precisely that the satellite is an extended body rather than a point particle. Then formula (6.1.1) is valid for each volume element dm of the satellite, and when extended over its whole volume V it takes the form

$$U = \mu \int_V \frac{dm}{r}. \tag{6.1.2}$$

Let R denote the distance from the center of attraction O^* to the center of mass O of the satellite (Figure 6.1). Consider a coordinate system $Ox'y'z'$ rigidly attached to the satellite. The distance ρ from O to the current point dm is then given by the formula $\rho = \left(x'^2 + y'^2 + z'^2\right)^{1/2}$, and the cosine of the angle between the vectors $\mathbf{R}$ and $\boldsymbol{\rho}$ is given by

$$\cos(\widehat{\mathbf{R}, \boldsymbol{\rho}}) = \frac{x'}{\rho}\gamma + \frac{y'}{\rho}\gamma' + \frac{z'}{\rho}\gamma''.$$

Here γ, γ', γ'' are the direction cosines of the vector $\mathbf{R}$ relative to the axes x', y', z', i.e., $\gamma = \cos(\widehat{\mathbf{R}, x'})$, $\gamma' = \cos(\widehat{\mathbf{R}, y'})$, $\gamma'' = \cos(\widehat{\mathbf{R}, z'})$. In these notations, r in formula (6.1.2) is given by

$$r = \sqrt{R^2 + 2R(x'\gamma + y'\gamma' + z'\gamma'') + x'^2 + y'^2 + z'^2}. \tag{6.1.3}$$

The integration in (6.1.2) is carried out with respect to the current coordinates x', y', z', and hence one can obtain the explicit dependence

$$U = U(R, \gamma, \gamma', \gamma''). \tag{6.1.4}$$

Thus, the force function (6.1.2) depends on the distance from the satellite to the attracting body as well as on the orientation of the satellite with respect to the direction "center of attraction–center of mass of the satellite."

One can show (see [6.1]) that the projections of the torques of gravitational forces can be written in the form

$$\left.\begin{aligned} M_{x'} = \gamma'' \frac{\partial U}{\partial \gamma'} - \gamma' \frac{\partial U}{\partial \gamma''}, \quad M_{y'} = \gamma \frac{\partial U}{\partial \gamma''} - \gamma'' \frac{\partial U}{\partial \gamma}, \\ M_{z'} = \gamma' \frac{\partial U}{\partial \gamma} - \gamma \frac{\partial U}{\partial \gamma'}. \end{aligned}\right\} \tag{6.1.5}$$

The concrete form of U and of the torques (6.1.5) depends, generally speaking, on the shape of the body and the distribution of the mass in its volume. This complicates the analysis. However, a simplification arises thanks to the fact that satellites have small dimensions compared with the distance R. Consequently, the quantities x'/R, y'/R, z'/R are small and the expression (6.1.3) can be expanded in a series of powers of these quantities. If in this expansion one retains only the terms of order less than or equal to two, then the integral (6.1.2) is readily calculated. Here one takes into account that the integrals $\int_V x'\,dm = \int_V y'\,dm = \int_V z'\,dm = 0$ (since the origin of the coordinate system $Ox'y'z'$ is the satellite's center of mass) and that the position of the axes x', y', z' in the satellite can be chosen so that

$$\int_V x'y'\,dm = \int_V x'z'\,dm = \int_V y'z'\,dm = 0.$$

The axes x', y', z' chosen in this manner are called the *principal central axes of inertia* of the satellite. Under the adopted simplifying assumptions the calculation of U by means of formula (6.1.2) yields

$$U \approx \frac{\mu m}{R} + \frac{\mu}{2R^3}(A+B+C) - \frac{3}{2}\frac{\mu}{R^3}(A\gamma^2 + B\gamma'^2 + C\gamma''^2). \tag{6.1.6}$$

Here m is the mass of the satellite and the constants

$$A = \int_V (y'^2 + z'^2)dm, \quad B = \int_V (z'^2 + x'^2)dm, \quad C = \int_V (x'^2 + y'^2)dm$$

are the *principal central moments of inertia* of the satellite. Accordingly, formulas (6.1.5) take the form

$$\left.\begin{aligned} M_{x'} = 3\frac{\mu}{R^3}(C-B)\gamma'\gamma'', \quad M_{y'} = 3\frac{\mu}{R^3}(A-C)\gamma\gamma'', \\ M_{z'} = 3\frac{\mu}{R^3}(B-A)\gamma\gamma'. \end{aligned}\right\} \tag{6.1.7}$$

2. Rotation of the Moon. Background material on stability theory

From antique times people have noticed that the face of Moon does not change. The Moon is always turned with the same side toward the Earth and people were not able to see the "back of its head" until October 1995, when an unmanned Soviet spacecraft photographed it.

This invariability of the Moon's position relative to the Earth means that the period of axial rotation of the Moon is identical to the period of its revolution around Earth. For simplicity we shall consider that the Moon's orbit is circular.

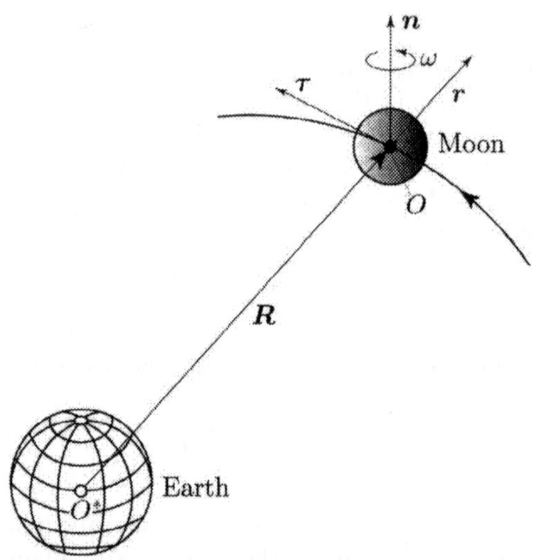

FIGURE 6.2. The orbital coordinate system

Let us introduce the orbital coordinate system $O\boldsymbol{\tau}\mathbf{nr}$ as follows (Figure 6.2). Place the origin of coordinates O in the center of mass of the Moon, and direct the **r**-axis along the geocentric radius vector **R** of the Moon, the $\boldsymbol{\tau}$-axis along the tangent to the circular orbit, and the **n**-axis along the normal to the Moon's orbital plane. Thus, the orbital coordinate system is a rotating system; the $\boldsymbol{\tau}$- and **r**-axes lie always in the Moon's orbital plane and rotate around the **n**-axis with the angular velocity ω of the orbital motion. (The linear velocity of the motion of the center of mass of the Moon in orbit is $V = \omega R$.)

The angular velocity of axial rotation of the Moon is also equal to ω and is directed along the vector **n**; the axial rotation takes place in the same direction as the orbital motion. Consequently, the Moon is always turned with the same side toward Earth and, as is readily understood, *it is at rest relative to the orbital coordinate system*. In other words, the Moon is in a *relative equilibrium* up to a slight rocking motion (*libration*).

Let us remark that the rotation of the Moon is described in the first approximation by the *laws of Cassini*, according to which the axis of rotation of the Moon is in relative equilibrium in the coordinate system that rotates together with the line of nodes of the Moon's orbit; moreover, the axis of rotation of the Moon is normal to the line of nodes and makes an angle of $6°41'$ with the normal to the Moon's orbital plane. Here we will neglect the motion of the line of nodes of the Moon's orbit. This allows us to replace the motion according to the laws of Cassini by a simpler motion, namely, relative equilibrium in the orbital coordinate system. (For more details on Cassini's laws see [6.14] and the exposition below.)

The equality of the angular velocities of revolution and axial rotation of a celestial body cannot be accidental, the more so as in the Solar system this apparently is not an isolated phenomenon. Recent astronomical observations tend to confirm that Jupiter's satellites Io, Europa, and Callisto, and Saturn's satellites Tethys, Rhea, Dione, and Iapetus are permanently turned with the same side toward the central planet (see [6.2], and also Astrophysical J., **165** (1971), p. 431). According to some data, even the satellites of Mars, Phobos and Deimos move in a similar fashion [6.2′], i.e., they are in relative equilibrium states.

For this behavior to be not accidental, there must exist some stabilizing factor that keeps the satellites (the Moon, for example) in a relative equilibrium position. This factor should also exert its influence on artificial Earth satellites.

Such a stabilizing factor is provided by the gravitational torques, defined by the gravitational potential (6.1.6). Let us show this.

Before we embark upon our analysis, let us give a number of definitions and – without proofs – theorems of A. M. Lyapunov's theory of stability. Proofs can be found, for example, in N. G. Chetaev's book [6.56].

The investigation of the stability of a motion $q(t)$ can be reduced, via a change of variables, to the investigation of the solution $x_i \equiv 0$ of a system of differential equations

$$\frac{dx_s}{dt} = X_s(x_i, t), \quad X_s(0,t) \equiv 0 \quad (i, s = 1, \dots, n). \tag{6.2.1}$$

Equations (6.2.1) will be called the *equations of the perturbed motion*; their particular solution $x_i \equiv 0$ is called the *unperturbed motion*. The right-hand sides of the equations (6.2.1) are usually required to be analytic functions.

DEFINITION 1 (A. M. LYAPUNOV). The unperturbed motion is said to be *stable* if for any arbitrarily small number $A > 0$ one can choose a number $\lambda > 0$ such that, for all initial perturbations $x_{10}, \dots, x_{n0}$ satisfying the condition $\sum_s x_{s0}^2 \leq \lambda$ and for all $t \geq t_0$, one has that $\sum_s x_s^2 < A$. In the opposite case the unperturbed motion is said to be *unstable*.

We wish to emphasize that the notion of stability that we used when we described the results of V. I. Arnold is weaker than the Lyapunov stability. There are two reasons for this:

1) Arnold requires the closeness of the perturbed and unperturbed *trajectories*, rather than the closeness, at each moment of time, of the positions of the points moving on these trajectories. In general, a conditionally-periodic trajectory that is stable in the sense of Arnold is not stable in the sense of Lyapunov.

2) In Arnold's results stability is established not for all initial conditions, but only for most of them. According to Lyapunov, a motion is unstable if one can find *at least one* perturbed trajectory that does not satisfy the definition of stability.

In the remaining part of this section we will deal only with Lyapunov stability.

Before Lyapunov, stability of motion was investigated mainly by linearizing the equations (6.2.1) and analyzing the solutions of the resulting linear equations. After Lyapunov's investigations it became clear that, generally speaking, the linear equations yield *only necessary* conditions for stability ("if the motion is stable, then those conditions are satisfied"). But the question "if these conditions are satisfied, is the motion stable?" cannot always be answered in the affirmative. The search for sufficient stability conditions is a very difficult one.

DEFINITION 2. Suppose that in the domain $t > t_0$, $\sum_s x_s^2 \leq H$, where $H \neq 0$, there is defined a function $V(x_s, t)$ whose values are all of the same sign or zero. Then V is called a *sign-constant function* (positive or negative).

DEFINITION 3. If the sign-constant function V does not depend on t and if the constant H in Definition 2 can be chosen small enough so that V vanishes only at the origin (i.e., only when $x_s = 0$ for all s), then V is called a *sign-definite function* (positive definite or negative definite).

EXAMPLES. In the case of two variables x_1, x_2, the function $V_1 = x_1^2 + x_2^2$ is positive definite, whereas the functions $V_2 = (x_1 + x_2)^2$ and $V_3 = x_1^2$ are only sign-constant (positive), since in addition to the origin V_2 vanishes on the set $x_1 = -x_2$, while V_3 vanishes on the set $x_1 = 0$. The function $V_4 = x_1^2 - x_2^2$ is not of constant sign.

DEFINITION 4. A function V that depends on t is said to be *sign-definite* if there exists a positive-definite function W which does not depend on t, such that one of the expressions $V - W$ or $-W - V$ is a positive function.

THEOREM 1 (A. M. LYAPUNOV). *If the differential equations* (6.2.1) *of perturbed motion are such that one can find a sign-definite function V whose derivative $V' = \sum_s \frac{\partial V}{\partial x_s} X_s + \frac{\partial V}{\partial t}$ by virtue of the equations of motion is a sign-constant function of sign opposite to the sign of V or is identically equal to zero, then the unperturbed motion is stable.*

The proof of this result (see, e.g., [6.65]) is based on the observation that, under the assumptions of the theorem, a trajectory is forever trapped inside a surface $V(x_s) \leq l$ that surrounds the origin.

Lyapunovs' theorem gives sufficient conditions for stability. A function V with the properties indicated in its formulation is called a *Lyapunov function*. There exists a rather rich arsenal of methods for constructing Lyapunov functions in problems of various classes. Unfortunately, however, no general algorithm for constructing Lyapunov functions for arbitrary problems is known.

Lyapunov's theorem will prove useful below, in Section 3 of this essay. Here we wish to discuss the interesting fact that, in some cases, this theorem allows one to settle the stability question by examining only the first approximation. But first we need some definitions.

A function $x(t)$ is said to be *bounded* if there is a constant $C > 0$ such that $|x(t)| < C$ for all $t \geq t_0$. A function whose absolute magnitude takes values larger that any arbitrarily large positive number is said to be *unbounded*. A bounded function is said to *decay* if $x(t) \to 0$ as $t \to \infty$.

LEMMA 1. *If $z(t) = x(t)e^{\lambda t}$ is a function that decays for $\lambda = \lambda_1$ and is unbounded for $\lambda = \lambda_2$, where λ_1 and λ_2 are real constants, then one can find a real λ_0 such that the function $z(t)$ with $\lambda = \lambda_0 + \varepsilon$ is unbounded for any positive value of ε and decays for any negative value of ε.*

DEFINITION 5. The number λ_0 is called the *characteristic exponent* of the function $x(t)$.

Now instead of the system (6.2.1) let us consider the system linearized about the origin:

$$\frac{dx_s}{dt} = p_{s1}x_1 + \cdots + p_{sn}x_n, \qquad s = 1, \ldots, n. \tag{6.2.2}$$

Here p_{sr} are real, bounded, continuous functions of t.

Let $x_1(t), \ldots, x_n(t)$ be a solution of the system (6.2.2). The *characteristic exponent* of this solution is defined to be the smallest of the characteristic exponents of the functions $x_1, \ldots, x_n$. As is known, one can always find n linearly independent solutions of the system (6.2.2). These will be characterized by a set of n characteristic exponents κ_s (each of which is finite, as one can verify). Let us denote their sum by

$$S = \sum_{s=1}^{n} \kappa_s .$$

Also, let μ denote the characteristic exponent of the function

$$\exp\left\{-\int \sum_{s=1}^{n} p_{ss}\,dt\right\}.$$

Then the following holds.

LEMMA 2. $S + \mu \leq 0$.

DEFINITION 6. The system (6.2.2) is said to be *regular* if $S + \mu = 0$, and *irregular* in the opposite case

The characteristic exponents allow us to analyze the qualitative behavior of solutions of the system (6.2.2) and draw conclusions about the stability of the unperturbed motion in the first (linear) approximation. This is important, because in the general case of variable coefficients $p_{sr}(t)$ we do not know how to integrate the system (6.2.2). Lyapunov showed, however, that in many cases the characteristic exponents of the first-approximation equations allow us also to decide about the stability or instability of the unperturbed motion in the *full system* (6.2.1) of differential equations of perturbed motion. More precisely, the following two assertions hold true.

THEOREM 2 (A. M. LYAPUNOV). *If the system* (6.2.2) *of differential equations of first approximation is regular and all its characteristic exponents are positive, then the unperturbed motion is stable.*

THEOREM 3. *If the system* (6.2.2) *of differential equations of first approximation is regular and if among its characteristic exponents there is at least that is negative, then the unperturbed motion is unstable.*

3. Stability of relative equilibrium in a gravitational field

The problem of the stability of a relative equilibrium of an artificial satellite in a gravitational field arose in the mid fifties of our impetuous century, but it has a long prehistory that goes back many centuries. In celestial mechanics the study of the stability of the relative equilibrium of the Moon and its oscillations (librations) goes back to Lagrange (1780). These classical investigations relied on the linearization of the equations governing small oscillations. Following the investigations of Lyapunov (1892) it became clear that in general the analysis of the linear equations does not settle the stability question. In 1959 (see [6.3]) the author of this book published the first results of a rigorous – in the sense of Lyapunov as well as in the sense of the mathematical setting of the problem – investigation, carried out already in 1956–57, of the question of the existence and stability of a relative equilibrium of a rigid body in a gravitational field. One of the basic results of this investigation is described below in a simplified form. For a more detailed analysis the reader is referred to [6.1] and [6.3].

If the Moon (or an Earth artificial satellite) moves in a circular orbit, then in the orbital coordinate system the Moon (satellite) is acted upon by gravitational, centrifugal, and Coriolis forces. The torques of the Coriolis force effect no work; the torques of the gravitational and centrifugal forces are defined by the potential (6.1.6) for gravitational forces and the analogous expression of the potential of centrifugal forces:

$$U_\beta = \frac{1}{2}\omega^2(A\beta^2 + B\beta'^2 + C\beta''^2). \tag{6.3.1}$$

Here β, β', β'' are the direction cosines of the normal $\mathbf{n}$ to the orbital plane relative to the axes x', y', z', i.e.,

$$\beta = \cos(\widehat{\mathbf{n}, x'}), \quad \beta' = \cos(\widehat{\mathbf{n}, y'}), \quad \beta'' = \cos(\widehat{\mathbf{n}, z'}).$$

Let us derive formula (6.3.1), for the benefit of the meticulous reader. The x'-component, say, of an elementary torque of centrifugal forces is

$$\begin{aligned} dM_{x'} &= y' f_{z'} - z' f_{y'} = \\ &= \{y'[(x'\alpha + y'\alpha' + z'\alpha'')\alpha'' + (x'\gamma + y'\gamma' + z'\gamma'')\gamma''] - \\ &- z'[(x'\alpha + y'\alpha' + z'\alpha'')\alpha' + (x'\gamma + y'\gamma' + z'\gamma'')\gamma']\}\omega^2 dm. \end{aligned}$$

Here $f_{y'}$ and $f_{z'}$ are the components of the elementary centrifugal force on the y'- and z'-axes, respectively, and α, α', α'' are the direction cosines of the unit vector $\boldsymbol{\tau}$ relative to the axes x', y', z'. Integrating over the whole volume and using the relation $\beta'\beta'' = -(\alpha'\alpha'' + \gamma'\gamma'')$, we get $M_{x'} = -\omega^2(C - B)\beta'\beta''$. A comparison with the analogous formulas (6.1.7) for the torque of the gravitational forces, defined by the force function (6.1.6), leads to the force function (6.3.1) of centrifugal forces.

Since the total potential energy does not depend explicitly on time (for a circular orbit), conservation of mechanical energy holds, i.e.,

$$T - U - U_\beta = H, \tag{6.3.2}$$

where T is the kinetic energy of the satellite in its relative rotation and H is the constant total energy of rotational motion of the satellite. If we observe that, in formula (6.1.6), $R = $ const and, moreover, $\mu/R^3 = \omega^2$, we can write the law of conservation of energy (6.3.2) in the explicit form (see [6.3])

$$\begin{aligned} \frac{1}{2}\left(A\overline{p}^2 + B\overline{q}^2 + C\overline{r}^2\right) &+ \frac{3}{2}\omega^2\left(A\gamma^2 + B\gamma'^2 + C\gamma''^2\right) - \\ &- \frac{1}{2}\omega^2\left(A\beta^2 + b\beta'^2 + C\beta''^2\right) = h. \end{aligned} \tag{6.3.3}$$

Here $\overline{p}$, $\overline{q}$, $\overline{r}$ are the components of the relative angular velocity of rotation of the satellite with respect to its principal axes x', y', z'.

In the relative equilibrium the relative angular velocity is equal to zero, and, as it turns out, the direction cosines can be taken equal to either 0 or 1 in absolute magnitude. For example, to the motion

$$\overline{p} = \overline{q} = \overline{r} = 0, \qquad \gamma = \gamma' = \beta = \beta'' = 0, \qquad \gamma'' = \beta' = 1, \tag{6.3.4}$$

there corresponds a relative equilibrium in which the satellite's z'-axis coincides with the direction of the radius vector ($\gamma'' = \cos\left(\widehat{z', \mathbf{r}}\right) = 1$), while its y'-axis coincides with the normal to the orbital plane ($\beta' = \cos\widehat{(y', \mathbf{n})} = 1$); moreover, the satellite's x'-axis coincides with the tangent to the orbit.

The motion (6.3.4) is referred to as the *unperturbed motion* and deviations from it are referred to as *perturbations*. The total energy (6.3.3) of the relative motion is preserved under arbitrary perturbations; the expression (6.3.3) is a first integral of the equations of motion (here we omit these equations; the complete equations of motion can be found in the book [6.1]).

Recalling the familiar relations between direction cosines:

$$\gamma''^2 = 1 - \gamma^2 - \gamma'^2, \qquad \beta'^2 = 1 - \beta^2 - \beta''^2, \tag{6.3.5}$$

one can recast (6.3.3) in the form

$$\begin{aligned}\frac{1}{2}\left(A\overline{p}^2 + B\overline{q}^2 + C\overline{r}^2\right) + \frac{3}{2}\omega^2\left[(A - C)\gamma^2 + (B - C)\gamma'^2\right] + \\ + \frac{1}{2}\omega^2\left[(B - A)\beta^2 + (B - C)\beta''^2\right] = h_0.\end{aligned} \tag{6.3.6}$$

Here h_0 is a new constant, determined by the initial data $\overline{p}_0$, $\overline{q}_0$, $\overline{r}_0$, γ_0, γ_0', β_0, β_0'. The left-hand side of expression (6.3.6) is a quadratic form in its variables. Let us require that this quadratic form be positive definite. All its coefficients will be positive provided that

$$B > A > C. \tag{6.3.7}$$

This immediately allows us to settle the stability question for the relative equilibrium (6.3.4). Indeed, since a sum of positive terms is larger than the sum of any number of its terms, under the assumption (6.3.4) relation (6.3.6) yields, for instance, the following bounds:

$$\left.\begin{aligned}\gamma^2 \le \frac{2h_0}{3\omega^2(A - C)} < \Delta^2, \qquad \gamma'^2 \le \frac{2h_0}{3\omega^2(B - C)} < \Delta^2, \\ \beta^2 \le \frac{2h_0}{3\omega^2(B - A)} < \Delta^2, \qquad \beta''^2 \le \frac{2h_0}{3\omega^2(B - C)} < \Delta^2.\end{aligned}\right\} \tag{6.3.8}$$

Here $\Delta^2 < 1$ is an arbitrarily preassigned number. It follows from (6.3.8) that one can always choose the initial data (h_0) so that the quantities $|\gamma|$, $|\gamma'|$, $|\beta|$, $|\beta''|$ will not exceed an arbitrarily small prescribed value $|\Delta|$, i.e., so that the perturbed motion will be arbitrarily close to the unperturbed one – the relative equilibrium. By the definition of Lyapunov stability, this means precisely that the unperturbed motion is stable.

Thus, inequalities (6.3.7) are sufficient conditions for the stability of the relative equilibrium of the satellite in a circular orbit. In order for inequalities (6.3.8) to hold, the quantity h_0 must be chosen so that

$$h_0 < \min\left\{\frac{3}{2}\omega^2(A - C)\Delta^2,\ \frac{1}{2}\omega^2(B - A)\Delta^2\right\}, \tag{6.3.9}$$

i.e., h_0 must be smaller than the smallest of the two quantities in the brackets. Notice that, by (6.3.9), for such a choice of h_0 the components of the angular velocity, $\overline{p}_0$, $\overline{q}_0$, $\overline{r}_0$, do not exceed a quantity of the order of ω. We are dealing here with very small quantities (for orbits of artificial satellites, $\omega \sim 0.05 \div 0.07$ degree/sec, and for the Moon $\omega \sim 0.00015$ degree/sec). In other words, the stabilizing action of the gravitational field can be discerned only for a very small value of the kinetic energy of rotation of the satellite around its center of mass.

The argument used above to establish the stability of the relative equilibrium does actually amount to the application of Lyapunov's stability theorem that we already are acquainted with. Recall its formulation: *if from the variables of the perturbed motion one can construct a positive-definite function V whose derivative $\dot{V}$ along the solutions of the differential equation is negative or identically equal to zero, then the unperturbed motion is stable.* In our case one can take for the function V the right side of equality (6.3.6); it is positive-definite if (6.3.7) holds. And since V is a first integral of the equations of motion, $\dot{V} \equiv 0$, hence stability holds.

This theorem of Lyapunov is a powerful tool for establishing stability in those cases in which there is no dissipation of energy in the system, and consequently asymptotic stability cannot hold. Many problems in the dynamics of space flight were investigated with respect to stability by means of this theorem or modifications thereof. A large number of such problems are described in V. V. Rumyantsev's book [6.4].

4. *Rhinogradentia* in orbit

It follows from conditions (6.3.7) that the stability of the relative equilibrium is ensured first and foremost thanks to the fact that the satellite (or the Moon) differs from a homogeneous ball. Indeed, in the case of a ball $A = B = C$ and to speak about the action of gravitational torques is meaningless.

Let us examine conditions (6.3.7) closer. The maximal moment of inertia B corresponds to the y'-axis of the satellite, which in the relative equilibrium, as one recalls, is directed along the normal $\mathbf{n}$ to the orbital plane. The minimal moment of inertia C corresponds to the z'-axis of the satellite, which in the relative equilibrium is directed along the radius vector of the orbit. As is known, in mechanics the mass geometry of a rigid body is characterized by its *central ellipsoid of inertia*, whose axes are directed along the principal central axes of inertia x', y', z' of the body and for which the lengths of the semiaxes, a, b, c, are expressed in terms of the principal central moments of inertia A, B, C through the formulas

$$a = \frac{k}{\sqrt{A}}, \quad b = \frac{k}{\sqrt{B}}, \quad c = \frac{k}{\sqrt{C}},$$

where k is a constant. Therefore, to the maximal moment of inertia there corresponds the minor axis of the ellipsoid of inertia, and conversely.

Summing up, we can formulate the following assertion, obtained by the author in 1956 and published in 1959 [6.3]:

For a relative equilibrium of a rigid body moving along a circular orbit in a Newtonian central force field to be stable it is sufficient that in the unperturbed motion the major axis of the central ellipsoid of inertia be directed along the radius vector of the orbit, the minor axis – along the normal to the orbital plane, and the middle axis – along the tangent to the orbit (Figure 6.3).

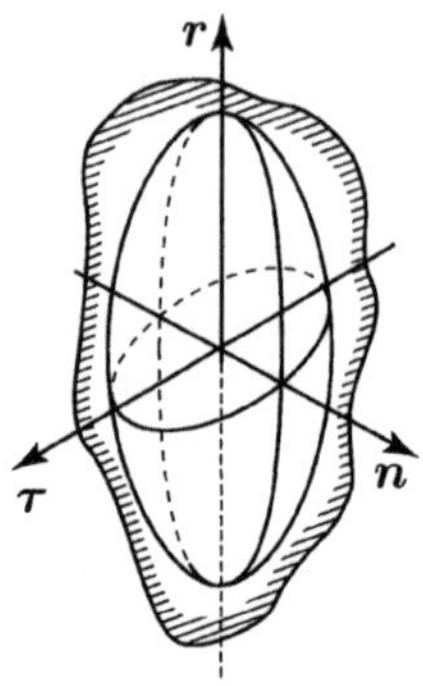

FIGURE 6.3. Stable position of the ellipsoid of inertia of a satellite in orbit

One can say about a satellite that if we wish to ensure its gravitational stability, then it must be fairly "squeezed" in the direction toward the orbital plane, and more stretched, "elongated" in the satellite–Earth direction.

Perhaps a "good" satellite design from this point of view is one whose shape is close to that of the well known animal *Rhinogradentia otopteryx volitans*, discovered by H. Stümpke and described in his monograph *Body Structure and Life of Rhinogradentia* [6.5] (see also the more accessible edition [6.6]).

5. Passive stabilization of artificial satellites

The possibility alluded to above of a gravitationally-balanced satellite design is of fundamental importance. Thanks to "gratis" gravitational energy one can keep the satellite as long as one wishes in a position in which the same side is turned toward Earth (like in the case of the Moon). The importance of this fact for meteorological observations, Earth's-surface photography, and other tasks can be hardly overestimated. In his survey [6.18], among various important achievements of mechanics, A. Yu. Ishlinskiĭ described also the aforementioned principle of stabilization of a satellite by means of gravitational forces.

In the practical implementation of such – as is referred to – *passive* stabilization system[1] it is important to reduce the initial, usually rather high, angular

[1] To distinguish it from active systems, in which stabilization is achieved by means of rocket engines, flywheels, and similar active devices, which requires an expenditure of some form of energy stored aboard the satellite

velocity of the satellite. The angular velocity must be confined to a very narrow range, in which gravitational capture of the satellite into the oriented position "works well." For this reason systems of gravitational stabilization must contain mechanisms for damping (reducing) angular velocities. For natural satellites (the Moon, the satellites of Jupiter and Saturn mentioned in the beginning of this essay), such a damping mechanism is provided by the torques of tidal friction forces, which in the course of millenary evolution have brought the rotational motion of the Moon and of several other celestial bodies to a "standstill" [6.2]. The damping of the initial rotations of artificial satellites is achieved by introducing some type of friction in the system, which dissipates the kinetic energy of rotation; to this end one resorts to elastic damping, magnetic damping, etc.

To build passive stabilization systems one can use not only the gravitational field, but also the stabilizing action of aerodynamic forces, of the Earth's magnetic field, and even of the Sun's radiation pressure.

In (the former) Soviet Union the first system of gravitational stabilization was proposed in 1956 by D. E. Okhotsimskiĭ [6.34]. In the literature one can find descriptions of the passive stabilization systems of a number of already launched satellites [6.7], [6.8]. A detailed description of the dynamics of passive stabilization systems is given in works of V. A. Sarychev (see, e.g., [6.9]).

6. Nonlinear oscillations

If a satellite is driven out of a relative equilibrium position, then it will execute oscillations in space about that position. The investigation of these oscillations is a difficult problem. A widely used method of analysis amounts to the following. Assume that the angular deviations from the equilibrium position (and also the angular velocities of the oscillations) are very small. Then instead of the exact equations of motion one can consider approximate equations, in which one retains only the small terms of order at most one. The solution of these linear equations of motion must remain bounded over the entire infinite time interval. Indeed, the presence of an unbounded solution would violate the a priori assumption that the oscillations are small. If for arbitrarily small initial deviations the bounded oscillations will also be arbitrarily small, then it would seem natural to conclude that the boundedness conditions of the linear oscillations are stability conditions of the original equilibrium position.

But, as we already mentioned, after the publication of Lyapunov's investigations it became clear that such "stability conditions," derived from the linearized equations, are by far not obligatorily "genuine" stability conditions. From the linearized equations one may conclude that the original unperturbed motion (equilibrium) is stable, while in fact this might not be the case. A beautiful illustration of this statement is our problem on spatial oscillations of a rigid body (satellite) in a gravitational field.

We shall examine the oscillations of a satellite about the position of relative equilibrium in a circular orbit. Then the conditions of boundedness of the linear oscillations reduce (see [6.1]) either to the already familiar conditions

$$B > A > C, \tag{6.6.1}$$

or to the set of conditions

$$\left.\begin{aligned} &B < C < A,\\ &\left[\frac{C}{A} + 3\left(\frac{B-C}{A}\right)\frac{C}{A} + \left(\frac{B}{A} - 1\right)\left(\frac{B-C}{A}\right)\right]^2 -\\ &\qquad - 16\frac{C}{A}\left(\frac{B-C}{A}\right)\left(\frac{B}{A} - 1\right) > 0. \end{aligned}\right\} \tag{6.6.2}$$

Inequalities (6.6.1), (6.6.2) are necessary conditions for the stability of the equilibrium position; in other words, if this position is stable, then (6.6.1), (6.6.2) hold. However, if conditions (6.6.1), (6.6.2) are satisfied, then the motion is not obligatorily stable. We have shown earlier that under conditions (6.6.1) (or, as we will say, "in region I") the relative equilibrium is actually stable, i.e., conditions (6.6.1) are sufficient conditions for stability.

We cannot say the same about conditions (6.6.2) (which define "region II"). To understand the character of the motion in that region (and also to study in more depth the motion in region I), it is necessary to examine the nonlinear equations governing the oscillations. Such nonlinear effects were studied (among others) by J. V. Breakwell and R. Pringle [6.10]. Let us describe briefly their method and the results of their analysis.

The function figuring in the left-hand side of equality (6.3.3) is the Hamiltonian of our problem, and after a number of transformations it can be reduced to the following normal form:

$$H = \pm\frac{1}{2}\left(\xi_1^2 + \omega_1^2\eta_1^2\right) + \frac{1}{2}\left(\xi_2^2 + \omega_2^2\eta_2^2\right) + \frac{1}{2}\left(\xi_3^2 + \omega_3^2\eta_3^2\right) + H^{(1)}(\xi_i, \eta_i). \tag{6.6.3}$$

Here η_i are angular variables, which describe the so-called normal oscillations, and ξ_i are the corresponding momenta. The explicitly written part of H (i.e., excluding $H^{(1)}$) describes precisely linear oscillations with frequencies ω_1, ω_2, ω_3 (which are expressible in terms of the satellite's moments of inertia). The plus [resp., minus] sign in (6.6.3) refers to region I given by (6.6.1) [resp., region II given by (6.6.2)].

The part $H^{(1)}$ of the Hamiltonian involves only terms of third and higher order in the variables ξ_i, η_i. If these terms are neglected, then the remaining part of the Hamiltonian can be used to obtain the linearized equations of motions. The solution of the latter can be written in the form

$$\left.\begin{aligned} \eta_1 &= \pm\frac{\sqrt{2\alpha_1}}{\omega_1}\sin\omega_1(\tau \pm \beta_1), & \xi_1 &= \sqrt{2\alpha_1}\cos\omega_1(\tau \pm \beta_1),\\ \eta_2 &= \frac{\sqrt{2\alpha_2}}{\omega_2}\sin\omega_2(\tau + \beta_2), & \xi_2 &= \sqrt{2\alpha_2}\cos\omega_2(\tau + \beta_2),\\ \eta_3 &= \frac{\sqrt{2\alpha_3}}{\omega_3}\sin\omega_3(\tau + \beta_3), & \xi_3 &= \sqrt{2\alpha_3}\cos\omega_3(\tau + \beta_3), \end{aligned}\right\} \tag{6.6.4}$$

Here τ is "time" and α_i, β_i are integration constants.

We note that the α_i are proportional to the squares of the amplitudes of oscillations. Moreover, $\eta_3(\tau)$ describes longitudinal oscillations (in the orbital plane), while $\eta_1(\tau)$ and $\eta_2(\tau)$ describe transversal oscillations of the satellite. Next, let us seek the solution of the exact equations of motions, corresponding to the full Hamiltonian (6.6.3), in the same form (6.6.4), where now α_i and β_i are no longer constants, but some functions of time subject to determination. This method of variation of constants leads to the differential equations

$$\frac{d\alpha_i}{d\tau} = -\frac{\partial H^{(1)}(\alpha_i, \beta_i, \tau)}{\partial \beta_i}, \qquad \frac{d\beta_i}{d\tau} = \frac{\partial H^{(1)}(\alpha_i, \beta_i, \tau)}{\partial \alpha_i}. \tag{6.6.5}$$

Here $H^{(1)}$ is understood as the term $H^{(1)}$ in (6.6.3), with η_i, ξ_i replaced by the expressions (6.6.4).

These equations, well-known in classical mechanics (see, e.g., [6.11]), are convenient for investigation because they are in canonical form. If $H^{(1)} \equiv 0$, then (6.6.5) immediately yields $\alpha_i = \text{const}$, $\beta_i = \text{const}$, which corresponds to the linear oscillations (6.6.4). For an arbitrary $H^{(1)}$ that does not vanish identically, the solution of equations (6.6.5) shows right away in explicit manner how the nonlinear oscillations differ from the linear ones.

If the oscillations do not differ very strongly from the linear ones, then $H^{(1)}$ is small and in order to investigate the equations (6.6.5) one can resort to an asymptotic method – the averaging method. Then, as it turns out, the mean value

$$\overline{H^{(1)}} = \lim_{T\to\infty} \frac{1}{T} \int_0^T H^{(1)}\, d\tau = 0,$$

provided only that the frequencies ω_i satisfy no resonance relations

$$\sum_{i=1}^{3} n_i \omega_i = 0,$$

where n_i are some integers (one of which can be zero). In other words, formulas (6.6.4) with constant α_i and β_i describe the exact motion with high accuracy outside of neighborhoods of any resonances. However, near resonances this is no longer true: the mean value of $H^{(1)}$ does not vanish.

Now let us retain in $H^{(1)}$ only the terms of order three. Then it turns out that the following third-order resonances are possible:

$$\text{(I)}\ \ \omega_3 = 2\omega_1, \qquad \text{(II)}\ \ \omega_3 = \omega_2 - \omega_1, \qquad \text{(III)}\ \ \omega_3 = \omega_2 + \omega_1. \tag{6.6.6}$$

There are no other third-order resonances. The motion in the neighborhood of the resonances (6.6.6) was investigated in the already cited work [6.4].

Let us consider, for instance, the type-(I) resonance from the list (6.6.6). Denote by $\varepsilon = \omega_3 - 2\omega_1$ the "detuning" parameter of the resonance relation (I).

Then in the ε-neighborhood of the type-(I) resonance from the list (6.6.6), equations (6.6.5) in the first approximation (in the sense of the averaging method) possess the following first integrals:

$$\alpha_2 = \text{const}, \qquad \beta_2 = \text{const}, \tag{6.6.7}$$

$$H^* = a\,\frac{2\omega_1}{\omega_2}\,\alpha_1^* \sqrt{\alpha_3}\,\cos\omega_3(\beta_3 \mp \beta_1^*) \mp \frac{\varepsilon}{\omega_3}\,\alpha_1^* = \text{const}, \tag{6.6.8}$$

$$C = \alpha_1^* \pm \alpha_3 = \text{const}, \tag{6.6.9}$$

where we used the notation

$$\alpha_1^* = \alpha_1 \frac{\omega_3}{2\omega_1}, \qquad \beta_1^* = \frac{2\omega_1\beta_1 \mp \varepsilon\tau}{\omega_3}.$$

In (6.6.8) the constant a depends only on the moments of inertia. Let us denote $\Phi = \omega_3(\beta_3 \pm \beta_1^*)$. Then from (6.6.8) and (6.6.9) it follows that

$$\cos\Phi = \frac{h_0 \pm \varepsilon_0\alpha_1^*}{\alpha_1^*\sqrt{\pm(C - \alpha_1^*)}}, \qquad \varepsilon_0 = \frac{\varepsilon}{2a\omega_1}. \tag{6.6.10}$$

Here h_0 is an integration constant, proportional to H^*.

Equation (6.6.10) allows us to construct integral curves in the (α_1^*, Φ)-plane and investigate the motion. Let us consider first region I, corresponding to the inequalities (6.6.1) (the plus sign in (6.6.10)). Figure 6.4 shows the integral curves for two cases: $|\varepsilon_0| > \sqrt{C}$ and $|\varepsilon_0| < \sqrt{C}$. As one can see, in the first case the oscillations of "amplitude" α_1^* that are initially small remain small. This is completely natural, since the inequality $|\varepsilon_0| > \sqrt{C}$ means that the motion is not close to a resonant one, and consequently is should differ only slightly from the linear case (i.e., from the case $\alpha_1^* =$const). On the contrary, if the detuning parameter $|\varepsilon_0|$ is sufficiently small ($|\varepsilon_0| < \sqrt{C}$, i.e., the motion is close to resonance), then, as seen in Figure 6.4. (b), an initially small value α_1^* grows significantly with time (provided only that $C \approx \alpha_3^0$ is not very small).

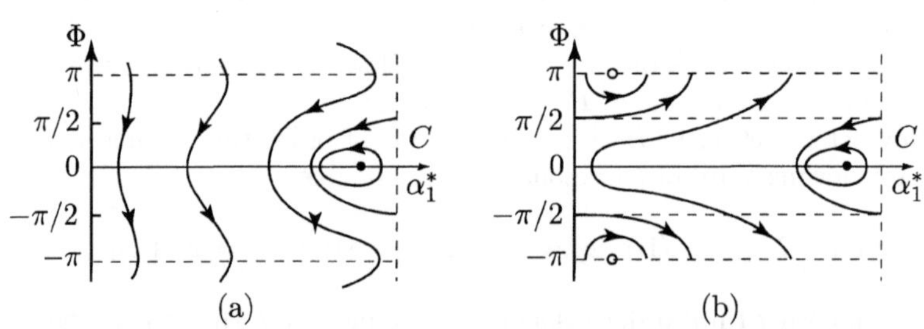

FIGURE 6.4. The amplitude-phase characteristic of the motion in region I (of sufficient conditions for stability): (a) nonresonance; (b) resonance

In other words, in the resonance case initial deviations from the equilibrium position in the *orbital plane* induce large deviations of the initially small angles of transverse oscillations. Due to the presence of the first integral (6.6.9), the oscillations in the orbital plane are "pumped" into transverse oscillations, then "pumped back," and so on. This phenomenon was first revealed by T. R. Kane [6.12] through numerical integration of the exact equations of motion. (The effect discovered by Kane was explained in subsequent work of Breakwell and Pringle [6.10].)

Figures 6.5 and 6.6, borrowed from [6.12], show the variation with time of the angle that describes the transverse oscillations of the satellite. The initial data are the same for both figures, but Figure 6.5 corresponds to the resonance case, while Figure 6.6 corresponds to the nonresonance case. While in the second case the amplitude does not exceed 0.5°, in the first case it reaches 20°!

Such is the effect of resonance. We should emphasize however that the existence of the resonant pumping of oscillations displayed above does not, of course, destroy the Lyapunov stability of the satellite's equilibrium position established earlier. Indeed, in both the resonance case and the nonresonance case, one can always choose h_0 in (6.3.6) and (6.3.8) so small that the amplitude of the oscillations will not exceed a preassigned value. The only difference is that in the resonance case this choice is considerably more restricted than in the nonresonance case.

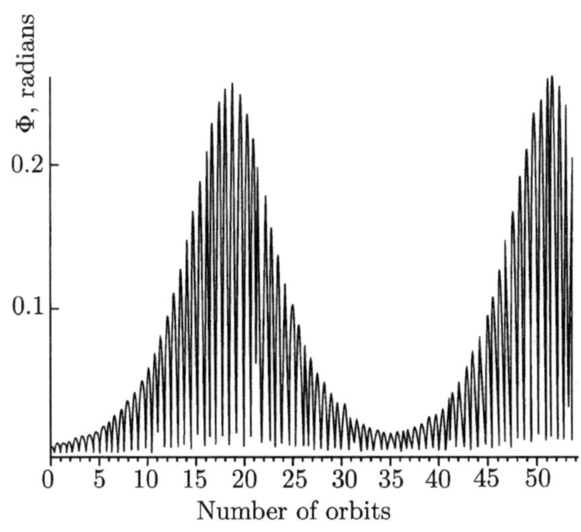

FIGURE 6.5. Transverse oscillations in a circular orbit: resonance

Now let us turn our attention to region II, where the necessary conditions for stability (6.6.2) hold. Here in formula (6.6.10) one has to take the minus signs. The corresponding "amplitude-phase portraits" of the motion are shown in Figure 6.7, (a) and (b). It turns out that, as above, for large enough deviations from reso-

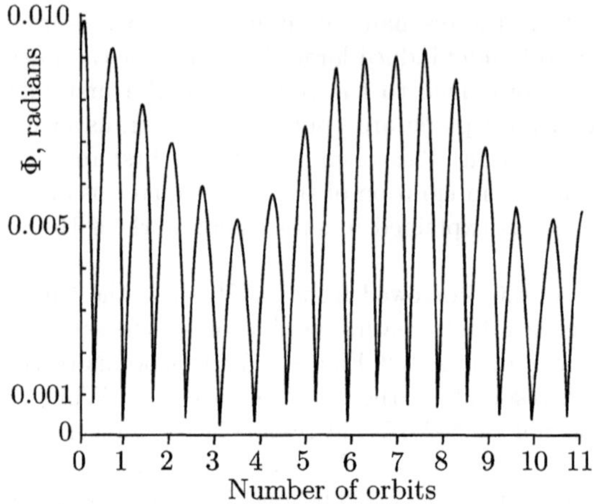

FIGURE 6.6. Transverse oscillations on a circular orbit: nonresonance

nance ($|\varepsilon_0| > \sqrt{|C|}$) the motion can remain bounded (Figure 6.7 (a)). However, in the neighborhood of the resonance ($|\varepsilon_0| < \sqrt{|C|}$), the motion becomes unbounded (Figure 6.7 (b))! This is the qualitative difference between a motion that satisfies the (necessary) conditions (6.6.2) and a motion that satisfies the (sufficient) conditions (6.6.1).

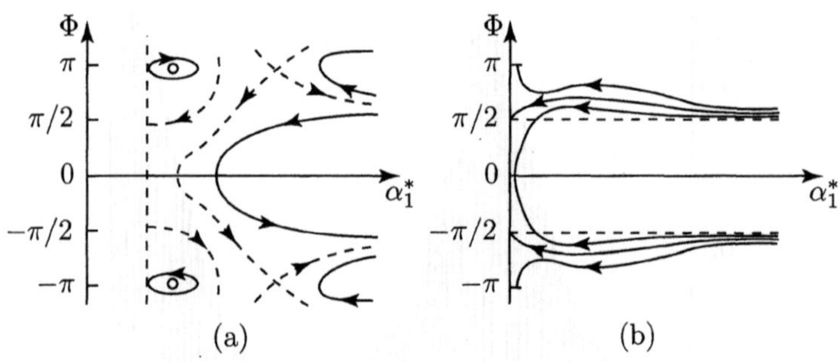

FIGURE 6.7. Amplitude-phase characteristic of the motion in region II (of necessary conditions for stability): (a) nonresonance, $|\varepsilon_0| > \sqrt{|C|}$; (b) resonance, $|\varepsilon_0| < \sqrt{|C|}$

The results of the above analysis, displayed in Figure 6.7, are best described as follows. Regardless of how small the frequency detuning $|\varepsilon| \neq 0$ is, one can always find initial conditions ($\sqrt{|C|} < |\varepsilon_0|$!) for which the motion is bounded; one fails to find such conditions only for $|\varepsilon_0| = 0$, i.e., for strict resonance. As a matter of fact, for strict resonance ($|\varepsilon_0| = 0$) the motion (in region II) is necessarily unbounded (in the approximation considered here). Moreover, due to the presence of the first integral (6.6.9), the longitudinal and transverse oscillations are simultaneously amplified. All this leads us to the conclusion that when only the necessary conditions for stability (6.6.2) are satisfied, the motion is indeed stable, provided that the frequencies satisfy no resonance relations, but in the case when resonance relations hold between frequencies stability may be lost and turn into instability.

Needless to say, this conclusion has only an approximate character (because in our analysis we resorted to approximations). However, recent mathematical investigations (see, e.g., [6.13], [6.14]) allow us to make this conclusion rigorous. It turns out that in the case where only the necessary conditions for stability are satisfied, when the leading terms of the Hamiltonian (6.6.3) form a quadratic form that is not positive-definite, the equilibrium position can be *unstable* if the frequencies of the linear oscillations satisfy strictly a resonance relation of odd order.

Thus, stability established by means of the linearized equations still requires verification (for instance, by constructing a Lyapunov function). Otherwise, this stability may prove to be deceiving, i.e., it may be lost at resonance. The analysis of the oscillations of a satellite about a relative equilibrium position demonstrate this quite convincingly.

Let us note that for practical problems of stabilization of satellites only region I (6.6.1) of sufficient conditions for stability is of importance. In such problems the introduction of dissipative forces turns stability into asymptotic stability. In region II (6.6.2) the dissipative forces destroy the stability and steer the satellite away from the equilibrium position satisfying the conditions (6.6.2).

7. Fast rotations

Gravitational "capture" of a satellite into a regime of oscillations about a relative equilibrium position is possible, as we already explained, only when the angular velocity of the satellite is rather small, because in that case the gravitational torques are relatively "weak."

But what happens when, as is usually the case, the initial angular velocity of the satellite is relative large? Then the action of the torques of external forces (for instance, gravitational) modifies the initial rotation of the satellite so little that in a crude approximation one can assume that this initial rotation remains unchanged, taking place by virtue of inertia. The inertial rotation of a rigid body around its center of mass is well studied in classical mechanics, under the name of *Euler-Poinsot motion*. However, the torques of external forces produce small, yet systematic, persistently acting perturbations. Due to these perturbations, small changes in the initial rotation of the satellite, accumulating over time, may result in a motion that differs considerably from the initial motion. This clearly leads to the following task: study the evolution of the Euler-Poinsot motion under the action of small torques of perturbing forces (in much the same manner as in the theory of orbits one studies the evolution of Keplerian orbits under the action of small perturbing forces). We note that the perturbing forces can be of other nature than gravitational.

The Euler-Poinsot motion (which will sometimes be referred to as the "unperturbed motion") is very simply described for the particular, yet practically important case when the satellite is a *dynamically-symmetric body*. This means that among the three principal central moments of inertia of the body two are equal. Assume, for example, that $A = B \neq C$. The axis corresponding to the moment of inertia C is called the *axis of dynamical symmetry* or the *longitudinal axis* of the satellite. In what follows we will deal precisely with this simple case. As is know from theoretical mechanics, the Euler-Poinsot motion of a dynamically-symmetric body is a *regular precession*. Let us describe this motion (Figure 6.8).

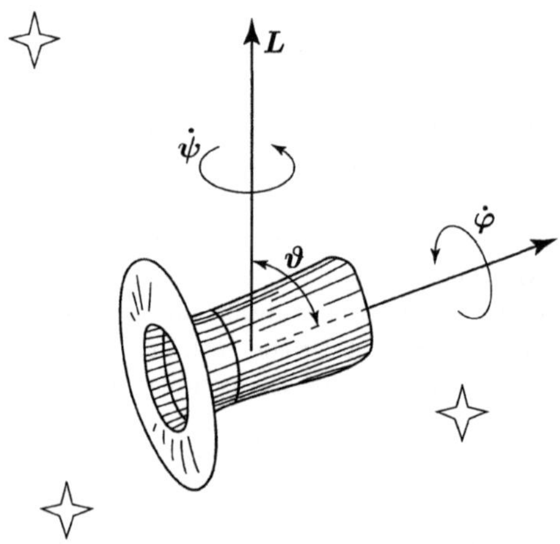

FIGURE 6.8. Regular precession

If p, q, r denote the components of the absolute angular velocity of the satellite relative to the principal central axes of inertia x', y', z', then the *angular momentum vector* $\mathbf{L}$ is the vector whose components with respect to the same axes are Ap, Bq, Cr. Here the longitudinal axis is z'. It turns out that in the Euler-Poinsot motion *the angular momentum vector* $\mathbf{L}$ *is constant in magnitude and direction in absolute space*: $\mathbf{L} \equiv \text{const}$. In a regular precession the component of the vector $\mathbf{L}$ along the satellite's longitudinal axis is constant: $Cr = Cr_0$. The symmetry axis of the satellite circles the angular momentum vector *at constant angular distance* ϑ *at a constant angular rate* $\dot{\psi}$; simultaneously, the satellite "spins" around its longitudinal axis at a *constant angular rate* $\dot{\varphi}$. The satellite "waltzes"[2] around the angular momentum vector like a tireless dancer spinning in a ballroom.

[2]The term "waltzing" in connection with the rotation of a satellite was coined some years ago by Nam Tum Po, then a graduate student at Leningrad State University.

The angle ϑ is called *nutation angle*, and the angular rates $\dot{\psi}$ and $\dot{\varphi}$ are called *angular rate of precession* and *angular rate of proper rotation* (*spin rate*), respectively. The position of the angular momentum vector in space can be specified by means of two constant angles: ρ and σ. Then the regular precession is described by six parameters:

$$L,\ \rho,\ \sigma,\ \vartheta,\ \psi,\ \varphi, \tag{6.7.1}$$

of which the first four are constant, while the last two change uniformly in time (with constant rates $\dot{\psi}$ and $\dot{\varphi}$).

Now let us "turn on" the action of the torques of perturbing forces. Then, generally speaking, the first four parameters (6.7.1) will no longer remain constant, and the dependence of ψ and φ on time will not be linear. The problem is then to determine precisely how the parameters (6.7.1) change (evolve) with time under the action of the perturbing torques. Our parameters (6.7.1) are analogous to the osculating elements in orbital dynamics, and pursuing this analogy we need first of all to derive differential equations "in the osculating elements" (6.7.1).

Consider, for example, the perturbations due to the gravitational potential (6.1.6). The parameter γ, γ', γ'' figuring in (6.1.6) must be expressed through the new variables (6.7.1). As it turns out, in the case $A = B$ the potential does not depend on φ, so that

$$U = U(\rho, \sigma, \vartheta, \psi, \omega_0 t). \tag{6.7.2}$$

The explicit dependence of U on time is a result of the motion of the satellite along its orbit with mean angular velocity ω_0 and, in the case of a noncircular orbit, also of the fact that the value of the current radius vector R, which appears in the expression (6.1.6) of U, is also a (periodic) function of time t: $R = R(\omega_0 t)$. The potential U is a 2π-periodic function of each of its arguments, i.e., for instance, $U(\psi + 2\pi) = U(\psi)$, $U(\omega_0 t + 2\pi) = U(\omega_0 t)$, and so on.

It turns out that in the present case the "equations in osculating elements" have the following form, derived by the author in 1962 [6.15], [6.1]:

$$\left.\begin{aligned}
&\frac{dL}{dt} = \frac{\partial U}{\partial \psi}, \qquad \frac{d\sigma}{dt} = \frac{1}{L\sin\rho}\frac{\partial U}{\partial \rho},\\
&\frac{d\rho}{dt} = \frac{1}{L\sin\rho}\left(\frac{\partial U}{\partial \psi}\cos\rho - \frac{\partial U}{\partial \sigma}\right),\\
&\frac{dL}{dt} = \frac{L}{A} - \frac{1}{L}\left(\frac{\partial U}{\partial \rho}\cot\rho + \frac{\partial U}{\partial \vartheta}\cot\vartheta\right),
\end{aligned}\right\} \tag{6.7.3}$$

$$\cos\vartheta = \frac{Cr_0}{L}. \tag{6.7.4}$$

These equations are valid not only in the case of gravitational perturbations, but also for arbitrary perturbations that possess a force function of the type (6.7.2). (Under certain assumptions, this type of perturbations cover the case of aerodynamic and magnetic torques or torques of radiation pressure.) The equation for φ

is omitted in (6.7.3), since the right-hand sides do not depend on φ and hence the equation for φ can be integrated separately (by a simple quadrature, after which one can integrate the system (6.7.3)).

Thus, the original problem has been reduced to the integration of the system (6.7.3) of four first-order differential equations (supplemented by the relation (6.7.4)).

In the absence of perturbing torques, $U \equiv 0$ and equations (6.7.3) yield $L = L_0$ (and then from (6.7.4) it follows that $\vartheta = \vartheta_0$), $\rho = \rho_0$, $\sigma = \sigma_0$, $\dot{\psi} = L_0/A \equiv \dot{\psi}_0$, which means that the unperturbed motion is simply a regular precession. We shall assume that small perturbations U are present; here "small" means that $L/A \gg (\max|U|)/L$. Then in equations (6.7.3) L, ρ, and σ are slow variables, while ψ is a fast variable in the sense employed in asymptotic methods, and in order to investigate these equations it is convenient to use the already familiar method of averaging with respect to the fast variable. Note that equations (6.7.3) actually contain two fast variables, ψ and $\tau = \omega_0 t$. However, ordinarily ψ changes at a rate of order $1 \div 10$ degree/sec, whereas τ changes at a rate of $0.05 \div 0.07$ degree/sec, and consequently one can readily assume that the variables ψ and τ are not in resonance (i.e., the frequencies $\dot{\psi}$ and ω are not commensurable) and carry out the averaging only with respect to ψ.

Denote

$$\overline{U} = \frac{1}{2\pi} \int_0^{2\pi} U \, d\psi. \tag{6.7.5}$$

Note that, thanks to the periodicity of U in ψ,

$$\frac{1}{2\pi} \int_0^{2\pi} \frac{\partial U}{\partial \psi} \, d\psi = \frac{1}{2\pi} [U(2\pi) - U(0)] = 0.$$

If one takes this into account, system (6.7.3) and equality (6.7.4) immediately yield the relations

$$L = L_0, \quad \cos\vartheta = \cos\vartheta_0, \tag{6.7.6}$$

$$\frac{d\psi}{dt} = \frac{L_0}{A} - \frac{1}{L_0} \left[\frac{\partial \overline{U}}{\partial \rho} \cot\sigma + \frac{\partial \overline{U}}{\partial \vartheta} \cot\vartheta \right], \tag{6.7.7}$$

and the canonical system of two differential equations

$$\frac{d\rho}{dt} = -\frac{1}{L_0 \sin\rho} \frac{\partial \overline{U}}{\partial \sigma}, \quad \frac{d\sigma}{dt} = \frac{1}{L_0 \sin\rho} \frac{\partial \overline{U}}{\partial \rho}. \tag{6.7.8}$$

Thus, the problem reduces to the integration of only two equations (6.7.8), after which once can use (6.7.7) to determine the angular rate of the perturbed precession (together with the small correction enclosed in brackets in (6.7.7)). The main conclusion derived from (6.7.6)–(6.7.8) is the following [6.16], [6.15]:

The perturbed motion of a dynamically-symmetric satellite in the potential force field (6.7.2) is a quasi-regular (i.e., almost regular) precession with modified

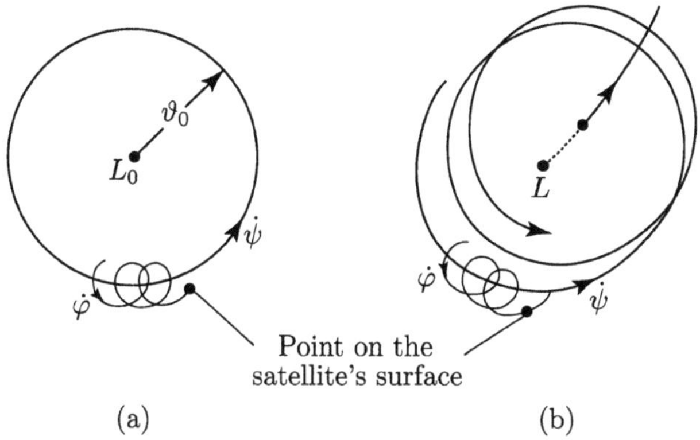

FIGURE 6.9. "Waltz figures." (a) unperturbed motion; (b) perturbed motion

angular rate of precession (6.7.7), which rotates the symmetry axis of the satellite by a constant angle ϑ_0 around the angular momentum vector $\mathbf{L}$ of constant length. *The main effect of the perturbations is that the direction of the vector* $\mathbf{L}$ *in space changes in time.* This change is governed by the equations (6.7.8).

To put it simply and briefly, the "waltz figure" of the satellite, displayed in Figure 6.8, becomes more complex only due to the slow variation of the direction of the vector $\mathbf{L}$. Figure 6.9 shows schematically the unperturbed as well as the perturbed "waltz figures."

The basic equations (6.7.8) of perturbed motion will be referred to here as the *evolution equations*.

8. Where the author slightly frightens the reader

Far, far away, on Lake Chad,
Gracefully strolls a giraffe.

N. Gumylev

I personally find the evolution equations (6.7.8) quite graceful, not so much because of their form, but because the entire perturbed motion does in fact follow from these two simple (and for that reason – beautiful) equations. But this beauty is deceptive. Indeed, before we can use the evolution equations we need to write in explicit form formula (6.7.2) for the force function U. That formula is far from simple – on the contrary, it is rather complicated. In his work, a researcher in mechanics must sometimes deal with an apparatus that is complicated as well lacking in gracefulness, which alas is necessary in order to reach the truth.

With the only purpose of bringing the reader closer to understanding this reality, we will write here the expression (6.7.2) in explicit form. It goes as follows:

if in (6.1.6) we retain the terms that are essential in the investigation of the motion around the center of mass and set $A = B$, then

$$U = \frac{3}{2} \frac{\omega_0^2}{(1-e^2)^3} (1 + e\cos\nu)^3 (A - C)\gamma''^2, \tag{6.8.1}$$

where

$$\begin{aligned}
\gamma''^2 &= \frac{1}{2} \sin^2\vartheta + \frac{1}{4} \sin^2\rho\,(3\cos^2\vartheta - 1) + \\
&+ \frac{1}{4} \sin^2\rho\,(3\cos^2\vartheta - 1)\cos 2(\nu - \sigma) + \\
&+ \sin\rho\,\cos\rho\,\sin\vartheta\,\cos\vartheta\,\sin\psi + \frac{1}{4}\sin^2\vartheta\,\sin^2\rho\,\cos 2\psi + \\
&+ \frac{1}{2}\sin\vartheta\,\cos\vartheta\,\sin\rho\,(1 + \cos\rho)\sin(\psi - 2\nu + 2\sigma) - \\
&- \frac{1}{2}\sin\vartheta\,\cos\vartheta\,\sin\rho(1 - \cos\rho)\sin(\psi + 2\nu - 2\sigma) - \\
&- \frac{1}{2}\sin^2\vartheta\,(1 + \cos\rho)^2\cos(2\psi - 2\nu + 2\sigma) - \\
&- \frac{1}{8}\sin^2\vartheta\,(1 - \cos\rho)^2\cos(2\psi + 2\nu - 2\sigma);
\end{aligned}$$

here the true anomaly ν is related to the dimensionless time $\tau = \omega_0 t$ via the formula

$$\tau = 2\,\mathrm{Arctan}\sqrt{\frac{1-e}{1+e}}\tan\frac{\nu}{2} - \frac{e\sqrt{1-e^2}}{1+e\cos\nu}\sin\nu. \tag{6.8.2}$$

In formulas (6.8.1) and (6.8.2), e denotes the eccentricity of the orbit and $\omega_0 = \mu^{1/2}a^{-3/2}$ denotes the mean motion in an orbit of semi-major axis a. The angles ρ and σ are chosen in a very concrete manner: ρ is the angle between the angular momentum vector $\mathbf{L}$ and the normal $\mathbf{n}$ to the orbital plane, while σ is the angle, measured in the orbital plane, from the radius vector of the orbit's pericenter to the projection of the vector $\mathbf{L}$ on the orbital plane.

I hope that, after being faced with formulas (6.8.1) and (6.8.2), the reader will emerge only "slightly frightened." Had I wished to frighten her or him badly, I would have written explicitly the expression of U for a three-axial ($A \neq B \neq C$) body.[3]

9. Explicit form of the perturbed motion

In order to derive from the expression (6.8.1) of U the expression for the quantity $\overline{U}$, which appears in the evolution equations (6.8.1), we simply need to discard all terms that contain trigonometric functions of ψ as factors. Indeed, these are precisely the terms that are "killed" in the averaging process (6.7.5). Carrying out

[3]The fast nonresonant rotational motions of a three-axial satellite in a gravitational field were studied by F. L. Chernous'ko [6.17].

this procedure, we see that $\overline{U}$ already looks much simpler, though an explicit dependence of $\overline{U}$ on time (through ν) still remains; due to this annoying circumstance the equations (6.7.8) are not integrable.

There are two ways out of this situation:

(1) a partial one: in the case of a circular orbit ($e = 0$), equations (6.7.8) can be reduce to an integrable form by means of the substitution $\varkappa = \nu - \sigma$ (see [6.1], [6.15]).

(2) a general one: for an arbitrary orbit one can consider τ as a fast variable and again apply the asymptotic method to equations (6.7.8), averaging the right-hand sides of the equations with respect to τ.

Let us follow the second way. Then in the equations (6.7.8) we must replace $\overline{U}$ by

$$\widetilde{U} = \frac{1}{2\pi}\int_0^{2\pi} \overline{U}\, d\tau = \frac{1}{(2\pi)^2}\int_0^{2\pi}\int_0^{2\pi} U\, d\psi\, d\tau.$$

To calculate this integral it is convenient to replace the variable τ by ν: indeed, $\overline{U}$ depends explicitly on ν, and

$$d\tau = \frac{d\tau}{d\nu}\, d\nu = \frac{(1-e^2)^{3/2}}{(1+e\cos\nu)^2}\, d\nu.$$

If we make this substitution in U, formula (6.8.1) yields

$$\widetilde{U} = \frac{3}{2}\frac{\omega_0^2}{(1-e^2)^{3/2}}(A-C)\left(\frac{1}{2}\sin^2\vartheta + \frac{1}{4}\sin^2\rho\,(3\cos^2\vartheta - 1)\right). \tag{6.9.1}$$

Replacing $\overline{U}$ in (6.7.8) by expression (6.9.1), we immediately get (see [6.16], [6.1])

$$\rho = \rho_0, \qquad \frac{d\sigma}{d\tau} = \frac{3}{4}\frac{(A-C)\omega_0}{L_0(1-e^2)^{3/2}}(3\cos^2\vartheta_0 - 1)\cos\rho_0. \tag{6.9.2}$$

The constancy of the angle ρ follows from the first of equations (6.7.8) because $\widetilde{U}$ does not depend on σ: the terms involving σ in (6.8.1) are all killed as a result of the two averaging steps. Thus, the perturbed motion of the angular momentum vector in the gravitational field is very simple. Namely, that vector undergoes a precession around the normal $\mathbf{n}$ to the orbital plane at a constant angular distance ρ_0 from $\mathbf{n}$ and with constant angular rate (6.9.2). Recall that the satellite itself "dances" a regular precession (6.7.6)–(6.7.7) around the angular momentum vector, where now in (6.7.7) we must replace $\overline{U}$ by $\widetilde{U}$. As it turns out, the angular rate of precession $\dot{\psi}$ remains strictly constant.

If the unperturbed motion is simply rotation around an axis of symmetry (a particular case of regular precession), then $\vartheta_0 = 0$ and $L = Cr_0$; this is precisely the case considered in classical celestial mechanics (precession of the Earth's axis under the influence of the attraction of the Sun and Moon). In this special case formulas (6.9.2) turn into the classical formulas for the angular rate of precession of the Earth's axis. It is know that the period of this precession is not too large and not too small – about 26,000 years. However, for artificial satellites the effects discussed are far more "ponderable, flagrant, and evident": the period of precession of the angular momentum vector, calculated by means of formula (6.9.2), is of only a few days.

10. Pegasus

Continuing our discussion, we note that over a duration of a few days, the orbit of the satellite itself changes rather notably, but up to this point we did not take that into consideration at all. As a matter of fact, in the first essay we have shown that due to the oblateness of the Earth, the satellite's orbital plane rotates (undergoes a precession) around the axis of the Earth with the angular rate

$$\frac{d\Omega}{d\tau} = -\frac{\bar{\varepsilon} R_{\mathrm{E}}^2}{p^2} \cos i \equiv K_\Omega \tag{6.10.1}$$

(this is the speed of the motion of the ascending node Ω of the orbit) and, moreover, the orbit rotates in its own plane with the velocity

$$\frac{d\omega_\pi}{d\tau} = -\frac{\bar{\varepsilon} R_{\mathrm{E}}^2}{2p^2}(5\cos^2 i - 1) \equiv K_\omega \tag{6.10.2}$$

(this is the velocity of the motion of the pericenter of the orbit, ω_π). In the foregoing analysis the perturbed motion (6.9.2) of the angular momentum vector turned out to be so simple only because we neglected the evolution of the orbit, described by formulas (6.10.1) and (6.10.2). A computation done, say, by means of formula (6.10.1) shows that a typical orbit of the satellite rotates in space by 360° in 10–12 days.[4]

The influence of the orbit evolution on the evolution of the rotation and orientation of a satellite was analyzed by us in the papers [6.16] and [6.19]; the theory constructed therein was subsequently developed in our book [6.1], taking into account various torques of forces acting on the satellite. An elegant improvement of this theory is due to S. I. Trushin [6.20]. Following the listed works, we write the equations governing the perturbed motion of the angular momentum vector in the form

$$\left.\begin{aligned} \frac{d\rho}{dt} &= K_\Omega \sin i \sin\Sigma, \\ \frac{d\Sigma}{dt} &= 2K_{\mathrm{g}} \cos\rho - K_\Omega(\cos i - \sin i \cot\rho \cos\Sigma). \end{aligned}\right\} \tag{6.10.3}$$

[4]The notation in formulas (6.10.1) and (6.10.2) is as follows: R_{E} is the equatorial radius of the Earth (6,371 km), p is the focal parameter of the orbit, i is the inclination of the orbit to the equator, and $\bar{\varepsilon} = 0.0016331$ is a dimensionless parameter, determined by the magnitude of Earth's oblateness.

Here the constant K_{g} is given by

$$K_{\mathrm{g}} = \frac{3}{8}\,\frac{(A-C)\omega_0}{L_0(1-e^2)^{3/2}}\,(3\cos^2\vartheta_0 - 1),$$

and K_Ω is the constant defined by formula (6.10.1). The variable ρ retains its previous meaning, while the new variable Σ is connected with the old one σ by

$$\Sigma = \sigma + \omega_\pi - \frac{\pi}{2}.$$

Let us mention here that for a triaxial ellipsoid ($A \neq B \neq C$) the evolution equations have exactly the same form (6.10.3), the only difference being that the constant K_{g} is given by a more complicated formula, found by F. L. Chernous′ko [6.17].

The meaning of equations (6.10.3) is quite simple. Consider the celestial sphere (Figure 6.10), and draw on it the celestial equator, its intersection with the orbital plane, and the trace of the angular momentum vector **L**.

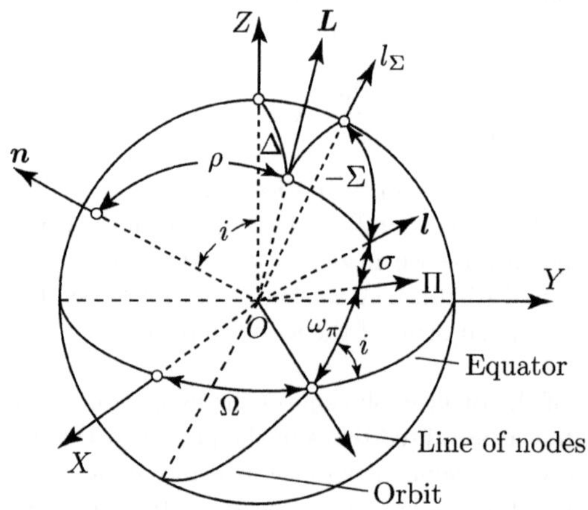

FIGURE 6.10. The relative position of the angular momentum vector and of the orbit: Z is the direction to the world's North pole, X is the direction to the spring (vernal) equinox point, **n** is the unit normal to the orbital plane, **L** is the direction of the angular momentum vector, $O\Pi$ is the direction of the radius vector of the perigee, **l** is the projection of **L** on the orbital plane, and l_Σ is the line from which the angle Σ is measured

Suppose that no gravitational torques (as well as no other torques) act on the satellite. Then the vector **L** remains fixed in absolute space (in the coordinate frame XYZ whose axes X and Z are directed toward the vernal equinox point and the world's North pole, respectively), but changes its position relative to the orbit as a result of the orbit's evolution. In other words, the angles ρ and σ, which define the position of the vector $\mathbf{L}$ with respect to the orbit, change due to the change in the angles Ω and ω_π that specify the position of the orbit. Figure 6.10 shows clearly that a change in the angle ω_π (motion of the orbit's perigee in its plane) cannot result in a change of the angle ρ between the normal $\mathbf{n}$ to the orbital plane and the direction of $\mathbf{L}$, and only the angle σ changes, namely, to an increment $\Delta\omega_\pi$ there corresponds the increment $\Delta\sigma = -\Delta\omega_\pi$.

It is therefore convenient to replace σ by a new angle, which does not depend on the motion of the perigee. As such an angle one can take, for example, the sum $\sigma + \omega_\pi$, which defines the angular position of the projection $\mathbf{l}$ of the vector $\mathbf{L}$ on the orbital plane relative to the line of nodes. This angle is indeed not affected by the motion of the perigee. However, even more convenient is to introduce the angle $\Sigma = (\sigma + \omega_\pi) - \pi/2$, and then measure the rotation of $\mathbf{L}$ starting from the "highest" point of the orbit (in Figure 6.10 there is shown the angle $-\Sigma$). Thus, the angles ρ and Σ are not affected by the motion of the perigee. However, they do change due to the motion of the orbit's node, and a calculation shows that the corresponding rates are

$$\left(\frac{d\rho}{d\tau}\right)_1 = K_\Omega \sin i \sin \Sigma,$$
$$\left(\frac{d\Sigma}{d\tau}\right)_1 = -K_\Omega(\cos i - \sin i \cot \rho \cos \Sigma),$$

where K_Ω is the angular rate (6.10.1) of the motion of the node. But the angle Σ changes also due to the influence of the torque of gravitational forces, at the rate (6.9.2). Adding first all the rates of change of the angle ρ, and then those of the angle Σ, we obtain the two evolution equations (6.10.3).

These equations can be reduced to canonical form with time-independent right-hand sides, from which one derives the first integral

$$K_\Omega(\cos i \cos \rho + \sin i \sin \rho \cos \Sigma) - K_g \cos^2 \rho = h_0; \qquad (6.10.4)$$

here h_0 is an integration constant. Equation (6.4.10) defines the family $\rho(\Sigma, h_0)$ of possible trajectories of the tip of the angular momentum vector. Naturally, for $K_\Omega = 0$ this family is a set of small circles $\rho = \rho_0$ (which $\mathbf{L}$ traces in its precession under the influence of the gravitational torques), while for $K_g = 0$ the family becomes a set of small circles $\Delta = \Delta_0$, where Δ is the angle between $\mathbf{L}$ and the Z-axis to the world's North pole (Figure 6.10) (indeed, in the absence of gravitational torques the vector $\mathbf{L}$ remains fixed in space, and so the angle Δ is constant). When both parameters K_Ω and K_g are different from zero, the family of trajectories (6.10.4) is considerably more complex. All the trajectories are closed and symmetric with respect to the big circle $\Sigma = 0, \pi$. As it follows from the form of the right-hand sides of equations (6.10.3), on these meridians the family of trajectories has fixed points (poles), with coordinates $\Sigma = 0, \pi$, $\rho = \rho_\pi$, where ρ_π satisfies the equation

$$K_g \sin 2\rho_\pi - K_\Omega \sin(\rho_\pi \pm i) = 0. \tag{6.10.5}$$

There are two or four such fixed points, depending on the values of i and of the parameter $\varkappa = -K_g/K_\Omega$. For example, if $|\varkappa| > 1$ there are always four fixed points, regardless of the value of i. More precisely (see [6.14]), there are two fixed points if

$$|2\varkappa|^{2/3} < \sin^{2/3} i + \cos^{2/3} i,$$

and four if the inequality sign is reversed.

Figure 6.11 exhibits an example of the dependence $\rho_\pi(i)$, obtained from equation (6.10.5), for a fixed value $\varkappa > 1$ (approximately equal to 1.1). The continuous [resp., dotted] lines show the set of values of ρ_π corresponding to stable [resp., unstable] poles of trajectories. Examining the vertical cross-section i =const on this figure we see that to each fixed $i \neq 0°, 90°$ there corresponds a family of trajectories with three stable poles and one unstable one. Moreover, one (stable) pole lies on the meridian $\Sigma = \pi$, while the remaining three poles lie on the meridian $\Sigma = 0$. The corresponding portrait of the trajectories is shown in Figure 6.12.

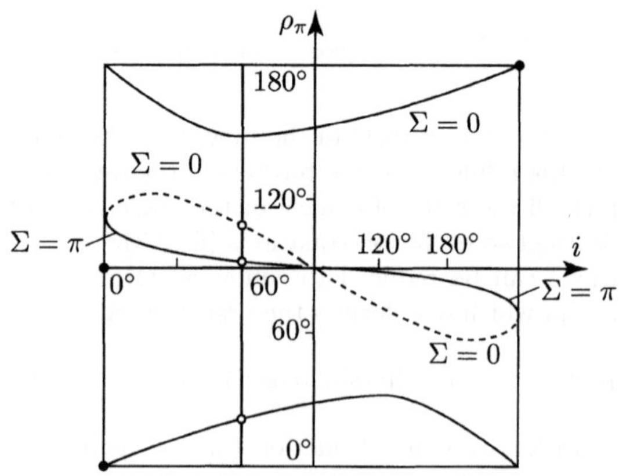

FIGURE 6.11. Dependence of the position ρ_π of a pole of a trajectory on the orbit inclination i for $\varkappa > 1$

Observing this family one notes that in the vicinity of the singular point 1 the motion is close to a "gravitational" one (of the type of a "spoilt" precession of the angular momentum vector around the normal to the orbital plane). However, in the vicinity of the singular point 4 the motion is qualitatively different: the tip of the vector **L** traces an elongated closed curve around the point 4, which lies on the

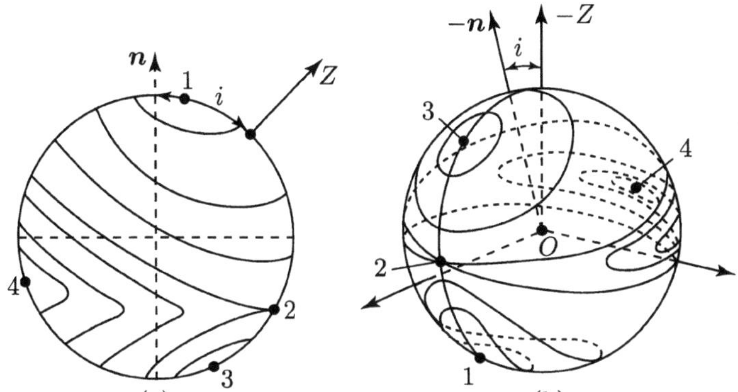

FIGURE 6.12. A family of trajectories of the tip of the angular momentum vector

meridian $\Sigma = \pi$ near the orbital plane. Furthermore, this kind of motion does not disappear for arbitrarily large (but finite) values of $|\varkappa|$. Regardless of how small the influence of the motion of the orbit's node (K_Ω) is compared with the influence of the gravitational torques (K_g), this small influence always generates a set of trajectories in the vicinity of the point 4 in Figure 6.12 that differ qualitatively from trajectories of gravitational type. For $|K_g| \gg |K_\Omega|$ the poles 1 and 3 lie very close to the points $\rho = 0$ and $\rho = \pi$, respectively, while the ρ-coordinates of the poles 2 and 4 are close to $\pi/2$. But the zone of "nongravitational" motion will nevertheless exist as a narrow "band-aid" near the orbital plane; only when $\varkappa \to \infty$ does the area of this zone tend to zero.

Such is the somewhat unexpected and manifestly nontrivial influence of the evolution of the orbit on the evolution of the motion around the center of mass. The reader can find more details on these effects in the book [6.1] (Ch. 1, §4, and also Ch. 8).

And now it is time to recall that the title of the present section refers to the famous winged horse – what is it doing here? But *Pegasus* is also the name of an American satellite for which the evolution of the motion around the center of mass fits beautifully into the theory just expounded. The character of this evolution was observed experimentally[5] and compared with the theory in the paper [6.21].[6] The strongest influence on *Pegasus* was exerted by precisely the gravitational torque and the motion of the node of its orbit; therefore, the evolution of its rotation is well described by equations of the type (6.10.3). In Figure 6.13, borrowed from [6.21], the continuous line shows the trajectory of the tip of the angular momentum vector of *Pegasus*, calculated from equations (6.10.3), while the small triangles indicate the observation data for that trajectory. We see that the points fit an exotic,

[5]The meaning of this statement will be explained in the 12th essay.

[6]The theory discussed in [6.21] repeats earlier results obtained in [6.1] and [6.17].

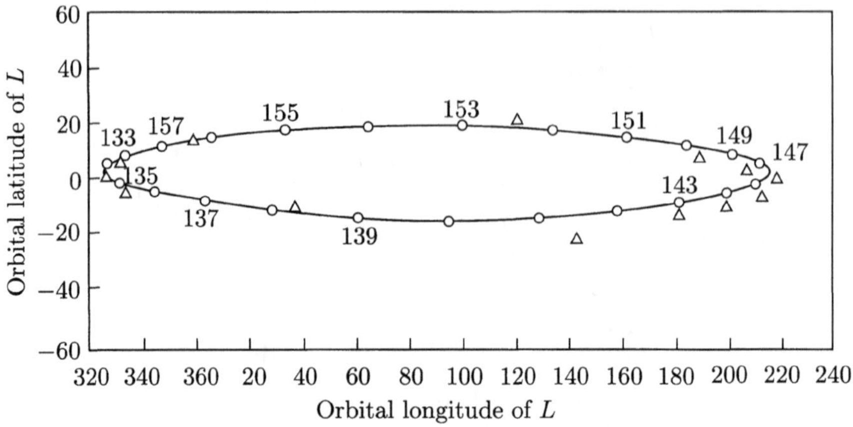

FIGURE 6.13. Trajectory of the tip of the angular momentum vector of the *Pegasus* satellite. The continuous line with dots shows computational results; the small triangles indicate observation data. The numbers along the curve show the time measured from the beginning of the year 1965

stretched along the orbital plane oval curve, which corresponds to the motion on one of the trajectories in the vicinity of the pole 4 shown in Figure 6.12.

11. Moon, Mercury, resonances ...

The equations (6.7.3) with the force function (6.8.1) admit a particular form, corresponding to the oscillations of the satellite in the orbital plane: $\rho = 0$, $\vartheta = 90°$, $\sigma = \sigma_0$. In this case (6.7.3) reduces to a system of only two equations:

$$\frac{dL}{dt} = \frac{\partial U}{\partial \psi}, \qquad \frac{d\psi}{dt} = \frac{L}{A}, \tag{6.11.1}$$

which in turn is equivalent to the single second-order equation

$$\frac{d^2\psi}{dt^2} - \frac{\partial U}{\partial \psi}\frac{1}{A} = 0, \tag{6.11.2}$$

where

$$U = -\frac{3}{4}\frac{\omega_0^2}{(1-e^2)^3}(1+e\cos\nu)^3(A-C)\cos(2\psi - 2\nu + 2\sigma_0). \tag{6.11.3}$$

If earlier we assumed that ψ is a fast variable, here this is not necessarily the case; in return, we will consider that the difference $A - C$ of the moments of inertia is small, so that the satellite is dynamically nearly spherical. This case is important for artificial satellites as well as for planets.

If $A - C = 0$, then the solution of equations (6.11.1) or (6.11.2) describes a uniform rotation:

$$\psi = \varpi\tau + \psi_0, \qquad \tau = \omega_0 t. \tag{6.11.4}$$

For $A - C \neq 0$, but small, one can seek a solution of equations (6.11.2) close to the uniform rotation (6.11.4) using an asymptotic method (with the small parameter taken to be the difference $A - C$ of the satellite's moments of inertia). Such an investigation was carried out by F. L. Chernous′ko in [6.22]. We will seek a solution of equations (6.11.1) in the form (6.11.4), where ψ_0 is a new variable.

It is actually more convenient to replace ψ_0 by another variable δ, related linearly to ψ_0:

$$2\delta = \pi + 2(\sigma_0 + \psi_0).$$

Denoting also $\varpi = m/2$, we have

$$\frac{\partial U}{\partial \psi} = \frac{\partial U}{\partial \delta}, \qquad \dot{\psi} = \frac{m}{2} + \dot{\delta}, \qquad \ddot{\psi} = \ddot{\delta},$$

where

$$U = \frac{3}{4}\frac{\omega_0^2}{(1-e^2)^3}(1 + e\cos\nu)^3(A - C)\cos(m\tau - 2\nu + 2\delta).$$

Now to obtain the asymptotic solution we need to average U with respect to τ and substitute the result in equations (6.11.1). However, for any m that is not an integer the result of the averaging is zero: the nonresonant rotation of the satellite does not differ from the uniform rotation. When m is an integer the function U is periodic in ν and then

$$\overline{U} = \frac{1}{2\pi}\int_0^{2\pi} U\,d\tau = \frac{1}{2\pi}\int_0^{2\pi} U\frac{d\tau}{d\nu}\,d\nu.$$

Averaging yields

$$\left.\begin{aligned} \overline{U} &= \frac{3}{4}\omega_0^2(A - C)\Phi_m(e)\cos 2\delta, \\ \Phi_m(e) &= \frac{1}{\pi(1-e^2)^{3/2}}\int_0^{\pi}(1 + e\cos\nu)[m\tau(\nu) - 2\nu]d\nu, \end{aligned}\right\} \tag{6.11.5}$$

where $\tau(\nu)$ is defined by formula (6.8.2). The equation of motion (6.11.2) becomes

$$\frac{d^2\delta}{d\tau^2} + \frac{3}{2}\frac{A - C}{A}\Phi_m(e)\sin 2\delta = 0. \tag{6.11.6}$$

One can show that equation (6.11.6) remains valid for a triaxial satellite as well, provided that in the denominator one replaces the moment of inertia A by the third moment of inertia B. Equation (6.11.6) contains substantial information, and we will now describe the basic properties and results that follow from it.

1. Equation (6.11.6) has the same form as the equation governing the oscillations of a physical pendulum.

2. It is integrable; a first integral – the energy integral – is given by

$$\frac{1}{2} A\dot{\delta}^2 + \frac{3}{2}\Phi_m(e)(A-C)\sin^2\delta = h_0. \tag{6.11.7}$$

This formula allows us to construct in the plane $(\delta, \dot{\delta})$ the *phase portrait* of the trajectories (i.e., the dependence $\dot{\delta} = \dot{\delta}(\delta, h_0)$ for different fixed values of h_0). This phase portrait is shown in Figure 6.14 below for the case when $\Phi_m(e)(A-C) > 0$.

3. By the definition of the angle δ, $\delta = 0$ (or $\delta = \pi$) means that at the moment of its perigee passage the satellite is oriented so that the axis corresponding to its moment of inertia C is directed along the radius vector of the perigee. The phase portrait in Figure 6.14 makes it clear that for $\Phi_m > 0$ [resp., $\Phi_m < 0$] the stable regime of rotation is that in which at perigee the direction in which the satellite is elongated coincides with the direction of the radius vector of the perigee ($A > C$) [resp., with the direction of the velocity vector ($A < C$)].

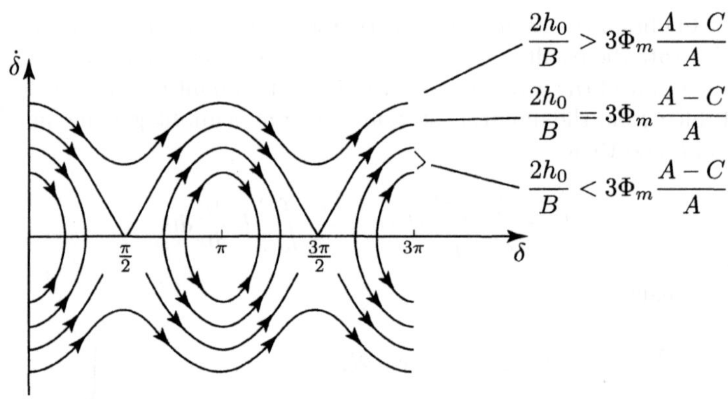

FIGURE 6.14. Phase portrait

4. The $(\dot{\delta}, \delta)$-phase plane is divided into a region of "rotational motion," where the angle δ increases (decreases) monotonically with time, and a region of "oscillatory motion," where the angle δ oscillates. The latter is called the *resonance zone*. Its width is given by

$$|\dot{\delta}|_{\max} = \sqrt{3\,\frac{A-C}{A}\,\Phi_m(e)}\,.$$

5. The arbitrary oscillations $\delta(\tau)$ in the resonance zone can be described in terms of elliptic functions of modulus

$$k^2 = \sin^2\delta_0 + \frac{A\dot{\delta}_0^2}{3(A-C)\Phi_m(e)} \leq 1,$$

so that

$$\sin\delta = k\,\mathrm{sn}\left(\sqrt{3\,\frac{A-C}{A}\,\Phi_m(e)}\,(\tau-\tau_0)+F_0, k\right), \qquad \max|\sin\delta| = k.$$

The dimensionless period of the oscillations is

$$T = \frac{4K(k)}{\sqrt{3\,\frac{A-C}{A}\,\Phi_m(e)}},$$

where $K(k)$ is the complete elliptic integral of the first kind.

6. The frequency of the small oscillations about the stable relative equilibrium position is given by

$$\omega_\delta = \sqrt{3\,\frac{A-C}{A}\,\Phi_m(e)}\,.$$

Now note that for a circular orbit ($e = 0$) $\Phi_m \equiv 0$ for all m, except $m = 0$. In the last case the period of axial rotation is exactly twice the period of orbital revolution. We already know that this is precisely how the Moon moves in its orbit.

In a circular orbit, apart from the Moon-type motions, no other stable resonant motions are possible. It is interesting to note that for $e = 0$ we have $\Phi_2(0) = 1$ and equation (6.11.6) is the *exact* equation of plane oscillations of a satellite about a relative equilibrium position, i.e., in this case equation (6.11.6) can be derived rigorously, without averaging. Then δ is the angle between the axis of inertia of the satellite and the current radius vector of its center of mass. As one can see, a Moon-type motion ($m = 2$) is also possible for an elliptic orbit. However, for an elliptic orbit there exist also infinitely many stable resonant motions, corresponding to different integer values of m. Among all possible resonant motions of interest are precisely the Moon-type motion ($m = 2$) and the Mercury-type motion ($m = 3$). Figure 6.15 shows the graphs of the functions $\Phi_2(e)$ and $\Phi_3(e)$ [6.22], [6.23].

For small values of e one can write (see [6.22], [6.23])

$$\Phi_2(e) = 1 - 5\,\frac{e^2}{2} + 13\,\frac{e^4}{16} - \cdots,$$

$$\Phi_3(e) = 7\,\frac{e}{2} - 123\,\frac{e^3}{16} + \cdots.$$

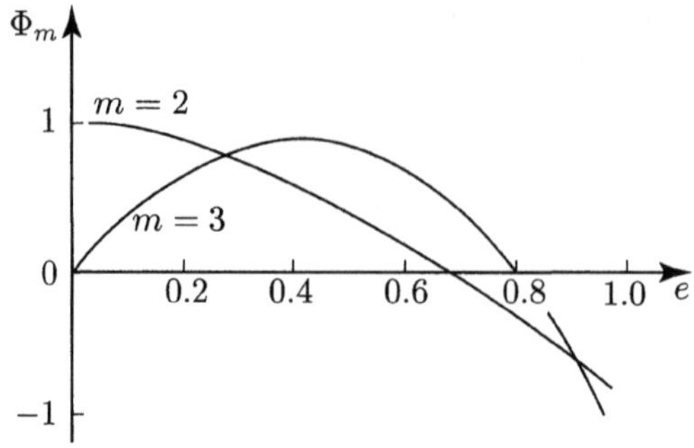

FIGURE 6.15. Graphs of $\Phi_2(e)$ and $\Phi_3(e)$

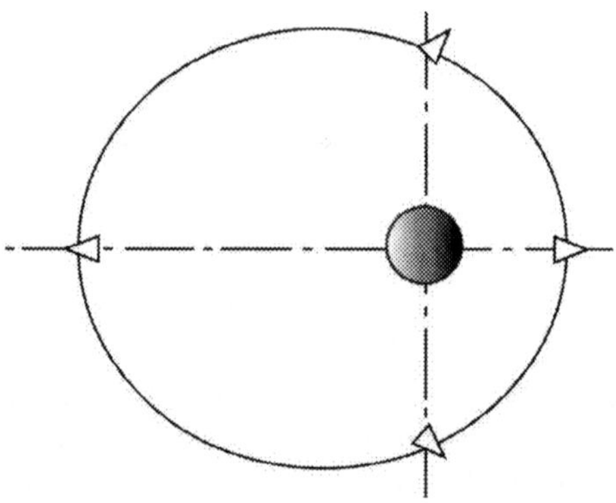

FIGURE 6.16. Moon-type rotation ($m = 2$)

The role of the resonances with $m > 3$ decreases in significance with the growth of m (at least for small eccentricities), because $\Phi_m(e) \approx e^{m-2}$.

Stable Moon-type rotations are depicted in Figure 6.16. For small eccentricities the satellite passes at perigee aligning its axis along the radius vector. However, for $e > 0.682$ the stable motion is that in which the satellite passes at perigee aligning itself in the direction of the velocity vector. The resonant rotation of Mercury was already mentioned in the 4th essay. Mercury's angular velocity of axial rotation is 3/2 larger than its orbital angular velocity ($\varpi = 3/2$). Mercury has a rather eccentric orbit ($e = 0.206$). For such an orbit $\Phi_3(e) \approx 0.7$, whereas $\Phi_1(e) \approx 0.9$. In other words, the width of the resonance zone for $m = 3$ ($\varpi = 3/2$) is of the same order as the width of the zone for $m = 2$ ($\varpi = 1$). In its evolution process Mercury, experiencing a reduction of its angular velocity under the influence of tidal forces, had a good chance of falling into the $m = 3$ resonance without reaching the resonance with the smaller angular velocity ($m = 2$), which

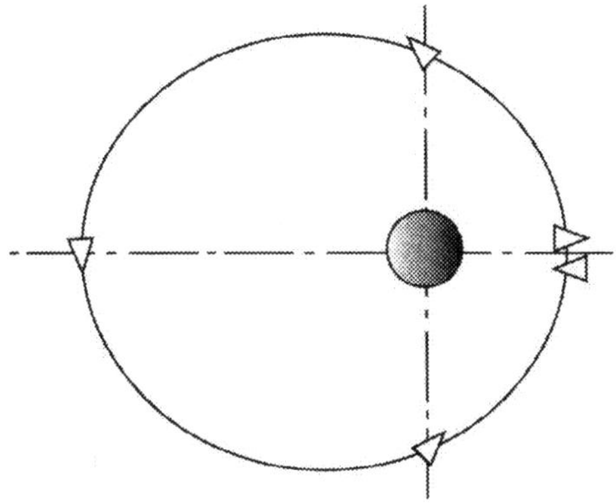

FIGURE 6.17. Mercury-type rotation ($m = 3$)

is exactly what happened in reality.[7] Figure 6.17 shows a Mercury-type rotation. At the pericenter the satellite is always aligned along the radius vector (with the "nose" and the "stern" alternately pointing toward the center of attraction), while at the apocenter it is aligned along the velocity vector (with the "nose" and the "stern" alternately pointing forward).

In this section we have examined briefly planar resonance motions in the gravitational field. As a mater of fact, resonant motions are more complicated due to the three-dimensional nature of the rotation of celestial bodies. Thus, for instance, the angular momentum vector of the rotational motion of Mercury makes an angle of about 90° degrees with the normal to its orbital plane [6.2], [6.27]. This reduces the width of the resonance zone compared with that obtained from the analysis of the planar case carried out here. A. P. Torzhevskiĭ [6.24 a, b] and A. P. Markeev [6.25] have conducted a long series of investigations on resonant rotations of satellites.

We can therefore conclude that the presence of stable resonant motions of celestial bodies is not accidental; rather, it is a consequence of the nature of gravitational forces.

12. Schiaparelli and others

Needless to say, the investigation of resonant motions is equally important in the theory of motion of natural and artificial celestial bodies. The classical example of 1 : 1 resonance (period of rotation equals period of revolution) is the motion of the Moon; but the same resonance plays a major role in systems of gravitational stabilization of artificial satellites. The 2 : 3 resonance was not used (and was not observed) for artificial satellites, but the discovery of this resonance in the motion

[7] It is highly interesting that the nature of the tidal forces did not allow Mercury to get stuck into a resonance with $m > 3$, but did welcome it in the resonance $m = 3$ [6.2]!

of Mercury is so instructive that is worth pausing here to tell briefly about the history of this discovery.[8]

For a long time it was considered that the rotation of Mercury around its axis is synchronized with its orbital motion, i.e., Mercury, like the Moon, rotates in a $1:1$ resonance. Then Mercury should almost exactly be turned permanently with the same side toward the Sun. This is precisely the way in which in 1889 the famous Italian astronomer Schiaparelli[9] interpreted the large number of observations he made on Mercury. Mercury's period of revolution around the Sun is known to be 88 days. Analyzing his observations, Schiaparelli concluded that Mercury's period of axial rotation also equals 88 days. For almost a century (till 1965) optical observations of different astronomers "confirmed" Schiaparelli's conclusion.

However, technological development (including space technology) made available over time essentially new, radar methods of planetary observation. Radar observations of Mercury in 1965 enabled the American scientists G. H. Pettengill, R. B. Dyce, and I. I. Shapiro [6.27], [6.28] to establish that the period of axial rotation of Mercury is not 88, but 59 ± 3 days. The Italian scientist G. Colombo (who worked for a long time in the United States) pointed out that Mercury's period of rotation, determined through radar measurements, is close to $\frac{2}{3}$ of Mercury's period of revolution ($\frac{2}{3}\,88 = 58.65$ days). He was the first to advance the idea that the rotation of Mercury is in resonance with its orbital motion (spin-orbit resonance), with the angular velocity of rotation and the angular velocity of revolution related as $3:2$. Of course, following these investigations the observation data collected by Schiaparelli and his successors have been reexamined. It was shown that their optical observations admit an alternative interpretation and that the rotation period of ~ 59 days is in excellent agreement with those observations!

Thus, Schiaparelli is at risk to be remembered in the history of science in the slightly sad role of a researcher that made two famous mistakes in the interpretation of his own observations: the synchronous rotation of Mercury, which does not hold, and the Mars canals, which apparently are also not there.[10]

Recounting the history of the determination of Mercury's period of rotation one cannot fail to mention how the period of rotation of Venus was actually determined. Because of the thickness of the Venusian atmosphere, optical observations allowed in principle to measure only the period of rotation of the upper layers of the atmosphere. Such observations showed that the atmosphere of Venus undergoes a retrograde rotation, which hinted at the (correct, as it later turned out) fact that the planet itself rotates retrogradely.

---

[8] For a bibliography on this subject the reader may consult the surveys [6.2] and [6.26].

[9] Schiaparelli became famous when in 1877, during the great opposition of Mars, he discovered a network of dark lines on Mars that he called "canals."

[10] To the utmost surprise of researchers, the photographs of the surface of Mars taken by the American space probes *Mariner* 4 in July 1965, *Mariner* 6 in July 1969, and *Mariner* 7 in August 1969 have shown that the surface of Mars resembles, of all things, the surface of the Moon. A considerable part of Mars' surface is covered by a system of craters similar to the lunar craters. At least some of the famous "canals" can be identified with chains of such craters [6.30].

As for the period of rotation of Venus itself, in the last century no progress has been achieved, except for stimulating the healthy curiosity of a number of researchers.[11]

The situation changed in 1962 after radar measurements were carried out in the (former) Soviet Union (by the group led by V. A. Kotel′nikov [6.31], [6.26]) and in the United States (by R. L. Carpenter [6.32] and R. M. Goldstein [6.33]). It was concluded that Venus rotates around it axis in opposite direction (retrogradely) with respect to its orbital motion and with a period of about 250 days. The inclination of its axis to the orbital plane is of $1^\circ.2$.

A bit later the value of the period of rotation was sharpened to 243.24 ± 0.1 days, which up to measurement errors coincides with the value of 243.16 days, for which in each inferior conjunction Venus is turned with the same side toward Earth.[12] P. Goldreich and S. J. Peale [6.2] were apparently the first who realized that this is the case. Thus, the axial rotation of Venus displays a totally phenomenal resonance, connected not only with the orbital motion of Venus itself, but also with the orbital motion of the Earth. In the time interval (of 583.92 days) between two inferior conjunctions, Venus executes five complete (retrograde!) rotations in its orbital coordinate system, and four complete rotations in the Earth's orbital coordinate system.

The theory of resonant rotational motions of Mercury and Venus is still in its initial stage of development [6.2], awaiting for inspired researchers.

13. Resonant rotations of celestial bodies and the generalized Cassini laws

The facts discussed above allow us to speak about the tendency toward synchronization of rotations as an objective law of nature.[13]

Systems of passive stabilization of satellites exploit this tendency.

Resonant rotations of celestial bodies are described by theories developed specially for this purpose. These theories do indeed confirm the synchronization of rotational motion as a law of nature. They allowed one to justify the empirical laws of rotation of the Moon, known as *Cassini's laws*, and to formulate generalized Cassini laws which govern, for instance, the rotation of Mercury.

The generalized Cassini laws describe a very subtle phenomenon of double synchronization (double resonance): first, between the axial rotation of a celestial body and its orbital motion, and second, between the motion of the axis of rotation of the body and the perturbed precession of the body's orbit.

---

[11]The author of this book, too, has a documented many-years record of this kind of curiosity: in 1948, while still a school student, he published (in the journal *Physics in the School*) a letter concerning the "theoretical" determination of the possible period of rotation of Venus, based on the freshly proposed cosmogonic theory of O. Yu. Shmidt.

[12]In the inferior conjunction the Sun, Venus, and Earth lie on the same line, with Venus and Earth on opposite sides of the Sun.

[13]The remaining part of this essay follows mainly the survey report made by the author in 1974 at the 4th *Satellite Dynamics* Symposium (Sao-Paolo, Brasil) [6.35].

The Moon has the double synchronization 1 : 1 and 1 : 1, while Mercury has the double synchronization 3 : 2 and 1 : 1.

The double synchronization described by the generalized Cassini laws can be successfully used for the passive stabilization of satellites, for example, for the double orbital stabilization of orbital space stations. We intend to survey below the theory of resonant rotations of celestial bodies that leads to the "*generalized Cassini laws,*" but not before we dedicate a few lines to the colorful personality of Giovanni Domenico Cassini.

At the time these lines were written we celebrated his 350th anniversary. G. D. Cassini (born June 8, 1625 in Perinaldo, deceased September 14, 1712 in Paris), Italian by birth, was a member of the Paris Academy and the first director – as well as the founder of a dynasty of directors – of the Paris Observatory. Representatives of this dynasty were long-lived (Giovanni Domenico died at 87, his son Jean at 79, his nephew César-François Cassini de Thury at 70, and the great-grandson Jean, the last director in the dynasty, at 97), worked intensely, and left the posterity a huge amount of their works, discoveries, and ... errors.

It is not easy to orient oneself in the maze of works written by the Cassini family (the dynasty had yet another member, a famous botanist, who was also an astronomer). As a result, in the subsequent literature Cassini's results were not always attributed to the, so to speak, local Cassini. Some discoveries of Giovanni Domenico were wrongly attributed to Jaques, while many of Jaques' errors were attributed to Giovanni Domenico. For instance, many sources mention that Jaques Cassini erroneously considered the Earth to be elongated at the poles, and not oblate. In other sources this error is attributed to Giovanni Domenico Cassini.

The most interesting figure of the dynasty was undoubtedly its founder. Apparently, he had a vivid, Italian, polemic personality. Born already after Kepler established his laws of planetary motion, G. D. Cassini nevertheless found necessary to offer his own ideas on the trajectories of celestial bodies. To this end he constructed certain curves of degree four. These curves, known in mathematics under the name of "Cassini ovals" or "Cassinoids," made Cassini's name immortal, but had nothing to do with the problem for which they were invented. A contemporary of Newton, G. D. Cassini refused to accept the validity of Newton's law of gravitation and, according to some accounts (difficult to verify!), he rejected even the teachings of Copernicus.

But it is not for these errors that G. D. Cassini's name is famous. He was an excellent observer and his discoveries continue to provide food for thought for theoreticians for already over 300 years. He discovered a gap in the Saturn's ring ("Cassini gap"). It now becomes clear that this gap is due to a resonance. He also discovered the main laws governing the rotation of the Moon ("Cassini's laws") – their subtle resonance structure was understood only recently. Saturn's satellites – Rhea, Tethys, Dione, Iapetus – were also discovered by G. D. Cassini. And again, in our times it was realized that these satellites constitute examples of resonant rotation. Thus, Cassini's name is directly or indirectly associated with some of the most actual problems in celestial mechanics – resonance problems.

Giovanni Domenico Cassini, being a prominent scientist of his times, had many students and was the founder of not only a "dynasty of directors," but also of a scientific school of mathematicians and astronomers. For instance, he supervised the studies of four Frenchmen who later, in 1685, were sent as missionaries by the king Louis the XIVth to China and joined the court of the emperor Kang Xi of the Qiang dynasty, to whom they provided advice and services. One of these four royal mathematicians and students of G. D. Cassini, Jean-François Gerbillon, in his quality of adviser and translator of the Chinese side, held negotiations with the ambassador F. A. Golovin in the town of Nerchinsk and took active part in the conclusion of the Treaty of Nerchinsk in 1689.

I feel attracted by the personality of Giovanni Domenico Cassini, a strange, full of contradictions figure in the history of science, with his subjective errors and objective progressive scientific contribution. His biography is little studied. But he lived in an interesting period; he was invited by Louis the XIVth to Paris and, possibly, shined at the court of the "Sun King." Incidentally, one of the duties assigned to the royal astronomer at the court was science education. Ladies of the court found it exciting to visit the Paris Observatory to watch the stars in the sky: "*Oh, I spent a delightful night in the company of Monsieur Cassini ...*"

But let us return to the problem of constructing a theory of resonant rotations of celestial bodies. Such a theory should be capable of discovering and using:

(a) a stabilizing factor, which keeps a celestial body in a resonant regime in a stable manner;

(b) a dissipative factor, which steers the body from an arbitrary initial rotation to capture into the resonant regime of motion.

A stabilizing factor is the torque of gravitational forces by which other celestial bodies act upon the body in question. Candidates for a dissipative factor are the tidal forces in the Solar System and other possible dissipative forces.

Let us describe successively the constructions that lead to a theory of resonant rotational motions of celestial bodies. Our departure point is the best known and studied case – the rotation of the Moon [6.36], [6.37].

In 1693 Cassini established the following laws governing the rotation of the Moon:

1. *The Moon rotates uniformly around an axis that remains fixed in the Moon's body; the period of axial rotation coincides with the period of orbital revolution around the Sun.*

2. *The Moon's equatorial plane has a constant inclination of* $1°32'.1$ *to the ecliptic.*

3. *The ascending node of the Moon's equator on the ecliptic and the descending node of the Moon's orbit on the ecliptic coincide at all times.*

We need to emphasize here that these laws are quite amazing. According to the third law, three planes – the ecliptic plane, the Moon's equatorial plane, and the Moon's orbital plane – always intersect along one straight line. This is even more amazing if we recall that two of these planes are in fact moving!

Figure 6.18 illustrates Cassini's laws. Since the Moon's orbit is inclined by an angle $i \approx 5°09'$ to the ecliptic, Cassini's second and third law imply that the angular velocity vector of the Moon's axial rotation, $\boldsymbol{\omega}$, is perpendicular to the line of nodes and makes an angle $\rho \approx 6°41'$ with the normal $\mathbf{n}$ to the Moon's orbital plane.

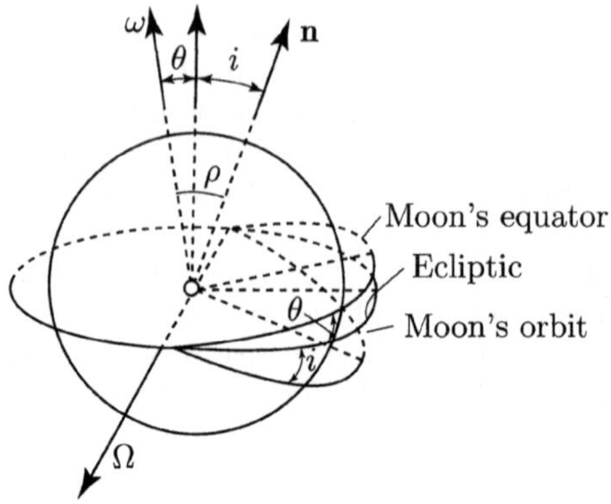

FIGURE 6.18. Regarding Cassini's laws

Starting with Lagrange and Laplace, Cassini's laws became the fundament of the theory of Moon's libration [6.38], [6.39]: researchers studied the linearized equations of motion, whose solution describes a motion that is close to the motion according to Cassini's laws. The classical linear theory is based on the empirical fact that the angle between the Moon's axis of rotation and the axis of the ecliptic is small, and so is the inclination of the Moon's orbit to the ecliptic.

A development of the classical studies was the quasi-linearized theory of the Moon's libration constructed by Sh. T. Khabibulin (in [6.40] and other works; see also the description of his theory in the book [6.41]). Khabibulin's theory encompasses a number of subtle effects manifest in the nonlinear oscillations of the Moon about its center of mass. Here our main interest will be not so much the oscillations of the Moon about the motion "according to Cassini," as another aspect of the problem, namely, the possibility of deriving from the equations of motions Cassini's laws themselves (and possible generalizations thereof), as laws of nature.

Cassini's laws have an empirical character; they do not satisfy the exact equations of motion, and as such they cannot be derived from the exact, nonlinear equations of motion.

The very fact that motion "according to Cassini" exist in nature testifies to the stability (in some sense) of this motion. One may expect that there exist approximate, yet sufficiently close to the exact ones, nonlinear equations of motion that admit stable "Cassini-type" motions as stationary solutions. A construction of such equations and their analysis is given in our works [6.14], [6.42]. An investigation on the same theme was carried out by M. L. Lidov and A. I. Neĭshtadt in their joint work [6.43]. Earlier the problem was partially studied by G. Colombo [6.44] and S. Peale [6.45].

The progress achieved in recent years in the investigation of the motion of bodies in a gravitational field (see, e.g., [6.1], [6.3], [6.16], [6.17], [6.19], [6.20], [6.22]. [6.24b]) allows one to construct nonlinear equations with the properties indicated above.

These equations are first-approximation equations in the sense of the asymptotic methods of nonlinear mechanics (the averaging method). Their construction takes into account in essential manner the factors that determine the rotation of the Moon: the rotation of Moon's orbital plane in space (secular motion of the node of Moon's orbit due to the perturbations produced by the Sun); the influence exerted on the Moon by the torques of the gravitational forces of Earth; resonance phenomena (commensurability of the Moon's period of axial rotation and orbital period). Asymptotic methods are applicable here thanks to the fact that the torques of gravitational forces are small because the Moon's ellipsoid of inertia is close to a sphere. Similar factors are taken into account in the case of Mercury.

A close nonlinear model of Cassini's laws is the stable relative equilibrium of a body in a circular unperturbed (i.e., fixed in space) orbit, described in sections 2–4 of the present essay (Figure 6.3).

This relative equilibrium represents the 1 : 1 synchronization between the period of axial rotation and the orbital period.

In a relative equilibrium one of the axes of the body (satellite) is stably directed toward the center of attraction. This fact plays a fundamental role in systems of gravitational stabilization of artificial Earth satellites. A satellite is stabilized in such a position that one of its sides faces the Earth permanently – "like the Moon," as one would usually add. This statement, however, does not quite reflect the actual state of affairs. Indeed, the relative equilibrium obeys the first of the three laws of Cassini, but fails to obey the other two.

The second and third of Cassini's laws are intimately connected with the evolution of the Moon's orbit under the influence of external perturbations (mainly those due to the Sun's attraction), specifically – with the fact that the Moon's orbital plane rotates. This effect amounts to the precession of the normal to the orbital plane around the normal to the ecliptic plane with the angular rate $K_\Omega = 1,934°08'31''$ over 36,525 ephemeris days (100 years) [6.36]. (Thus, the period of the node's motion is $\sim$ 18.6 years.) The meaning of the second and third of Cassini's laws is that during this precession the Moon's axis of rotation, the normal to the ecliptic plane, and the normal to the Moon's orbital plane always lie in the same plane, without coinciding with one another (Figure 6.18). This is where the difference between Cassini's laws and the "law of relative equilibrium" resides. In the latter, the axis of rotation of the satellite (Moon) coincides with the normal to the orbital plane. The deviation of the axis of rotation of the real Moon from the normal to the orbital plane is quite notable: $6°41'$.

According to Cassini's second and third laws, the plane that contains the Moon's axis of rotation and the normal to the Moon's orbital plane is always normal to the line of nodes, and hence it rotates with the angular rate K_Ω around the normal to the ecliptic plane. Thus, the Moon's axis shadows the motion of the normal to the Moon's orbital plane.

In other words, yet another $1:1$ synchronization takes place here, namely, between the period of motion (precession) of the Moon's axis in space and the period of precession of the Moon's orbit.

Such a motion of the Moon's axis translates into the statement that the axis remains at rest in a coordinate system which rotates with the angular rate K_L around the normal to the ecliptic plane.

During the period 1958–1965 the author has investigated the fast motion of a satellite under the influence of gravitational torques, with allowance for the evolution of its orbit due to external perturbations [6.16], [6.19], [6.1]. The method of research was to derive and analyze certain evolution equations of motion which are good approximations of the exact equations and provide a sufficiently accurate qualitative and quantitative picture of the motion. The results of our investigations are briefly presented in Section 10 of this essay. We have shown that the satellite must perform an unperturbed motion around the constant-in-magnitude angular momentum vector, while that vector moves slowly in space. The trajectories of the tip of the angular momentum vector are closed in the rotating coordinate system with one of the axes collinear to the line of nodes and another axis – the axis of rotation – coinciding with the normal to the orbital plane. The shape of such a trajectory is shown in Figure 6.12.[14]

We wish to emphasize that the fact that the trajectories of the tip of the angular momentum vector are closed in the coordinate system that rotates together with the orbit reflects the tendency toward synchronization of the rotation of the orbit and that of the angular momentum vector. In the papers [6.16], [6.19], [6.1] this phenomenon was discovered in the situation when not only gravitational, but also magnetic and aerodynamic effects are involved.

A particular case of the motion described above is the axial rotation of a body around a principal central axis of inertia, aligned in the direction pointing to the singular point 3 in Figure 6.12(b) (or to some other stationary point). The character of the motion of the tip of the angular momentum vector in the vicinity of this direction (Figure 6.12(b)) shows clearly that for this particular motion the position of the axis of inertia in question is stable. Moreover, *this particular motion*

[14]The papers [6.16], [6.19] and [6.1] deal with the case of a dynamically symmetric body; however, the results can be readily extended to the case of a triaxial body, thanks to a paper published in 1963 by F. L. Chernous'ko [6.17]. The latter is devoted to the study of the fast rotation of a triaxial satellite in a gravitational field. The evolution of the satellite's orbit was not taken into account. The effects discovered in [6.16], [6.17], [6.19], and [6.1] were rediscovered in some later papers of other researchers [6.44], [6.46]–[6.48].

satisfies – in a generalized sense – the second and third of Cassini's laws, a fact indicated by G. Colombo in 1966 [6.44]. In the same paper Colombo advanced and debated the idea of the applicability of generalized Cassini laws to the motion of celestial bodies other than the Moon (Mercury and Iapetus, for instance).

But by the very method of investigation indicated in our papers [6.16], [6.19], and [6.1], the motion cannot obey the first of Cassini's laws. The evolution equations were constructed by taking advantage of the presence of a small parameter, namely, the ratio of the orbital angular velocity to the axial angular velocity of the body. Hence, in this formulation the second velocity is considerably larger than the first, and hence cannot be equal to it, as required by Cassini's first law.

However, the small parameter required in the construction of evolution equations can be also introduced in a different way. Namely, we can use the fact that the ellipsoid of inertia of the Moon is close to a sphere, i.e., the moments of inertia of the Moon are close to one another. The evolution equations for this case were derived in 1963 by F. L. Chernous'ko [6.17] (neglecting the evolution of the orbit). They allow us to consider arbitrary values of the angular velocity (and not only resonant ones), including values close to the orbital angular velocity. Incorporating the influence of the orbit's evolution in these equations following [6.16], [6.19], and [6.1], we are led to the same qualitative type of motion as above: the satellite performs an unperturbed (with some stipulations) motion around the constant-in-magnitude angular momentum vector, which in turn traces in space the trajectories shown in Figure 6.12. A particular case of this motion is as follows: the angular momentum vector coincides with a principal central axis of inertia of the satellite and at the same time with the direction toward the stationary point in Figure 6.2, and the satellite's angular velocity of axial rotation is strictly equal to its orbital angular velocity. This motion *formally* satisfies all three of Cassini's laws.

We used the word "formally" because for the indicated particular motion the nonresonance condition, which was part of our setting, is violated. While being stable with respect to the second and third of Cassini's laws, this motion is not stable with respect to the first law: any change in the axial angular velocity is preserved, as a result of which the Moon, though perhaps slowly, would eventually turn with the other side toward the Earth.

Cassini's first law establishes the *resonant* nature of the motion of the Moon: the axial and orbital angular velocities of the Moon are related as 1 : 1. Resonant and near-resonant motions have their specific properties, and the evolution equations governing such motions have a more complex structure than the ones governing nonresonant motions. The constructions indicated above do not take into account the resonant character of the phenomenon considered.

In his 1963 paper [6.22], which will be discussed in Section 11 of this essay, Chernous'ko derived and studied a useful equation that describes planar resonant rotations of artificial and natural celestial bodies whose ellipsoid of inertia is nearly spherical. Later, in 1968–69, resonant motions of a satellite in a gravitational field in three-dimensional space were studied by A. P. Torzhevskiĭ [6.24]. According to his investigations, the perturbed motion of the satellite is no longer as simple as in the nonresonant case. Nevertheless, there exist stationary motions in which the satellite rotates around a central principal axis of inertia (the angular momentum vector is constant in magnitude and is directed along that axis of inertia), and the motion of the angular momentum vector relative to the fixed (nonevolving) orbit is similar to its motion in nonresonant cases relative to such an orbit. Furthermore, the motion is stable (under certain conditions) against perturbations of the axial rotation, a fact that was not grasped in previous analyses of nonresonant cases. Among the stable stationary motions studied by Torzhevskiĭ there is a case, suitable for our study, of $1:1$ resonance between the orbital and axial frequencies for a body whose principal moments of inertia are close to one another.

Combining the constructions of the aforementioned works [6.1], [6.15]–[6.17], [6.19], [6.22], [6.24] we arrive naturally at "motions according to Cassini," which are stable (in the sense of the evolution equations) with respect to all three of Cassini's laws. In the limit in which the angular rate of precession of the orbit tends to zero we must obtain a relative equilibrium together with conditions for the stability of the relative equilibrium described in [6.3], and also in [6.1].

Let us add that for the Moon's orbit the drift of the perigee is considerably larger than the drift of the node. The perigee rotates at an angular rate $K_\omega = 4,069°02'03''$ over 36,525 ephemeris days (the period of the perigee's motion is ~ 8.85 years). According to a result of S. I. Trushin [6.20] (1970), one can expect that the drift of the perigee does not destroy the stationarity of the motion, at least if we retain only the principal terms of the perturbing potential.

The ideas discussed above were realized in the author's papers [6.14] and [6.42].

Without entangling the reader in an exposition of the formal constructions, we will describe next some results of the investigations carried out in [6.14] and [6.42]. For details the reader is referred to the surveys [6.49], [6.50] or to the author's book [6.51].

The evolution (averaged) equations of motion constructed in [6.14] and [6.42] admit, depending on the values of the parameters, either two or four types of stationary motions. In each of these motions the ascending node of the orbit coincides with either the ascending, or the descending node of the satellite's equator, depending on the values of the parameters. All these motions may be termed "motions according to Cassini." One distinguishes direct motions, in which the angle ρ_0 between the angular momentum vector and the normal to the orbital plane is near 0 (these are genuine "motions according to Cassini"), retrograde motions (ρ_0 close to π), and "toppled-over" motions (ρ_0 close to $\pi/2$). In a direct [resp., retrograde] motion the axial rotation takes place in the same direction as the orbital revolution [resp., in the opposite direction].

Generalized Cassini laws. Let us assume that a rigid body has a nearly-spherical triaxial ellipsoid of inertia and that the center of mass of the body moves around the center of attraction with mean angular velocity ω_0 in an orbit that differs from a Keplerian orbit only by a constant precession of the normal to the orbital plane around a direction fixed in space and a constant rotation of the semi-major axis of the orbit in its plane. The body is acted upon by gravitational torques generated by the center of attraction.

Then among the possible motions of the body there are motions which in the first approximation (in the sense of asymptotic methods of nonlinear mechanics) satisfy the following *generalized Cassini laws* (see [6.14], [6.42], [6.51]):

1. The body rotates uniformly around a central principal axis of inertia; the angular velocity of this rotation is close to one of the resonance values: ω_0 (for Moon) or $\frac{3}{2}\omega_0$ (for Mercury).[15]

2. The axis of angular rotation of the body and the normal to the orbital plane make a constant angle ρ_0. For fixed values of the parameters of the body and of its orbit there are two or four possible values of the angle ρ_0.

3. The axis of angular rotation of the body, the normal to the orbital plane, and the axis of precession of the orbit lie in one plane.

Furthermore, the following "law of constant phase" holds:

4. Each times the body passes at the pericenter of its orbit one of the principal axes of inertia orthogonal to the axis of rotation, and the radius vector of the pericenter lie at equal angular distances from the line of nodes.

Let us explain these laws in more detail. To this end we introduce the following notations: A, B, C are the principal central moments of inertia; i is the constant inclination of the orbit to the ecliptic plane; K_Ω and K_ω are the constant angular rates of the motion of the node and of the pericenter of the orbit, respectively, as a result of perturbations from other celestial bodies. Also, let L_0 denote the magnitude of the angular momentum vector in the stationary motion. The angular velocity of the stationary motion is then equal to L_0/B and is given by the formula

$$\frac{L_0}{B} = \varpi + K_\omega + K_\Omega \cos(i \mp \rho_0), \qquad \text{where} \quad \varpi = \frac{m}{2}\,\omega_0, \qquad m = 2, 3.$$

This is precisely Cassini's first law (in a sharpened form): the axial angular velocity is equal to the resonant value – the orbital angular velocity for the Moon, and 3/2 of the orbital angular velocity for Mercury (with a small correction due to the motion of the pericenter and of the node of the orbit).

Cassini's third law is also slightly generalized: the coinciding nodes of the orbit and of the equator can have different as well as identical designations.

[15] Other stationary motions are also possible.

The angle ρ_0 figuring in the formulation of the generalized second law of Cassini is expressed in terms of the parameters of motion as follows. Denote

$$\alpha = \frac{3}{8}\frac{\omega_0^2 I}{L_0 K_\Omega}(A - C),$$

$$\beta = \frac{3}{8}\frac{\omega_0^2}{L_0 K_\Omega}\{[3B - (2 + I)C] + [B - (2 + I)A]\},$$

$$\chi = \frac{\beta}{\sqrt{\sin^2 i + (\alpha + \cos i)^2}},$$

and

$$\cos\rho^* = \frac{\alpha + \cos i}{\sqrt{\sin^2 i + (\alpha + \cos i)^2}}, \qquad \sin\rho^* = \frac{\sin i}{\sqrt{\sin^2 i + (\alpha + \cos i)^2}}.$$

Here $I = \Phi_m(e)$ depends only on the eccentricity (Figure 6.15). The resonance $1 : 1$ (Moon) corresponds to $m = 2$, while the resonance $3 : 2$ (Mercury) corresponds to $m = 3$. Then ρ_0 is determined from the equation

$$\cos\rho^* \mp \sin\rho^* \cot\rho_0 + \chi\cos\rho_0 = 0. \tag{6.13.1}$$

This equation was considered already in 1965 in the book [6.1], in the study of nonresonant rotation of an artificial satellite, with allowance for the evolution of its orbit and a number of perturbing factors (gravitational, aerodynamic, and magnetic). A particular case is equation (6.10.5). Let us mention that the analysis carried out in [6.1] covers a number of results of later works [6.44], [6.46]–[6.48].

Under the condition:

$$\cos^{2/3}\rho^* + \sin^{2/3}\rho^* < \chi^{2/3}$$

there exist four solutions of equation (6.13.1); in particular, when $|\chi| > 2$ this condition holds for all ρ^*.

If the opposite inequality

$$\cos^{2/3}\rho^* + \sin^{2/3}\rho^* > \chi^{2/3}$$

holds, then equation (6.13.1) has two solutions; in particular, for $|\chi| < 1$ the last inequality holds for all ρ^*. Let us remark that in the problem considered here the assumption of small values of $|\chi|$ (i.e., $|\chi| < 1$) means that the influence of the orbit's evolution dominates the influence of the gravitational torques, while for large values of $|\chi|$ the situation is reversed.

As for the fourth law, it has no analogue among Cassini's original laws. However, the fourth law plays a role of principle. Namely, it shows that to resonant motions there correspond completely determined fixed phase relations – in the present case, between the angular motion of the radius vector of the orbit and the angular motion of the principal central axes of inertia of the body. In the case of the Moon (resonance 1 : 1) this law simply means that one of the Moon's axes of inertia, in its motion, "shadows" the radius vector of the orbit.

The generalized Cassini laws reflect a double resonance in the rotation of celestial bodies: first, between axial rotation and orbital revolution (spin–orbit resonance), and second, between the precession of the orbit and that of the body's axis of rotation. The Moon [resp., Mercury] displays a double resonance 1 : 1 and 1 : 1 [resp., 3 : 2 and 1 : 1].

Example: the Moon. In the case of the Moon we take (see [6.36]): $B = 0.88836978$, $A = 0.88800195$, $C = 0.88781798$ (in units of $10^{35}\,\mathrm{Kg{\cdot}m^2}$); $i = 5°08'43''$. The period of rotation of the node of the Moon's orbit is about 18.6 years ($K_\Omega = 1934°08'31''$ over 100 years, i.e., 36,525 ephemeris years). Therefore, $|K_\Omega|/\omega_0 = 0.00398324$. The period of the motion of the Moon's perigee is about 8.85 years ($K_\omega = 4069°02'03''$ over 100 years).

In the formulas of this section we may consider that $L_0 = B\omega_0$. Using the above data we calculate

$$\chi \approx -0.218, \qquad \cos\rho^* \approx 0.970,$$

i.e., the condition for the existence of only two stationary motions is oversatisfied. One can show that in the stationary motion we are looking for ($\rho < \pi/2$) the ascending node of the Moon's equator and the descending node of the Moon's orbit coincide. Substituting the numerical values given above in the coefficients of the equation (6.13.1) (where we take $I = 1$), and calculating the root ρ of that equation close to $\rho = 0$, we obtain the value $\rho = 6°39'42''$. This yields the value $\rho_0 - i = 1°30'59'' \approx 1°31'$ for the inclination angle of the Moon's axis to the ecliptic. Observation data (Cassini's second law) give for that angle the value $1°32'1''$. The good agreement between the calculated and observed values testifies to the high accuracy of the model of motion adopted above.

Stability. Equation (6.13.1) has either two roots, corresponding to stable positions of the body's axis of rotation, or four roots, three of which correspond to stable positions of the axis, and one to an unstable position. Let us denote the unique possible unstable value of ρ_0 by ρ_0^N. We are listing the main results of the stability analysis of the (1 : 1) resonant motion carried out in [6.14] and [6.42].

The following are necessary – and almost sufficient – conditions for stability:

1. The body rotates around the shortest axis of its ellipsoid of inertia.
2. The radius vector of the orbit shadows the longest axis of inertia.
3. The inclination ρ_0 of the axis of rotation to the normal to the orbital plane satisfies the double inequality $(3 + 2\sqrt{6})/15 < \cos\rho_0 < 1$.
4. $\rho_0 \neq \rho_0^N$.

For the Moon all these four conditions are satisfied.

As follows from the evolution equations, the motion of the axis in the neighborhood of its stationary position is conditionally-periodic with two frequencies: $\omega_1 \sim \sqrt{\varepsilon}$ and $\omega_2 \sim \varepsilon$, where ε is the adopted order of smallness of the coefficients in the right-hand sides of the equations of motion. Therefore, generally speaking the curves traced by the axis of the body on the celestial sphere are not closed.

We have already mentioned that Jupiter's satellites Io, Europe, and Callisto (J I, J II, and J IV) and Saturn's satellite Iapetus (S VII), and also, possibly, Saturn's satellites Rhea, Tethys, and Dione, and Mars's satellites Phobos and Deimos rotate, like the Moon, in 1 : 1 resonance. Hence, the theory constructed above applies to them as well. One may guess that tidal effects have steered these satellites, as they did with the Moon, towards stationary motions "according to Cassini" (in the generalized sense of the present essay).

In [6.14] and [6.42] we have also derived the following sufficient conditions for the stability of a resonant motion of Mercury type (3 : 2 resonance):

1. $e \neq 0$.
2. The body rotates around the shortest axis of its ellipsoid of inertia.
3. The motion of the longest axis of the ellipsoid of inertia obeys the fourth generalized Cassini law.
4. $\rho_0 \neq \pi$, $\rho_0 \neq \rho_0^N$.

Let us mention yet another fact: for fixed values of the principal moments of inertia, the width of the resonance zone, i.e., of the "stability domain," is a quantity of order $\omega_0\sqrt{\Phi_m}$; very small values of Φ_m would mean that the probability of capture into resonance is very small. It turns out that, at least from this point of view, Mercury's chances to find itself in the 3 : 2 resonance are not small compared with the Moon's chances to find itself in the 1 : 1 resonance. This follows from the relation

$$\Delta = \frac{(\Delta_{\max})_{\mathrm{Mer}}}{(\Delta_{\max})_{\mathrm{Moon}}} = \frac{T_{\mathrm{Moon}}}{T_{\mathrm{Mer}}}\sqrt{\frac{\Phi_3}{\Phi_2}} = \frac{28}{88}\sqrt{\frac{0.714}{0.993}} \approx 0.25,$$

where $\Delta_{\max}$ denotes the width of the resonance zone for Mercury and for Moon, respectively.

Remarks on the resonant rotation of Venus. The question of the theoretical explanation of the resonant rotation of Venus, described in Section 5 of the 4th essay, was discussed by P. Goldreich and S. J. Peale in [6.2]. As we already mentioned, the angular velocity of rotation of Venus, ϖ, is connected with the angular velocity of orbital revolution, ω_{V}, and the angular velocity of orbital revolution of Earth, ω_{E}, by the resonance relation

$$\varpi = 4\omega_{\mathrm{V}} - 5\omega_{\mathrm{E}}. \tag{6.13.2}$$

Each time Venus reaches its shortest distance to Earth, it is turned toward the Earth with the same side. In [6.2] it was shown that a stable resonance of the indicated type is indeed possible and that for the resonance (6.13.2) in which we are interested the influence of the Earth dominates, while the influence of the Sun is negligibly small in the mean. Goldreich and Peale [6.2] considered only the model of planar rotation of Venus in the field of gravitational torques generated by the Sun and by the Earth and derived for its description the following approximate (in the sense of asymptotic methods) equation:

$$\ddot{\varkappa} + \frac{3}{2}\frac{A-C}{B}\frac{fM_\oplus}{b^3}\left[k_1(p) + k_2(p)\right]\sin 2\varkappa = 0. \tag{6.13.3}$$

Here $\varkappa$ denotes the angular deviation of the motion from the strictly resonant motion, A, B, C are the moments of inertia of Venus, $M_\oplus$ is the mass of the Earth, b is the semi-major axis of the Earth's orbit, and f is the constant of gravitation. Also, k_1 and k_2 are numbers depending on the "resonance number" p, with k_1 [resp., k_2] reflecting the influence of the gravitational torque of the Earth [resp., of the Sun]; this influence on the resonant motion in question manifests itself only in the mean. Thus, for the resonance (6.13.2) it turns out that $p = -5$ (the minus sign is due to the "retrograde" rotation of Venus) and $k_1 = 2.513$, $k_2 = 1.78 \cdot 10^{-2}$. As one can see from (6.13.3), the purely resonant rotation ($\varkappa \equiv 0$) can indeed be stable, and this is almost exclusively due to the influence of the Earth ($k_1 \gg k_2$).

However, one can also give some arguments against the resonance in question actually taking place.

(1) The resonance (6.13.2) is in no way superior to many other neighboring resonances for which the width of the resonance zone is of the same order of magnitude. Hence, it is difficult to explain why the fact that Nature favors this particular resonance is not an accident.

(2) The relative width of the resonance zone of Venus, Δ_{V} (compared with the width of the resonance zone of the Moon, Δ_{Moon}, or of Mercury, Δ_{Mer}), is negligibly small:
$$\Delta = \frac{(\Delta_{\max})_{\mathrm{V}}}{(\Delta_{\max})_{\mathrm{Moon}}} \approx 0.225 \cdot 10^{-3}.$$

(3) The analysis carried out in [6.2] leaves us slightly dissatisfied because it neglects the strong influence of the Sun. Indeed, in the mean the influence of the Sun is negligibly small; but the amplitude of the periodic perturbations due to the Sun (which are averaged out in [6.2]) may turn out to be quite significant, because the maximal value of the Sun's gravitational torque is 10^6 times larger than the cooresponding value for the Earth.

A more careful analysis of this question [6.50] reveals the following. In order for the periodic solar perturbations to not drive the motion out of the resonance zone it is necessary that the ratio of the principal moments inertia of Venus obey the bound
$$\frac{A-C}{B} \lesssim 2.5 \cdot 10^{-5}.$$

The values of the moments of inertia of Venus are not known. However, recalling that for the Moon the similar ratio is about $5 \cdot 10^{-4}$, we conclude that the above requirement probably lies at the boundary of the reasonable bounds. Therefore, from this point of view, too, the investigated resonance in the motion of Venus lies at the boundary of what is possible.

Let us add that, according to the study made in [6.2] and [6.50], tidal phenomena cannot explain the capture of Venus into the resonance (6.13.2) (in contrast to the case of Moon or of Mercury). Venus rotates retrogradely (in opposite direction) with respect to its orbital motion. In [6.50] it is shown that the tidal torque generated by the Sun has a tendency of "turning back" Venus, i.e., forcing it to shift from retrograde to direct rotation. Therefore, in order to explain the capture of Venus into the resonance (6.13.2) we need to devise other models of dissipative effects. This is a fourth argument that does not speak in favor of the existence of the resonance (6.13.2) being a law of nature (rather than an accident).

We see that the resonance in the rotation of Venus is not so indisputable as the resonances observed in the rotation of the Moon and of Mercury. If it nevertheless exists, then all the arguments given above would fully justify the assertion that the resonant motion of Venus is a truly amazing natural phenomenon. The amazing, up to now not fully explained resonant rotation of Venus apparently shows that the tendency toward synchronization in nature lies beyond the grasp of our present-day knowledge.

14. Tendency toward synchronization of rotational motion in complex gravitational fields. Lunar-solar precession and nutation of the Earth's axis

To this point we have considered the rotation of a body under the dominating influence of the gravitational field of a single center of attraction. However, in nature one may encounter situations when one has to take into account the influences, of comparable magnitudes, of a larger number of centers of attraction. A classical example of such a situation is the perturbation of the motion of the Earth's axis, to which the gravitational attraction of the Moon and the Sun bring comparable contributions.

The perturbed motion of Earth – i.e., lunar-solar precession and nutation – is not resonant in the same sense that the motions of the Moon and Mercury are. It is therefore the more so remarkable that a tendency toward synchronization is indeed manifest in the motion of Earth as well.

Let us introduce some notations: A and C are respectively the longitudinal and the transversal moments of inertia of Earth, m_{Moon} and m_{E} the mass of the Moon and of the Earth, respectively, e_{Moon} the eccentricity of the Moon's orbit, e_{E} the eccentricity of the Earth's orbit, ϑ_0 the angle between the Earth's axis of rotation and its angular momentum vector (to a high order of accuracy, $\vartheta_0 \approx 0$), L_0 the length of the angular momentum vector of the Earth, and $\dot{\Omega}$ the angular rate of precession of the node of the Moon's orbit. Also, let ω_{E} [resp., ω_{Moon}]

denote the mean angular velocity of the orbital motion of the Earth [resp., Moon]. Further, let us introduce the constant parameters

$$K_C = \frac{3}{8}\frac{A-C}{L_0}(3\cos^2\vartheta_0 - 1)\frac{m_{\rm Moon}}{m_{\rm Moon}+m_{\rm E}}\frac{\omega_{\rm Moon}}{(1-e_{\rm Moon}^2)^{3/2}},$$

$$K_\theta = \frac{3}{8}\frac{A-C}{L_0}(3\cos^2\vartheta_0 - 1)\frac{\omega_{\rm E}^2}{\omega_{\rm Moon}(1-e_{\rm E}^2)^{3/2}},$$

$$K_\Omega = \frac{\dot{\Omega}}{\omega_{\rm Moon}},$$

and construct the function

$$\Phi = -L_0K_C\cos^2\rho - L_0K_\theta\cos^2\Delta + L_0K_\Omega\cos\Delta,$$

where ρ is the angle between the angular momentum vector and the normal to the Moon's orbital plane and Δ is the angle between the same vector and the normal to the ecliptic plane (the Moon's orbit is inclined to the ecliptic by a constant angle i). By the evolution equations of motion [6.51], the angles ρ and Δ depend on time. However, those equations admit the first integral

$$\Phi = \Phi_0 = \text{const}, \qquad (6.14.1)$$

which has the following fundamental consequence:

All possible trajectories of the tip of the angular momentum vector are closed in a coordinate system that rotates together with the Moon's orbital plane.

Apparently, this conclusion – and the integral (6.14.1) – were not established in the classical theory of the lunar-solar precession and nutation of the Earth's axis. It undoubtedly is a manifestation of the tendency toward synchronization of the precession of the Earth's axis and the precession of the Moon's orbit. Indeed, the sets of closed trajectories shrink to singular points that correspond to steady motions. These steady motions can in turn be interpreted as further generalizations of Cassini's laws to the case when several centers of attraction (two, in the present case) exert their influence. In the process of evolution under the action of dissipative factors the motion must converge toward such a steady motion, in which the Earth's axis undergoes a precession with an angular rate equal to the angular rate of precession of the Moon's orbit. Of course, the real situation is far from the steady motion just described (the period of precession of Moon's orbit is 18.6 years, whereas that of the Earth's axis is 26000 years). But here we are interested only in what is possible in principle in such situations. A manifestation of the tendency toward synchronization is the actual nutation of the Earth's axis with a period of 18.6 years.

For details the reader is referred to the book [6.51].

15. A model of tidal phenomena and capture into resonant rotation

We have shown above that the gravitational torques acting on a celestial body can ensure a stable resonant motion, described by generalized Cassini laws. However, the initial data of the motion may lie outside the resonance zone; in fact, the probability that the initial data fall into the resonance zone is small. Hence, there must exist a mechanism which guarantees a sufficiently high probability of the motion being captured – in the process of its evolution – into the resonance zone, when the initial data lie outside that zone. Such a mechanism is apparently provided by tidal phenomena. The theory of tidal phenomena is complex and depends on many factors. Here we will confine ourselves to the consideration of some simple models of tidal phenomena [6.50], [6.51].

Let us consider a spherical, not totally rigid planet which moves in the Sun's gravitational field. As a consequence of the differences in the forces of attraction exerted by the Sun on the closest, the central, and the farthest (relative to the Sun) regions of the planet, tidal bulges form on the planet. The mass m of these bulges is proportional to k/r^3, where r is the distance to the Sun [6.2], [6.52]. Due to the rotation of the planet with respect to the orbital coordinate system with relative angular velocity ω, the tidal bulges shift by some angle δ with respect to the "planet's-center-of-mass–Sun" axis. (The Sun–planet system adopted here is just a typical choice; all considerations apply also to the pair Moon–Earth or satellite–planet). This immediately results in the usual gravitational torque

$$M \sim -\frac{m}{r^3}\sin 2\delta = -\frac{k\sin 2\delta}{r^6}, \qquad k > 0. \tag{6.15.1}$$

Let us denote by $\mathbf{e}_\omega$ the unit vector of the direction of the planet's relative angular velocity and by $\mathbf{e}_r$ the unit vector of the direction planet–center of attraction. Consider the plane $\mathbf{e}_\omega$, $\mathbf{e}_r$. The component of $\mathbf{e}_\omega$ along $\mathbf{e}_r$ does not contribute to the formation of tidal bulges; relative to $\mathbf{e}_r$ the bulges are pulled by the component $\mathbf{e}_\omega^1 = [\mathbf{e}_\omega \times \mathbf{e}_r] \times \mathbf{e}_r$, which is normal to $\mathbf{e}_r$ and lies in the plane $\mathbf{e}_\omega$, $\mathbf{e}_r$. Consequently, the axis of a tidal bulge lies in the plane normal to $\mathbf{e}_\omega^1$ and turns in that plane (under the action of the gravitational torque) toward $\mathbf{e}_r$. This means that the tidal torque $\mathbf{M}$ is directed along the vector $\mathbf{e}_\omega^1$. Using (6.15.1), we obtain the following model formula for the tidal torque:

$$\mathbf{M} = \frac{k\sin 2\delta}{r^6}\,[\mathbf{e}_\omega \times \mathbf{e}_r] \times \mathbf{e}_r\,. \tag{6.15.2}$$

This formula defines the tidal torque provided that the dependence of the angle δ on time and on the parameters of the problem is known. This dependence is in turn determined by the interaction of the forces of viscous friction in the planet's interior and the forces of attraction, and also by the angular velocity ω of relative rotation of the planet.

Two analytic models of the dependence $\delta(\omega)$ are usually employed [6.2], [6.52]:

$$\sin 2\delta \sim 2\delta \sim k_1\omega, \qquad k_1 = \text{const}, \tag{6.15.3}$$

and

$$\sin 2\delta = \sin 2\delta_0, \qquad \delta_0 = \text{const}. \tag{6.15.4}$$

The linear-in-angular-velocity model (6.15.3) is valid for a restricted range of angular velocities; however, it appears that in all real situations one can indeed confine ourselves to that range. The signature model (6.15.4) covers a wider range, but is crude for small values of ω and δ. Let us describe briefly the effect produced by the tidal torque, referring the reader for details to [6.51].

An analysis of the averaged equations of rotation of the planet under the action of the tidal torque (6.15.2)–(6.15.3) shows that the rotation tends toward an axial rotation around a principal axis of inertia.

Let $\varpi = L/(A\omega_0)$ be the ratio of the planet's absolute angular velocity of rotation to its angular velocity of orbital revolution ω_0, and let ρ denote the angle between the angular momentum vector $\mathbf{L}$ and the normal to the orbital plane.

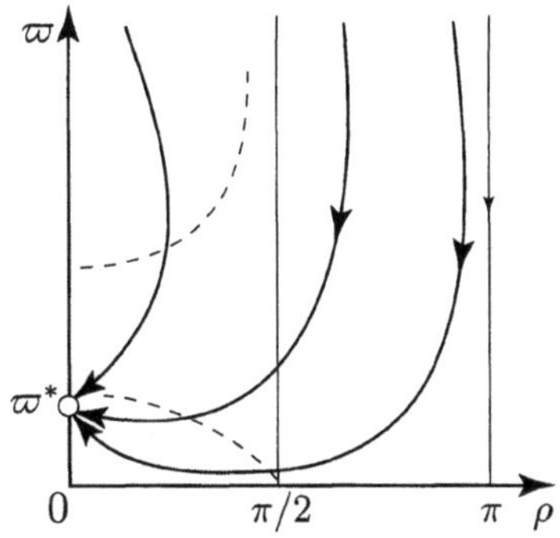

FIGURE 6.19. Evolution of the rotation parameters under the action of tides

The integral curves $\varpi(\rho)$ are schematically shown in Figure 6.19 [6.51]. In the limit the motion tends to the unique stationary point

$$\rho = 0, \qquad \varpi = \varpi^* = \frac{1 + \frac{15}{2}e^2 + \frac{45}{8}e^4 + \frac{5}{16}e^6}{(1-e^2)^{3/2}\left(1 + 3e^2 + \frac{3}{8}e^4\right)}. \tag{6.15.5}$$

Therefore, the tidal torque tends to steer the motion toward a direct rotation around the normal to the orbital plane with the angular velocity given by (6.15.5). We see that (in the absence of other torques) the tidal torque necessarily steers the motion from retrograde rotation ($\rho \sim \pi$) to direct rotation, a fact that is of interest for the theory of rotation of Venus. Apparently, the retrograde rotation of Venus cannot be explained as capture under the action of tidal torques.

No less interesting is the fact that the limit angular velocity depends on the eccentricity of the orbit. The graph of the function $\varpi^*(e)$ is shown in Figure 6.20. We see that for $e = 0$ one has $\varpi^* = 1$, that is, in a circular orbit the limit rotation is in 1 : 1 resonance (relative equilibrium). G. Colombo [6.44] and I. I. Shapiro

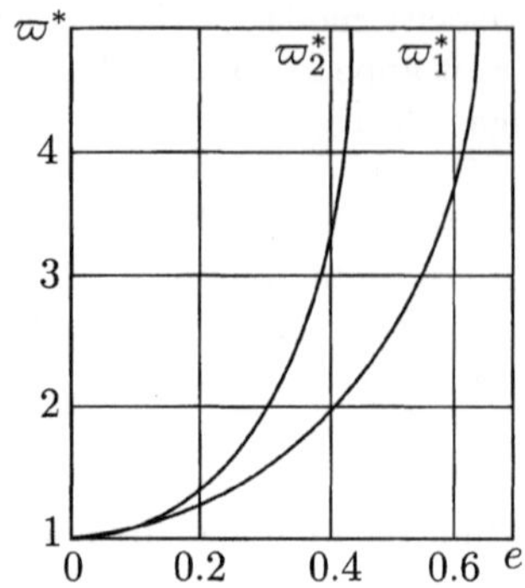

FIGURE 6.20. Dependence of the limit values of the dimensionless angular velocity of rotation of a planet on the eccentricity of its orbit

[6.52] studied the planar motion of Mercury from this point of view and discovered that precisely a value $e \approx 0.2$ of the eccentricity of Mercury's orbit can justify its capture into the 3 : 2 resonant rotation. In the model (6.15.3), $\varpi^* = 3/2$ when $e \approx 0.28$; other models may lower the required value of e. For instance, under the assumption (6.15.4), the limit value is $\varpi^* = 3/2$ if $e \approx 0.23$. More generally, to the spectrum of resonant values of ϖ^* there corresponds a spectrum of values of the eccentricity e for which the given value ϖ^* is attained in the limit under the action of tidal forces [6.50] (see Table 6.1).

ϖ^*		1	1.5	2.0	2.5	3.0	3.5
e	Model (6.15.3)	0	0.284	0.392	0.465	0.519	0.561
	Model (6.15.4)	0	0.219	0.29	0.34	0.365	0.385

TABLE 6.1.

In Table 6.1 the second row is calculated by means of formula (6.15.5), and the third row by means of the corresponding formula for the model (6.15.4):

$$\varpi^* = 1 + \frac{19}{2}e^2 + \frac{461}{24}e^4 + \frac{119703}{720}e^6 + \cdots . \tag{6.15.6}$$

Formula (6.15.6) is approximate – it contains only the principal terms of the series expansion in powers of e. Formula (6.15.5) is exact.

If in addition to the tidal phenomena one takes into account the gravitational torques studied above, formulas (6.15.5) and (6.15.6) acquire a somewhat different meaning. Namely, the orbits with the eccentricities displayed in Table 6.1 ensure that the probability P of capture into the corresponding resonance is maximal

($P \approx 1$).[16] If the eccentricity is close to one of the values indicated in Table 6.1, the probability of capture, though smaller than 1, is nonetheless large. Under the assumption (6.15.3), the calculation of the probability of capture of the Moon ($e \approx 0.05$) [resp., Mercury ($e \approx 0.2$)] into the resonance 1 : 1 [resp., 3 : 2] yields $P \approx 0.6$ [resp., $P \approx 0.5 \times 10^{-4}$].

In other words, the probability of capture of Venus into resonant rotation by tidal forces is negligibly small. We should mention also that the calculation of the probability of capture was carried out for the planar model of motion. For Venus such an estimate is quite problematic because, as we saw above, in the presence of tidal forces the planar rotation is unstable with respect to three-dimensional perturbations.

As for the Moon and Mercury, they are not affected by this problem because of their direct rotation; other models of tidal effects may increase the estimate of the capture probability (for Mercury – by an order of magnitude [6.2]).

16. Magnetic and magneto-gravitational stabilization

Among all possible systems of passive stabilization of satellites the inexperienced reader will find the magnetic system the most easily to understand and "familiarly looking." And to no surprise, since from childhood we are used with the stabilization of a compass needle. Indeed, the "compass-needle effect" is used in the practice of space flight to stabilize satellites with respect to the Earth's magnetic field, wherever stabilization is needed. On a light satellite it suffices to mount a magnet of constant strength with a pencil to ensure that the satellite will reliably "track" a field line of the Earth's magnetic field. Such tracking is provided by the torque of magnetic forces

$$\mathbf{M} = \mathbf{I} \times \mathbf{H}, \tag{6.16.1}$$

where $\mathbf{I}$ is the *magnetic moment vector*, or *magnetization vector*, which characterizes the degree of magnetization of the satellite, and $\mathbf{H}$ is the strength vector of the Earth's magnetic field. The torque (6.16.1) tends to "turn" the axis of the satellite along which the vector $\mathbf{I}$ is directed toward the vector $\mathbf{H}$, achieving in this way the stabilization of the satellite with respect to $\mathbf{H}$.

However, if one goes into details one finds that things are not so simple because of the complex structure of the Earth's magnetic field. The simplest approximation of the magnetic field – the so-called *dipole field* – is described by the formula

$$\mathbf{H} = \frac{\mu_E}{r^3}\left[3(\mathbf{k}_E \cdot \mathbf{e}_r)\mathbf{e}_r - \mathbf{k}_E\right]. \tag{6.16.2}$$

Here $\mathbf{k}_E$ is the unit vector of the axis of the magnetic dipole (for the sake of simplicity, one assumes that $\mathbf{k}_E$ is directed along the Earth's axis of rotation, pointing North), $\mathbf{e}_r$ is the unit vector in the direction of the current radius vector

[16] The probability of capture is defined as the ratio of the phase volume of the set of initial conditions leading to capture to the volume of the domain of phase space under consideration.

of the orbit (whose magnitude equals r), and the constant μ_E is the magnetic moment of the dipole (for Earth, $\mu_E \approx 8 \cdot 10^{25}$ oersted× cm^3). Formula (6.16.2) shows that the torque (6.16.1) depends in a complicated manner on the position of the satellite; moreover, the vector **H** varies nonuniformly along the orbit, which also complicates the equations of motion.

Let us demonstrate this on the example of motion in a circular polar orbit. In such an orbit there exists a planar (in the orbital plane) motion of the satellite and this motion is described by the following equation (see [6.1]):

$$\frac{d^2\varphi}{du^2} + \alpha\sqrt{1+3\sin^2 u}\,\sin\varphi = \frac{6\sin 2u}{(1+3\sin^2 u)^2}, \qquad \alpha = \frac{I_0\mu_E}{B\mu}. \tag{6.16.3}$$

Here φ is the angle between the *magnetization axis* of the satellite and the vector **H**. The magnetization axis, i.e., the axis of the satellite's magnetization vector **I** ($|\mathbf{I}| = I_0 = $const) is by assumption fixed in the satellite and lies permanently in the orbital plane. Further, u is the argument of the latitude (measured from the equator), and the expression of the dimensionless constant α includes, in addition to the indicated quantities, the satellite's moment of inertia B with respect to the axis normal to the orbital plane and the gravitational constant μ.

Equation (6.16.3) shows first of all that an exact orientation ($\varphi = 0$) with respect to the geomagnetic field cannot be achieved, because of the presence of the right-hand side of the equation (which arises due to the nonuniform variation of the vector **H** along the orbit). But one may hope that it is possible to construct a periodic solution of equation (6.16.3) describing periodic oscillations of the satellite's axis about the magnetic field line (i.e., about the vector **H**). As an examination of the coefficients of equation (6.16.3) reveals, the period of these oscillations is necessarily π (because the vector **H** returns to the initial direction after the satellite completes half of an orbital revolution).

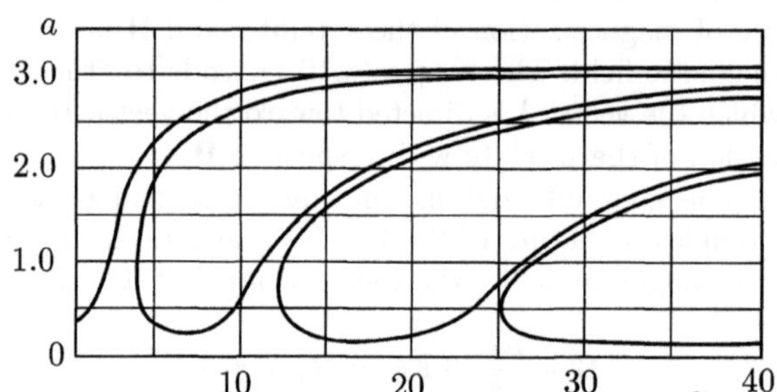

FIGURE 6.21. Dependence of the amplitudes of periodic oscillations of a satellite on the magnetization parameter

The results of numerical calculations of periodic solutions of (6.16.3) are shown in Figure 6.21 (these results, obtained by the author, were first published in a paper by his postgraduate student and disciple A. A. Khentov [6.53]). The figure displays the dependence of the amplitude a of a periodic solution on the single parameter of the problem, α_*; therefore, it may be referred to as the "phase-amplitude characteristic." We see immediately that there is not a unique periodic solution. The number of periodic solutions grows unboundedly as $\alpha_* \to \infty$. On the lower segments of all branches of our curve (Figure 6.21), magnetic stabilization is indeed achieved: the amplitudes of oscillations are small (and in addition the corresponding motions are stable). However, near certain critical ("resonant") values of α the amplitudes are not small, and so if we wish to ensure stabilization of the satellite we must avoid those critical values. They can be easily estimated based on the following considerations. In the region of small amplitudes, $\sin\varphi \approx \varphi$; the right-hand side of equation (6.16.3) can be expanded in a Fourier series $\sum_{k=1}^{\infty} b_k \sin(2ku)$, and similarly for the coefficient of φ in the left-hand side:

$$\alpha_* \left(\frac{a_0}{2} + \sum_{k=1}^{\infty} a_k \cos(2ku) \right);$$

a good approximation for $a_0/2$ is $\sqrt{5/2}$. If we seek a periodic solution of the linearized equation (6.16.3) in the form $\varphi = \sum_{k=1}^{\infty} A_k \sin(2ku)$, then for the coefficients A_k we obtain in the first approximation

$$A_k \approx \frac{b_k}{\left(\sqrt{\dfrac{5}{2}}\,\alpha_* - 4k^2 \right)},$$

from which it follows that the resonant (critical) values of α_* are given by $\alpha_* = 8k^2/\sqrt{10} \approx 2.5k^2$, $k = 1, 2, \ldots$, i.e., we obtain the sequence

$$\alpha_* = 2.5;\ 10;\ 22.5;\ 40;\ 62.5;\ 90;\ 122.5;\ 160;\ 202.5;\ 250;\ \text{and so on.}$$

This sequence is well "perceived" also in Figure 6.21 (in [6.53] the reader will find somewhat more accurate values of the numbers in this sequence). Let us add that on all periodic solutions $\sin\varphi \to 0$ as $\alpha_* \to \infty$, i.e., the amplitude of the solution tends either to π or to 0, which is also quite obvious in Figure 6.21.

In all cases, in absolute space the satellite completes exactly two rotations around its axis over each period of orbital revolution (2 : 1 resonance).

But what is the situation when the orbit of the satellite is not polar? Recalling the compass needle, we can predict with confidence that here, too, stabilization is possible; and this confidence provides already enough justification for writing down the equations of motion and integrating them.

However, in his scientific life the author has more than once reached the conclusion that in research the "pondering" phase is often more valuable and fruitful than the "work" phase, which includes writing equations, calculating their solutions, and so on. Pondering on a problem, turning it around this and that way,

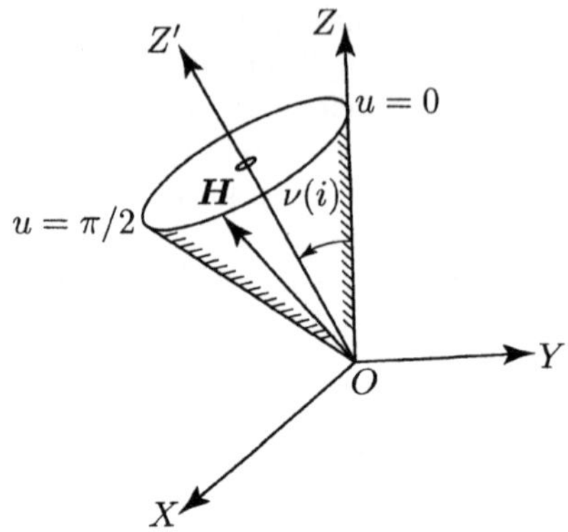

FIGURE 6.22. The magnetic-strength cone (**H**-cone)

you arrive (if you are successful) to simple geometrical and mechanical considerations, which make the problem transparent, the plan of action clear-cut, and the expected effects predictable. In the description of the problem of magnetic stabilization we will stop precisely at the pondering phase, without touching upon the "work" stage, with its cumbersome mathematical apparatus.

With this goal in mind, let us consider first the behavior of the vector **H**, given by formula (6.16.2), along the orbit of an artificial Earth satellite. We introduce the König coordinate system $OXYZ$ (Figure 6.22) with the origin at the center of mass of the satellite, with the axis OZ collinear with the axis of rotation of the Earth, and with the direction of the axis OX collinear with the direction from the center of the Earth to the ascending node of the orbit. In the motion of the satellite in its orbit the vector **H** (6.16.2) describes in the system $OXYZ$ a conical surface (the **H**-*cone*), which closes after half of an orbital revolution of the satellite. Indeed, (6.16.2) shows that $\mathbf{H}(\mathbf{e}_r) = \mathbf{H}(-\mathbf{e}_r)$. One of the generatrices of this cone coincides with the axis OZ, and the cone itself is symmetric with respect to the plane YZ (Figure 6.22). Let T_H denote the period of motion of the vector **H** along the **H**-cone, and let T_0 be the orbital period of the satellite. Introduce the frequencies $\omega_H = 2\pi/T_H$ and $\omega_0 = 2\pi/T_0$. Then $\omega_H = 2\omega_0$. Let u be the argument of the latitude of the satellite's center of mass and i be the inclination of the satellite's orbit to the equator of Earth. The magnitude of the vector **H** is then equal to $H = \mu_E r^{-3}\sqrt{1+3\sin^2 i \sin^2 u}$. When the satellite is in the ascending node of the orbit ($u = 0$), the magnitude of **H** takes the minimal value (with respect to u), $H_{\min} = \mu_E/r^3$, and the vector **H** itself is directed along the axis OZ. For $u = \pi/2$ the length of the vector **H** has the maximal value (with respect to u), $H_{\max} = \mu_E r^{-3}\sqrt{1+3\sin^2 i}$, and maximally deviates from the axis OZ, making with it an angle 2ν, $\nu \le \pi/2$, where (see [6.54])

$$\tan\nu = \frac{3}{2}\,\frac{\sin 2i}{1 - 3\sin^2 i + \sqrt{1+3\sin^2 i}}\,. \tag{6.16.4}$$

Note also that the tip of the vector **H** traces a circle that lies in a plane parallel to the satellite's orbital plane. For an equatorial orbit the **H**-cone degenerates into a line, while for a polar orbit it unfolds into plane. We have thus described the main properties of the **H**-cone.

Let us call the axis Z' that lies inside the **H**-cone in the plane ZY and makes with the axis OZ the angle ν given by formula (6.16.4) the **H**-*cone axis*. Next, let us construct a circular cone whose axis coincides with the **H**-cone axis and whose angle at the vertex is 2ν, and call it the ν-*cone*. As it turns out, the ν-cone is almost identical to the **H**-cone. Let $\varkappa$ denote the angular deviation of the vector **H** from the **H**-cone axis. Then $\varkappa = \nu$ when $u = 0$ and $u = 2\pi$, while $\varkappa > \nu$ for any other value of u in the interval $0 < u < \pi$. In other words, the circular ν-cone lies entirely inside the **H**-cone, sharing with the latter two diametrically opposite generatrices. However, the difference $\varkappa - \nu$ is always small and does not exceed $1°11'$ (for $i \approx 52°6'$; for orbits with other inclination i the maximal value of $\varkappa - \nu$ is even smaller [6.54]). Thus, the **H**-cone is indeed very close to the circular ν-cone.

Here, however, we should point out that the velocity of motion of the vector **H** along the **H**-cone is not constant; the nonuniformity in the rotation of **H** is quite important. Nevertheless, the fact that the **H**-cone is close to a circular cone allows us to propose the following method of constructing and describing the motions of a satellite stabilized with respect to the vector **H**.

It is known that in the general case the unperturbed motion of a dynamically symmetric satellite around its center of mass is a regular precession: the symmetry axis of the satellite rotates uniformly around the stationary angular momentum vector, lying at a constant angular distance from the latter. Assume that the satellite is magnetically neutral. Let us spin the satellite into a regular precession such that its axis of symmetry rotates uniformly with frequency ω_H along the surface of the ν-cone. To this end the angular momentum vector must be directed along the axis of the ν-cone, and the angle between the satellite's axis and the axis of the ν-cone must be equal to ν.

Clearly, such a regular precession forces the satellite's axis to "track" the magnetic strength vector **H**. Twice over a precession period the axis of the satellite coincides with **H**, while in the intervals between these moments it deviates from **H**, attaining the maximal deviation twice. The reason for the deviation of the satellite's axis from the vector **H** is that the latter rotates nonuniformly, whereas the former rotates uniformly.

Thus, the axis of the satellite executes periodic oscillations about the vector **H** with a period equal to half the satellite's orbital period. The maximal value Δ of the amplitude of these oscillations depends only on the inclination of the orbit. Specifically, the largest Δ is attained on a polar orbit ($i = 90°$) and equals $19.5°$. (In Figure 6.21, for $\alpha_* = 0$ the value of the amplitude a is exactly $19.5°$). On other orbits Δ is smaller – it decreases monotonically to $0°$ as i decreases to $0°$ [6.54].

However, the periodic regime of tracking **H** just described is unstable due to the instability (with respect to the precession angle) of the precession of the neutral satellite. Now let us "switch on" the magnetic moment **I** of the satellite, aligning it along the symmetry axis. Then there arises the magnetic torque (6.16.1). It has two useful functions: it stabilizes the tracking process of the vector **H** by the satellite's axis and, for sufficiently large values of $I_0 = |\mathbf{I}|$, it reduces the amplitude of the oscillations between the satellite's axis and **H**. The corresponding stable periodic regimes can be constructed by taking as generating solution the regular precession described above of the satellite's axis along the ν-cone with a precession frequency equal to $\omega_H = 2\omega_0$.

At this point, armed with an understanding of the essential issues, we could have decided to proceed further and derive the equations of motions. And indeed, the author's ideas described above were taken to the phase of analytic investigations, computation formulas, graphs, and numbers by his postgraduate student and disciple A. A. Khentov [6.54].

If the magnetization of the satellite is small, then the periodic deviations from the **H**-stabilization regime may not only fail to decrease, but may actually increase. This is already plain in Figure 6.21 (for small α_* the amplitude of the oscillations is larger than for $\alpha = 0$). True, the periodic motions become stable, but what is the benefit of that if the amplitude reaches 30° or more?

A good way of comprehending the nature of these large deviations and finding tools to fight them back it to approach the problem from the point of view of perturbation theory. The apparatus of perturbation theory does indeed apply for small values of the magnetic torques.

As we recall, the perturbed motion of the satellite is described as the superposition of the perturbed motion of the angular momentum vector and the motion of the satellite with respect to that vector. Among such motions there are stationary resonant regimes, in which the angular momentum vector **L** is at rest and remains constant in magnitude and the axis of the satellite rotates around **L** with the frequency $\omega_H = 2\omega_0$. Further, among these stationary motions there are some which behave approximately like the periodic motions considered above. Unfortunately, the stationary direction of the vector **L** in these motions does not coincide (and very notably so!) with the axis of the ν-cone (**H**-cone). As a result, the satellite strongly deviates from **H**.

This is a solemn moment: could one perhaps "improve" the stationary direction of the vector **L** using other forces than magnetic ones?

As a matter of fact, among the resonant motions of a satellite in a *gravitational* field there exists a stable motion in 2 : 1 resonance, in which the axis of the satellite completes two rotations over one revolution of the satellite in orbit. This allows one in principle to use the "gravitational" 2 : 1-resonant motion in order to ensure that the axis of the satellite will track the magnetic strength vector **H**. "Switching on" the magnetic moment does not spoil the resonant precession of the satellite's axis, and in fact improves the stationary position of the vector **L** thanks to the "play" between the magnitudes of the magnetic and gravitational torques.

The motion designed in the indicated manner depends on two free parameters: the angle at the vertex of the precession cone and the ratio of the magnitudes of the magnetic and gravitational torques. Choosing the values of these parameters in a suitable manner one can ensure that the precession cone and the ν-cone will coincide.

Furthermore, the combined influence of the two stabilizing factors should enlarge the stability zone compared with the zone achieved by any of them separately.

Thus, a satellite can be stabilized with respect to the Earth's magnetic field by exploiting in principle the joint influence of magnetic and gravitational torques. This idea of magneto-gravitational stabilization with respect to the Earth's magnetic field was proposed and investigated in the joint paper [6.55] of the author and Khentov.

Apparently, similar effects may play a definite role also in the rotation of some natural artificial satellites of planets if those planets and satellites possess their own magnetic fields. We are already familiar with the idea that a number of natural satellites (the Moon, some satellites of Jupiter and Saturn, the two satellites of Mars) rotate in a 1 : 1 resonance regime. Mercury rotates in a 3 : 2 resonance regime. The motion of these celestial bodies obeys the generalized Cassini laws, which are a consequence of the action of gravitational forces – and only such forces – on a body. But the influence of magnetic torques may result in the existence of 2 : 1 resonance regimes of rotation of natural celestial bodies!

Favorable conditions for such regimes arise for those satellites which have their own magnetic field and at the same time revolve around planets which have a magnetic field. It is possible that some of the satellites of Jupiter or Saturn satisfy these requirements. At present very little is known about the magnetic fields of satellites of planets in the Solar System (except for the Moon), and no 2 : 1 resonance regimes of rotation of natural celestial bodies were discovered. But if you ever hear about such a discovery, remember this book!

Additional comments for this translation

A. On the rotation of Venus. The phenomenal rotation of Venus has stimulated a series of works [6.2], [6.51], [6.57], [6.58], devoted to the construction of a resonance theory of this rotation. These investigations have pointed at the difficulties of the realization of such a resonance, in particular, at the small width of the resonance zone and the small probability of capturing Venus into resonance. An explanation of the retrograde rotation of Venus was found in its cosmogonic origin [6.69] according to the new cosmogonic theory of T. M. Eneev and N. N. Kozlov [6.60], in the influence of its extremely dense atmosphere [6.61], and in other factors.

According to present-day data [6.62], the period of rotation of Venus is of

$$243.022 \pm 0.006 \quad \text{terrestrial days}$$

and the parameter $(A - C)/B$ in equation (6.13.3) equals

$$(8.19 \pm 0.33) \cdot 10^{-6}.$$

This value of the inertia parameter satisfies the condition $(A-C)/B \leq 2.5 \cdot 10^{-5}$ under which the motion will not leave the resonance zone due to the influence of solar perturbations (p. 191). However, this is where the optimism of the enthusiast supporter of the resonance theory is cut short. Accurate computations (see [6.63], [6.64], first announced in [6.65]) by a refined numerical method using the indicated values of the period of rotation and inertia parameter have demonstrated that the rotation of Venus lies outside the resonance zone, and hence is not resonant. The possibility that the rotation of Venus is not resonant was also indicated earlier [6.66]–[6.69].

According to the computations reported in [6.63]–[6.65], the width of the resonance zone is extremely small, namely, (in terms of the dimensionless angular velocity) it is about $\Delta_1 \approx 4.35 \cdot 10^{-5}$. Moreover, the true angular velocity of Venus is separated from the center of the resonance zone by a quantity $\Delta_2 \approx 1.4 \cdot 10^{-3}$, which is two orders of magnitude larger than half of the width of the resonance zone.

In conclusion let us note that to the resonant rotation of Venus would correspond the resonance number $p = -5$, which means that in the time interval between two consecutive inferior conjunctions with Earth Venus performs exactly 5 turns in its retrograde rotation with respect to the current Sun–Venus radius vector. The closest resonance zones with a width of the same order as the $p = -5$ zone have resonance numbers $p = -4.5$ and $p = -5.5$. And although, as it turned out, the true period of rotation of Venus lies outside the $p = -5$ resonance zone, it clearly is attracted to the latter, being separated from the $p = -4.5$ and $p = -5.5$ zones by "distances" that are higher by many orders of magnitude (more precisely, by quantities of order 10^3) than the "distance" to the $p = -5$ resonance zone. This may mean something, but then again it may not. The question of the nature and origin of the retrograde rotation of Venus still remains open at the time when these comments to the translation were written. And for the time being, in their new book [6.70] devoted to resonant rotation of celestial bodies, Beletskiĭ and Khentov had to set up the chapter on the rotation of Venus roughly as follows: "§ 4. Resonant rotation of Venus. § 5. Refutation of the resonance theory of rotation of Venus."

B. On the relationship between orbital and rotational motion in gravitational fields. The stability of the relative equilibrium of a satellite in orbit (p. 152) and other effects in the rotational motion of celestial bodies considered in the 6th essay assumed no underlying relationship between orbital motion and rotation (spin–orbit problem). The orbital motion is given independently of rotation – for example, by Kepler's laws.

Incidentally, such a relationship does exist, and is described for instance in [6.1], [6.11], [6.71], and becomes notable for bodies of extended (nonnegligible) dimensions; it can become an essential factor for tethered orbital systems [6.72], for example. Later, in the 9th essay, we will describe the "gravity flyer effect," which is based on the fact that the orbit of an extended body in a Newtonian central force field is, rigorously speaking, not Keplerian.

The author's joint paper with O. N. Ponomareva [6.71] (see also [6.64]) investigates interesting effects of the relationship between orbital and rotational motions in gravitational fields. Here are two instructive results obtained in that paper.

1. Consider the motion of a "dumbbell" consisting of two identical masses m connected by a massless rod of length l, which moves in the gravitational field of a homogeneous spherical body of mass M_0. Then there exist a motion of the dumbbell such that the orbit of its center of mass is circular (of radius R_0) and the dumbbell itself is aligned with the current radius vector of the orbit. In the setting of the "restricted problem" adopted above in this essay, such an orbit is considered to be Keplerian and, by the result formulated on p. 152, Figure 6.3, the "along-the-radius" position of the dumbbell is stable with respect to perturbations of its angular motion.

However, in the general setting this is no longer the case: a sufficiently long dumbbell will be unstable. As shown in [6.71], for a dumbbell of small mass ($m/M_0 \sim 0$) such a motion is stable with respect to angular and orbital (planar) perturbations if

$$\frac{l}{2R} < \sqrt{3} - \sqrt{2} \approx 0.318,$$

but is unstable if this inequality is not satisfied! This instability is essentially orbital – it catastrophically changes the orbit. (An amusing detail: in [6.71] instead of $\sqrt{3} - \sqrt{2}$ the authors had the number $\sqrt{5 - \sqrt{24}}$. Can you see right away that these two numbers are the same?)

2. Consider the motion of a spherical body of mass M_0 and a body of mass M whose triaxial ellipsoid of inertia is nearly spherical. Then there exists a planar circular motion in which the distance R_0 between the centers of mass of the bodies M_0 and M remains constant and the principal central axes of inertia of the body M are directed along the radius vector of the circular orbit, the tangent to the orbit, and the normal to the orbital plane, respectively. Denote the moment of inertia of the body M relative to the axis normal to the orbital plane by B. Then if the condition $B/(MR_0^2) < 3 \cdot M_0/(M + M_0)$ is satisfied the motion is stable when the ellipsoid of inertia is stretched along the tangent to the orbit (as in the result described on p. 152, Figure 6.3). The condition written above means that the body M has bounded dimensions ($B \sim Ml^2$, where l is a typical dimension of the body M).

If the opposite inequality holds (the body M is sufficiently large), then the motion in which the body M is stretched along the radius vector is *not* stable! But here is an interesting fact: the so-called "gyroscopic stability" of the motion sets in when the body M is stretched perpendicularly to the radius vector! It is convenient to treat this case as the gyroscopic stability of a small satellite of mass M_0) in *diurnal orbit* around a large planet (the Earth) of mass M and with triaxial ellipsoid of inertia. A diurnal satellite is positioned in a stable (gyroscopically stable) manner on the imaginary prolongation in space of the smallest of the two axes of the planet's ellipsoid of inertia that lie in the orbital plane.

Thus, two problems – that of gravitational stabilization of the relative equilibrium of the Moon (or of an artificial satellite) and that of the gyroscopic stabilization of a diurnal satellite of Earth – pass into one another by a continuous change of parameters. In the setting of the restricted formulation the relationship between these two problems is overlooked.

C. On the stability of relative equilibrium in a gravitational field and the passive stabilization of an artificial satellite. In the author's paper [6.3] the result giving sufficient conditions for stability of relative equilibrium was proved not only for the restricted formulation considered in sections 3 and 4 of the present essay, but also in the general case, when the interaction between the translational and rotational motion of the satellite is taken into consideration. At meeting on the presidium of the Academy of Sciences of USSR on September 14, 1956, Mstislav Vsevolodovich Keldysh read a report (then secret). Here is an excerpt from the stenographic notes of that report ([6.37], pp. 238–239):

One has to say that many specialists in mechanics already for a long time consider that in the mechanics of a rigid body all is done and no interesting problems remain to be solved. As it turns out, the following new interesting problem was solved in connection with satellites.

It was shown that any body that flies in the gravitational field of Earth has only three equilibrium positions relative to the radius vector that points from the center of the Earth to the center of mass of the body, and among these three positions only one is stable. In order for stability to hold, it is necessary that the largest [resp., smallest, middle] axis of inertia of the body be directed along the indicated radius vector, i.e. vertically to the Earth's surface [resp., normal to the trajectory, along the trajectory].

... This interesting problem of the mechanics of a rigid body was solved by a very young member of the Department of Applied Mathematics, V. V. Beletsky.

At this point the editors of the volume [6.73] referred to above made the following comment: *What Keldysh had in mind in this particular instance was the fundamental theorem on the equilibrium of a rigid body relative to its center of mass in a central gravitational field, proved for the first time by V. V. Beletsky. The first scheme of gravitational stabilization of a satellite using damping was proposed by D. E. Okhtsimskiĭ already in 1954, before Beletsky proved his theorem.*

D. E. Okhotsimskiĭ himself dated his first scheme of gravitational stabilization in 1956 (see [6.34]), which agrees with other sources (for instance, [6.9]).

Seventh Essay

In a Spiral to Space

Imagine a dazzling, elastic spiral,
whirling away into the sky!...

A. Voznesenky, *Triangular pear*

1. Low thrust

In various laboratories of the world scientists design and develop reactive engines in which the reactive jet is generated by accelerating charged particles in an electric field (ion engines, operating on the principle of interaction of an electric current with a magnetic field, plasma engines, and other types of engine [7.1]). The reactive thrust produced by such engines, unlike that produced by engines that use a chemical fuel, is very small and can impart a spaceship an acceleration of only several mm/sec^2. In other words, the reactive acceleration is thousands of times smaller than the acceleration of gravity at the Earth's surface ($g_0 = 9.81$ m/sec^2). Clearly, with such an acceleration one cannot escape the gravitational field of Earth. To that end one needs accelerations larger than g_0, which are generated by rocket engines that use chemical fuel.

But once a spaceship is brought to an orbit around the Earth, a low-thrust engine may become quite suitable. Low-thrust engines can operate continously for a long time. This allows one to achieve a slow (due to the smallness of the thrust), but prolonged evolution of the orbit of the spaceship. The spaceship may, without hurrying, turn after turn, move along an unwinding spiral around the Earth till the moment it reaches escape velocity, when it will leave the vicinity of Earth and fly away to other planets. Such a maneuver can prove convenient, since it requires a smaller mass expenditure than a flight that uses only chemical jet engines. This would allow one to send in an interplanetary flight a larger useful payload.

A trajectory of a spaceship that initially flies on a closed orbit around a planet, and then accelerates to escape (parabolic) velocity is termed an *escape trajectory*. Some properties of spiral escape trajectories are considered in this essay.

2. Escape parameters and paradoxes

Here we will confine our study to the special, yet important case when the thrust of the jet engine is directed all the time along the tangent to the orbit and the resulting acceleration $\mathbf{f}_d$ is constant in magnitude. We will use the following notation: r is the distance from the center of attraction to the spaceship (whose mass,

as is customary in all such cases, is considered to be concentrated at the spaceship's center of mass) and V_d is the speed of the spaceship; the thrust vector $\mathbf{f}_\text{d}$ is assumed to be collinear with the velocity vector $\mathbf{V}_\text{d}$; finally, α denotes the angle made by the vectors $\mathbf{r}$ and $\mathbf{V}_\text{d}$. Figure 7.1 also shows the polar angle φ. Further, let r_0 be an arbitrarily fixed distance of the spaceship to the center of attraction (for instance, the initial distance), and g be the acceleration of gravity at distance r_0. It is convenient to introduce the dimensionless variables

$$\rho = \frac{r}{r_0}, \qquad V = \frac{V_\text{d}}{\sqrt{g r_0}}, \qquad \tau = \sqrt{\frac{g}{r_0}}\, t. \tag{7.2.1}$$

Here τ and t are the dimensionless and the dimensional current time, respectively.

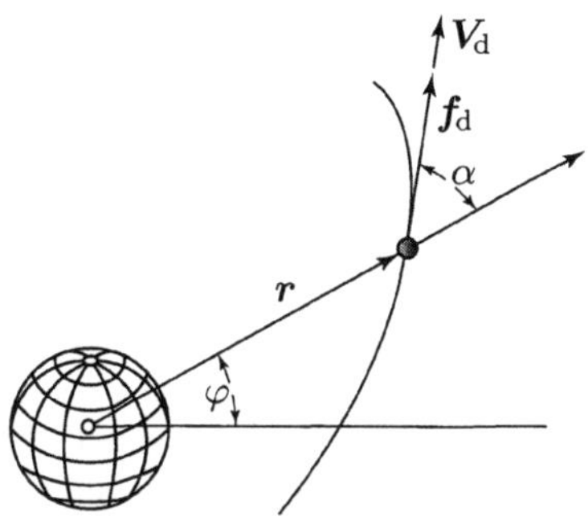

FIGURE 7.1. Trajectory parameters

Let us introduce also the dimensionless reactive acceleration

$$f = \frac{f_\text{d}}{g}. \tag{7.2.2}$$

Further, following the paper of V. V. Beletsky and V. A. Egorov [7.2], let us write the equations of motion in the variables V, ρ, α in the form

$$\left.\begin{aligned} &\frac{dV}{d\tau} = f - \frac{\cos\alpha}{\rho^2}, \qquad && \frac{d\rho}{d\tau} = V\cos\alpha, \\ &\frac{d}{d\tau}\cos\alpha = \frac{\sin^2\alpha}{\rho V}\left(\frac{V^2}{2} + h\right), \quad && h = \frac{V^2}{2} - \frac{1}{\rho}. \end{aligned}\right\} \tag{7.2.3}$$

The first of these equations is the equation of motion in the projection on the tangent to the trajectory, with f being the reactive acceleration and the second term in the right-hand side being the projection of the acceleration due to the Newtonian force of attraction. The second equation simply describes the evolution of the radial component of the spaceship's velocity. Finally, the third equation governs the motion projected on the direction normal to the trajectory. This equation contains the constant h, which represents the dimensionless mechanical energy of motion.

To completely determine the trajectory we need to write also the equation for the polar angle φ:

$$\frac{d}{d\tau}(\rho^2\dot{\varphi}) = f\rho\sin\alpha. \tag{7.2.4}$$

This equation can be integrated by a quadrature once the solution of the system (7.2.3) is known.

The system (7.2.3) is closed, and its integration together with (7.2.4) yields a complete description of orbits and of the law governing the motion. Unfortunately, the integration cannot be carried out in analytic, closed form.

To analyze the motion qualitatively it is convenient to write also the differential equation for the osculating elements of the orbit: the energy h and the focal parameter p. These equations are readily obtained by differentiating with respect to time the identities $h = (V^2/2) - 1/\rho$ and $p = \rho^2 V^2 \sin^2\alpha$. Substituting into the resulting equations the expressions of $\dot{V}$, $\dot{\rho}$ and $\dot{\alpha}$ given by (7.3.2), we obtain

$$\frac{dh}{d\tau} = fV, \qquad \frac{dp}{d\tau} = \frac{2fp}{V}. \tag{7.2.5}$$

These equations show immediately that the energy h and the focal parameter p are monotonically increasing. The energy h is connected with the magnitude of the semi-major axis of the orbit by the relation $h = -1/(2a)$; here, of course, $a > 0$, and hence $h < 0$ (until escape velocity is reached, and then $h = 0$). The fact that h is monotonically increasing means that the semi-major axis a is monotonically decreasing. Since the parameters p and a increase simultaneously, it follows that the orbit "swells up," i.e., its dimensions grow with time. It is clear that for small values of the acceleration f, one turn of the orbit differs only slightly from the Keplerian ellipse that would be obtained by setting $f = 0$. But since $da/d\tau > 0$ and $dp/d\tau > 0$, this ellipse increases in size with each turn of the spiral, so that *our spaceship moves around Earth along an unwinding spiral.*

An example of such a spiral (obtained by numerical integration of the equations of motion (7.2.3) and (7.2.4)) is shown in Figure 7.2. The computation was carried out using a rather large value $f = 0.01$ to reduce computing time; nevertheless, the orbit shown in Figure 7.2 possesses all the main qualitative features characteristic also to orbits with smaller values of f. It was assumed that the initial orbit was circular (for a circular orbit $V = 1$, $\rho = 1$, and $h = -0.5$).

As one can see in Figure 7.2, after completing several revolutions around the Earth on an unwinding spiral, the spaceship begins to move along a finite segment – a smooth arc of trajectory, which no longer winds around the Earth; on this segment the ship attains escape velocity (or, equivalently, h becomes equal to zero). After that, the spaceship moves away from the Earth, building up more and more energy.

Let us analyze in an approximate manner the properties of motion that follow from equations (7.3.2). Suppose that the initial orbit is circular and that f is very

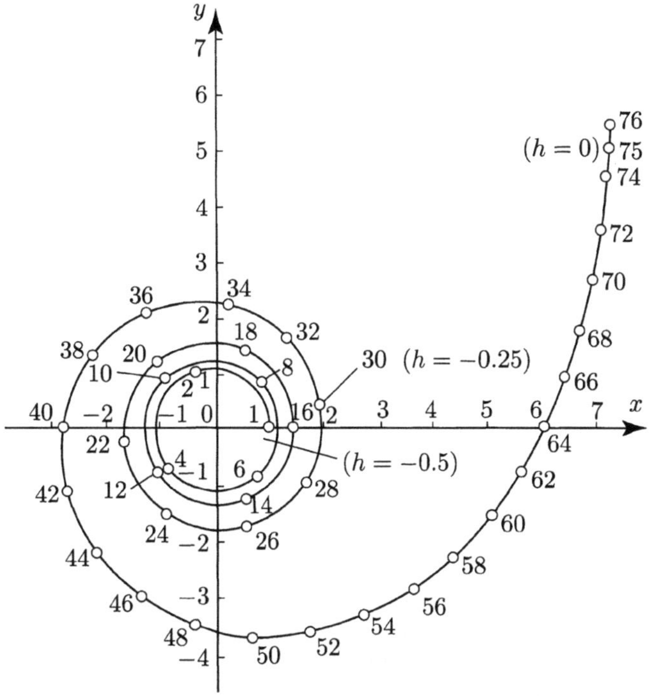

FIGURE 7.2. Example of escape trajectory starting from a circular orbit. Dimensionless flight time is indicated along the trajectory (∘)

small. Then the spaceship makes a large number of spiral turns around the Earth, and each such turn can be approximately regarded as a circular orbit (at least for a sufficiently long segment of orbit at the initial stage of the motion). This is precisely the approximation we are going to adopt (its legitimacy is confirmed by numerical computations). If each turn of orbit spiral is (approximately) a circular Keplerian orbit, then necessarily

$$V^2 = \frac{1}{\rho}.$$

But in this case

$$h = \frac{V^2}{2} - \frac{1}{\rho} = -\frac{V^2}{2}.$$

Substituting this value of h in relation (7.2.5) for the derivative of h, we obtain

$$\frac{dV}{d\tau} = -f \tag{7.2.6}$$

(and then from the first equation (7.2.3) it follows that $\cos\alpha = 2f\rho^2$).

Equation (7.2.6) is remarkable because it plainly displays the following apparent paradox: if a satellite is acted upon by a force directed parallel to its velocity vector in the direction of motion, then the velocity of the satellite will *decrease* (on the average); this reduction in velocity takes place as if the force applied changed its direction into the opposite one and pushes the satellite backward! Of course, there is no "paradox" here: the explanation is that the energy of the motion, and hence the semi-major axis of the orbit, grow; but according to Kepler's third law, an increase in the semi-major axis of the orbit results in an increase in the period of revolution of the satellite around the Earth, i.e., in a decrease in its average velocity. This is exactly what formula (7.6.2) says.

In the first essay we mentioned the aerodynamic "paradox" in the motion of a satellite. Namely, under the influence of atmospheric drag the average velocity of the satellite increases. The nature of the aerodynamic paradox is identical to that of the "paradox" considered here. The atmospheric drag force is tangent to the orbit of the satellite, but points backward with respect to the direction of motion. As a result, the energy of motion decreases and, together with it, so does the semi-major axis of the orbit; hence, the period of revolution also decreases, i.e., the average velocity of the satellite's orbital motion increases. This increase can also be described by formula (7.2.6), where now we must take $f < 0$. More precisely, in this case $f = -c\rho_{\text{atm}}V^2$, where c is a constant parameter and ρ_{atm} denotes the density of the atmosphere.

Now let us go back to our problem of escape of a spaceship under the action of a constant acceleration tangent to its orbit.

The decrease in the average velocity of the satellite continues for as long as $h < 0$. However, since the energy h increases monotonically, it must necessarily reach the value zero in a finite time and then continue to grow in the positive range. This follows from the observation that for $h = 0$ we have $V \neq 0$ and, by (7.2.5), $dh/dt \neq 0$. But if $h > 0$, then from the third of the differential equations (7.2.3) we conclude that $\cos\alpha$ increases monotonically. Moreover, $\cos\alpha \to 1$ as $\tau \to \infty$. Then the second equation in (7.2.3) shows that $p \to \infty$ as $\tau \to \infty$, while the first equation in (7.2.3) shows that $dV/d\tau \approx f > 0$ as $\tau \to \infty$, since in that equation the term $\cos\alpha/\rho^2$ becomes arbitrarily small when $\tau \to \infty$. Thus, we may consider that the average orbital velocity decreases [resp., increases] monotonically for $h < 0$ [resp., $h > 0$]. Hence, near the value $h \approx 0$ the velocity has an absolute minimum and $dV/d\tau = 0$.

This conclusion can be used to obtain approximate values of the orbital parameters at the end of the escape segment. We are most of all interested in precisely the moment at which the spaceship attains escape velocity ($h = 0$). At what distance from Earth does this happen? After how much time? What is the spaceship's velocity at that time? The condition $dV/d\tau =$ for $h = 0$ helps us answer these questions. Specifically, using this condition and the first equation in (7.2.3) we obtain

$$\rho_* = \sqrt{\frac{\cos\alpha_*}{f}}, \tag{7.2.7}$$

where the asterisk indicates that the parameters are evaluated at the end of the escape segment. The value $\cos\alpha_*$ is unknown for the moment.

Since the escape velocity V_* satisfies the condition $V_*^2 = 2/\rho_*$, (7.2.7) yields

$$V_* = \sqrt[4]{\frac{4f}{\cos\alpha_*}}. \tag{7.2.8}$$

To calculate the escape time τ_*, i.e., the time needed to reach the escape velocity, we proceed as follows. Integrating the first of equations (7.2.5) we obtain

$$h_* - h_0 = f\overline{V}\tau_*, \tag{7.2.9}$$

where $\overline{V}$ is the mean value of the velocity on the escape segment. One can put $\overline{V} = (V_0 + V_*)/2$, where V_0 is the initial value of the velocity and V_* is the escape velocity. But since $h_* = 0$, $h_0 = -1/2$, $V_0 = 1$, and V_* is defined by formula (7.2.8), we deduce from (7.2.9) that

$$\tau_* = \frac{1}{f\left(1 + \sqrt[4]{\dfrac{4f}{\cos\alpha_*}}\right)}. \tag{7.2.10}$$

To complete the calculation of the orbital parameters at the endpoint of the escape segment it remains to determine the angle α_*. An exact formula was obtained in [7.2][1], but for crude calculations one can manage without it. We are interested in evaluating with relatively small accuracy the parameters at the terminal point of the escape segment. To this end we will use the fact that $\cos\alpha_* \to 1$ when $\tau \to \infty$ and approximate $\cos\alpha_*$ by its limit value 1. Then formulas (7.2.7), (7.2.8), and (7.2.10) acquire the following simplified forms:

$$\rho_* = \frac{1}{\sqrt{f}}, \qquad V_* = \sqrt[4]{4f}, \qquad \tau_* = \frac{1}{f\left(\sqrt[4]{4f}\right)}. \tag{7.2.11}$$

As a comparison with results obtained by numerical integration of the exact equations of motion shows, the error in the determination of the escape parameters introduced by the above approximation is only of 5–15%. Such an accuracy is fully satisfactory for crude calculations, preliminary orbit selection, and so on. Once a desired variant of orbit is selected with the help of the approximate formulas (7.2.11), one can always calculate that variant and variants close to it by integrating numerically the equations of motion.

To pass to dimensional values of the escape parameters we need to substitute the values of ρ_*, V_*, and τ_* calculated according to (7.2.11) in formulas (7.2.1).

[1]That formula is: $\sin\alpha_* = \frac{1}{2}\left(\sqrt[4]{4f}\right)^{2\sqrt{f}/(2\sqrt{f}-1)}$.

Let us mention also that $V = dS/d\tau$, where S is the (dimensionless) distance traversed by the spaceship. Substituting this expression for V in the first of equations (7.2.5) and integrating, we obtain the exact relation

$$h - h_0 = f(S - S_0), \tag{7.2.12}$$

from which one can readily determine the distance S. Thus, for an acceleration $f = 10^{-4}$, a spaceship starting from a circular orbit ($h_0 = -1/2$) traverses up to the moment when it attains escape velocity (i.e., when $h = 0$) a (dimensionless) distance $S = 0.5 \cdot 10^4$. But $S = S_d/r_0$, where r_0 denotes the dimensional radius of the original orbit and S_d is the dimensional distance traversed. If $r_0 = 7,000$km, we get $S_d = 35$ million km.

Some results of the calculation of escape parameters of spiral trajectories are given below in several tables. All data shown in these tables concern the moment when escape velocity is reached. The asterisk indicates values calculated by means of the approximate formulas (7.2.11); the same parameters without an asterisk were obtained by numerical integration of the exact equations of motion.

f	τ_*	τ	δ_τ	ρ_*	ρ	δ_ρ	V_*	V	δ_V	α	n
10^{-2}	69.1	75	0.078	10.00	8.939	0.118	0.447	0.473	0.072	37.5	4
$5 \cdot 10^{-3}$	145	157	0.074	14.14	12.438	0.137	0.375	0.401	0.063	36	8
10^{-3}	799	856	0.066	31.62	27.846	0.135	0.251	0.268	0.068	33	39
$5 \cdot 10^{-4}$	1651	1758	0.061	44.72	39.506	0.132	0.211	0.225	0.066	32.5	79
10^{-4}	8761	9192	0.047	100.00	87.715	0.139	0.141	0.151	0.066	31.5	398

TABLE 7.1. Escape from a circular orbit

Table 7.1 also gives an estimate of the accuracy of the computation done using the approximate formulas: δ_ρ, δ_V, and δ_τ are the relative errors of the computations done using these formulas (for instance, $\delta_\rho = |(\rho_* - \rho)/\rho|$, and similarly for δ_V and δ_τ). The last column of Table 7.1 shows the numbers of turns of the escape segment traversed by the spaceship up the moment when it attains escape velocity. In Table 7.1 all parameters are dimensionless, whereas tables 7.2–7.4 show some dimensional characteristics of the escape segment.

f	f_d, mm/sec^2	t, days	r, km	V_d, km/sec
10^{-2}	89.5	0.75	56,630	3.65
$5 \cdot 10^{-3}$	44.7	1.5	82,970	3.10
10^{-3}	8.95	8.5	185,760	2.07
$5 \cdot 10^{-4}$	4.5	17.5	263,545	1.74
10^{-4}	0.9	92	585,150	1.17

TABLE 7.2. Escape near Earth. The initial altitude of the circular orbit above the surface of Earth is $H_0 = 300$ km

f	f_{d}, mm/sec^2	t, days	r, km	V_{d}, km/sec
10^{-2}	26	2	110,580	2.68
$5 \cdot 10^{-3}$	13	2	153,870	2.275
10^{-3}	2.6	21.5	344,483	1.52
$5 \cdot 10^{-4}$	1.3	44	488,730	1.58
10^{-4}	0.26	232	1,085,120	0.86

TABLE 7.3. Escape near Earth. The initial altitude of the circular orbit above the surface of Earth is $H_0 = 6000$ km

f	f_{d}, mm/sec^2	t, days	r, km	V_{d}, km/sec
10^{-2}	31.6	0.93	33,000	1.61
$5 \cdot 10^{-3}$	15.8	2	45,920	1.37
10^{-3}	3.16	11	102,810	0.91
$5 \cdot 10^{-4}$	1.6	22	145,860	0.77
10^{-4}	0.3	115	323,840	0.51

TABLE 7.4. Escape near Mars. The initial altitude of the circular orbit above the surface of Earth is $H_0 = 300$ km

An examination of the tables leads us to the conclusion that for an acceleration of the order of 1 mm/sec^2 the escape time is equal to about 100 days for motion around the Earth and about 300 days for motion around Mars (assuming that the escape begins at an altitude of 300 km above the planet's surface). The same durations are required to brake the spaceship from parabolic velocity to the velocity with which the spaceship will move on a circular orbit with an altitude of 300 km. If the initial orbit has an altitude of 6,000 km, then using an acceleration of 1 mm/sec^2 the escape velocity is attained after about 40 days. (Calculations for higher altitudes of the initial orbit (of the order of 6,000 km) are of importance because such orbits lie outside the Earth's radiation belt. The energy spent to bring the spaceship to such an altitude can be compensated by reducing the weight of the antiradiation shield).

Let us mention one more instructive formula. As Table 7.1 shows, the number n of turns of the escape trajectory (starting from the initial orbit) is very well approximated by the formula

$$n = \frac{0.04}{f}. \tag{7.2.13}$$

This formula was obtained by analyzing numerical results; attempts to derive it analytically have failed.

Let give also (Figure 7.3) graphs (obtained by numerical integration) of the dependence of the parameters of an escape trajectory on time. Note that the velocity V changes in time precisely as we inferred from the qualitative analysis of the equations of motions: in the mean the velocity decreases until escape velocity is attained. As it turns out, when escape velocity is attained (i.e., when the energy

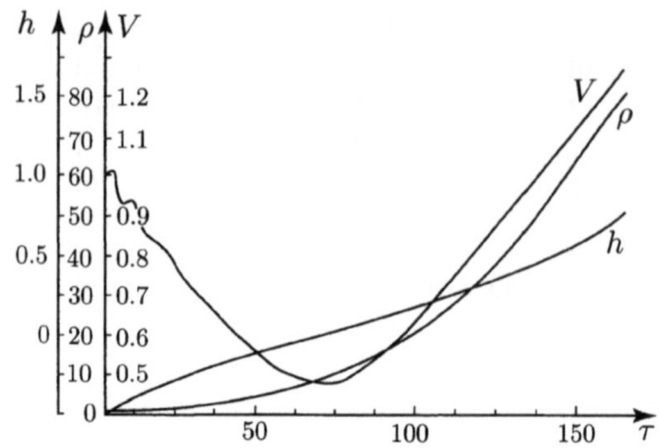

FIGURE 7.3. Time-dependence of the parameters of an escape trajectory from a circular orbit (dimensionless variables); $f = 0.01$

h attains the value zero) the velocity has an absolute minimum; after that moment the velocity increases monotonically. At the same time, Figure 7.3 exhibits details of the behavior of an escape trajectory that were not revealed by the approximate analysis. First of all, one observes the presence of small oscillations about the monotonically decreasing mean value (on the segment of acquisition of escape velocity). Generally speaking, the variation of the trajectory enjoys no monotonicity property. If we start the escape from a circular orbit, then the application of a reactive acceleration transforms the osculating orbit into a weakly elliptic one; and, as we know, on an elliptic orbit the velocity and the position vector change periodically. These oscillations, superposed on the mean monotone variation of the parameters, yield the picture shown in Figure 7.3 (the oscillations in ρ are not seen there because of their scale).

3. A monotone escape spiral

One can, however, construct a whole class of trajectories for which all parameters change in a monotone manner [7.3]. To do this, let us write the full system of equations in osculating elements for our problem:

$$\left.\begin{aligned} \frac{dp}{d\tau} &= f\frac{2p}{V}, & \frac{de}{d\tau} &= f\frac{2(e+\cos\nu)}{V}, \\ \frac{d\omega}{d\tau} &= f\frac{2\sin\nu}{Ve}, & \frac{d\nu}{d\tau} &= \frac{\sqrt{p}}{\rho^2} - f\frac{2\sin\nu}{Ve}. \end{aligned}\right\} \tag{7.3.1}$$

Here, as in the first essay, e is the osculating eccentricity of the orbit, ω is the osculating longitude of the perigee, and ν is the true anomaly. The quantities ρ and V are expressed in terms of the introduced variables by the formulas

$$\rho = \frac{p}{1+e\cos\nu}, \qquad V = \sqrt{\frac{1}{p}\left(1+e^2+2e\cos\nu\right)}. \tag{7.3.2}$$

Recall that our variables are dimensionless.

Now let us perform the change of variables

$$V = (2f)\overline{V}, \qquad \rho = (2f)^{-1/2}\overline{\rho}, \qquad p = (2f)^{-1/2}\overline{p}, \qquad \tau = (2f)^{-3/4}\overline{\tau}. \quad (7.3.3)$$

Then in the variables $\overline{V}$, $\overline{\rho}$, $\overline{p}$, $\overline{\tau}$ the equations (7.3.1) and (7.3.2) will have the same form as if in (7.3.1) we would put $2f \equiv 1$ and replace V, p, ρ and τ by $\overline{V}$, $\overline{\rho}$, $\overline{p}$ and $\overline{\tau}$, respectively. Therefore, once we determine some trajectory using (7.3.1) with $2f = 1$, we obtain, according to (7.3.3), a whole family of trajectories with arbitrary values of f.

As for the trajectories with $2f = 1$, we will follow D. E. Okhotsimskiĭ [7.3] and seek a solution of equations (7.3.1) such that when $\tau \to -\infty$ one has that $e \to 0$, $p \to 0$, and also $\cos\nu \to 0$ (henceforth the bar over symbols will be omitted). It turns out that there is only one solution with the indicated properties. That solution is in a certain sense universal, because it does not depend in any way on the preassigned initial conditions, corresponding to a nonzero distance to the center of attraction. The parameters of this solution vary monotonically until escape velocity is attained. The osculating ellipse that this solution describes is the closer to a circle the smaller its dimensions are. Any initial orbit corresponding to this solution will be elliptic, with a determined small eccentricity (provided the dimensions of the orbit are not too large).

To the universal trajectory described above there corresponds, via the substitution (7.3.3), a family of trajectories enjoying the same property that $e \to 0$, $p \to 0$, $\cos\nu \to 0$ as $\tau \to -\infty$; let us add here that the solution is constructed so that $e = 1$ when $\tau = 0$.

Let us describe the final results of calculation of a monotone escape spiral [7.3]. It is convenient to divide the spiral into three characteristic segments: the first segment, corresponding to very small values of p, is described by an asymptotic solution in the form of series in powers of p; the second segment is determined by numerical integration (with the initial data found using the asymptotic formulas for the first segment); and the third segment, corresponding to large values of p, is described by series in powers of $1/p$. Escape velocity ($e = 1$) is attained on the second segment.

It turns out that on the first segment the solution is represented by series in powers of p^4:

$$\left.\begin{aligned} &e = p^2 \sum_{k=0}^{\infty} \varepsilon_k p^{4k}, \qquad \cos\nu = p^2 \sum_{k=0}^{\infty} \nu_k p^{4k}, \\ &\omega = -3.40492 + p^{-2} \sum_{k=0}^{\infty} \varphi_k p^{4k}, \\ &\tau = 1.34571 + p^{-1/2} \sum_{k=0}^{\infty} \tau_k p^{4k}. \end{aligned}\right\} \quad (7.3.4)$$

A key step in obtaining these formulas is the transformation of a part of equations (7.3.1) to the form

$$\frac{dp}{de} = f_e(p, e, \cos\nu), \qquad \frac{d\cos\nu}{dp} = f_\nu(p, e, \cos\nu).$$

The coefficients ε_k, ν_k, φ_k and τ_k of the expansions (7.3.4) are given in Table 7.5.

k	ε_k	ν_k	φ_k	τ_k
0	$1.0 \cdot 10^0$	$1.0 \cdot 10^0$	$-0.5 \cdot 10^0$	$-2.0 \cdot 10^0$
1	$-6.0 \cdot 10^0$	$-3.0 \cdot 10^1$	$2.75 \cdot 10^0$	$4.285714 \cdot 10^{-1}$
2	$3.53 \cdot 10^2$	$3.177 \cdot 10^3$	$-4.835417 \cdot 10^1$	$-5.75 \cdot 10^0$
3	$-5.0216 \cdot 10^4$	$-6.528080 \cdot 10^5$	$4.292169 \cdot 10^3$	$3.604946 \cdot 10^2$
4	$1.271935 \cdot 10^7$	$2.162289 \cdot 10^8$	$-8.098106 \cdot 10^5$	$-5.110699 \cdot 10^4$
5	$-5.007378 \cdot 10^9$	$-1.051549 \cdot 10^{11}$	$2.555165 \cdot 10^8$	$1.281759 \cdot 10^7$
6	$2.829338 \cdot 10^{12}$	$7.073346 \cdot 10^{13}$	$-1.204903 \cdot 10^{11}$	$-5.010652 \cdot 10^9$
7	$-2.173519 \cdot 10^{15}$	$-6.303206 \cdot 10^{16}$	$7.935154 \cdot 10^{13}$	$2.819974 \cdot 10^{12}$

TABLE 7.5.

k	ε'_k	τ'_k	φ'_k	ν'_k
0	—	—	$-1.30465 \cdot 10^0$	$1.53298 \cdot 10^0$
1	$-4.58000 \cdot 10^{-1}$	$9.16000 \cdot 10^{-1}$	0	0
2	$6.80189 \cdot 10^{-1}$	$-2.04057 \cdot 10^0$	$-7.74093 \cdot 10^{-2}$	$-7.40216 \cdot 10^{-1}$
3	$-8.45172 \cdot 10^{-1}$	$3.38069 \cdot 10^0$	$-3.20210 \cdot 10^{-1}$	$5.44150 \cdot 10^{-1}$
4	$8.85247 \cdot 10^{-1}$	$-4.42624 \cdot 10^0$	$1.05876 \cdot 10^0$	$-3.77374 \cdot 10^{-1}$
5	$-4.86283 \cdot 10^{-1}$	$2.91770 \cdot 10^0$	$-2.39421 \cdot 10^0$	$-1.23027 \cdot 10^{-2}$
6	$-9.64784 \cdot 10^{-1}$	$6.75349 \cdot 10^0$	$4.44336 \cdot 10^0$	$8.58759 \cdot 10^{-1}$
7	$4.43942 \cdot 10^0$	$-3.55154 \cdot 10^1$	$-6.81926 \cdot 10^0$	$-2.41946 \cdot 10^0$
8	$-1.10624 \cdot 10^1$	$9.95619 \cdot 10^1$	$7.74454 \cdot 10^0$	$4.80913 \cdot 10^0$
9	$2.11182 \cdot 10^1$	$-2.11182 \cdot 10^2$	$-2.45367 \cdot 10^0$	$-7.51642 \cdot 10^0$
10	$-3.13216 \cdot 10^1$	$3.44538 \cdot 10^2$	$-1.89947 \cdot 10^1$	$8.38035 \cdot 10^0$
11	$2.90194 \cdot 10^1$	$-3.48233 \cdot 10^2$	$7.31602 \cdot 10^1$	$-1.84542 \cdot 10^0$
12	$1.76253 \cdot 10^1$	$-2.29129 \cdot 10^2$	$-1.79535 \cdot 10^2$	$-2.32322 \cdot 10^1$
13	$-1.71363 \cdot 10^2$	$2.39908 \cdot 10^3$	$3.41319 \cdot 10^2$	$8.44939 \cdot 10^1$
14	$5.27883 \cdot 10^2$	$-7.91825 \cdot 10^3$	$-4.90991 \cdot 10^2$	$-2.00619 \cdot 10^2$

TABLE 7.6.

For large values of p (on the third segment of the trajectory) the expansions have the form

$$\left.\begin{aligned} e &= 0.76649p + \sum_{k=1}^{\infty} \varepsilon'_k p^{-k}, \qquad \cos\nu = \sum_{k=1}^{\infty} \nu'_k p^{-k}, \\ \omega &= -1.5708 + p^{-1} \sum_{k=0}^{\infty} \varphi'_k p^{-k}, \\ \tau &= -1.611 + \sum_{k=0}^{\infty} \tau'_k p^{-k}. \end{aligned}\right\} \tag{7.3.5}$$

The coefficients ε'_k, ν'_k, φ'_k and τ'_k are given in Table 7.6. Finally, Table 7.7 gives values of the functions $e(p)$, $\cos\nu(p)$, $\omega(p)$ and $\tau(p)$ for intermediate values of p $(0.10 \leq p \leq 20.0)$. It also displays the values of twice the energy,

$$2h \equiv \overline{h} = \frac{e^2 - 1}{p}.$$

The behavior of the quantities e, $\cos\nu$, ω and τ as functions of p is also shown in Figures 7.4. Further, Figure 7.5 shows how the velocity V, the energy $\overline{h}$, and the distance ρ vary with time.

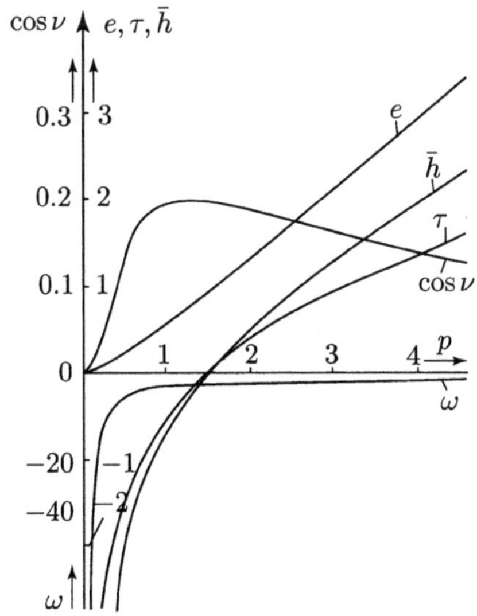

FIGURE 7.4. Behavior of osculating elements of Okhotsimskiĭ's universal spiral

p	e	$\cos\nu$	ω	τ	$\bar{h}$
0.10	0.009994	0.009970	−53.3820	−4.97871	−9.99900
0.12	0.014382	0.014313	−38.0922	−4.42754	−8.33161
0.14	0.019556	0.019382	−28.8661	−3.99908	−7.14013
0.16	0.025503	0.025128	−22.8710	−3.65360	−6.24594
0.18	0.032207	0.031475	−18.7539	−3.36729	−5.54979
0.20	0.039646	0.038332	−15.8022	−3.12493	−4.99214
0.22	0.047793	0.045597	−13.6115	−2.91623	−4.53507
0.24	0.056615	0.053158	−11.9390	−2.73398	−4.15331
0.26	0.066076	0.060904	−10.6313	−2.57297	−3.82936
0.28	0.076137	0.068730	−9.58789	−2.42927	−3.55073
0.30	0.086757	0.076539	−8.74073	−2.29994	−3.30824
0.32	0.097896	0.084250	−8.04234	−2.18264	−3.09505
0.34	0.109512	0.091795	−7.45886	−2.07556	−2.90590
0.36	0.121566	0.099118	−6.96556	−1.97723	−2.73673
0.38	0.134020	0.106179	−6.54409	−1.88645	−2.58431
0.4	0.146838	0.112949	−6.18056	−1.80226	−2.44610
0.45	0.180027	0.110190	−5.39184	−1.61006	−2.15020
0.5	0.215315	0.142001	−4.93070	−1.45562	−1.90728
0.55	0.251470	0.153636	−4.50292	−1.31363	−1.70321
0.6	0.288902	0.163344	−4.20657	−1.19254	−1.52756
0.7	0.365541	0.178381	−3.73868	−0.98117	−1.23769
0.8	0.443994	0.188657	−3.41258	−0.80437	−1.00359
0.9	0.523513	0.195430	−3.17267	−0.65205	−0.80659
1.0	0.603641	0.199643	−2.98886	−0.51783	−0.63562
1.2	0.764699	0.202929	−2.72567	−0.28806	−0.34603
1.4	0.925942	0.202030	−2.54612	−0.09406	−0.10188
1.6	1.08688	0.198833	−2.41563	0.07546	0.11332
1.8	1.24734	0.194388	−2.31640	0.22720	0.30881
2.0	1.40726	0.189299	−2.23832	0.36543	0.49019
2.2	1.56666	0.183923	−2.17524	0.49306	0.66110
2.4	1.72556	0.178472	−2.12318	0.61211	0.82398
2.6	1.88401	0.173075	−2.07947	0.72406	0.98058
2.8	2.04205	0.167807	−2.04223	0.83004	1.13213
3.0	2.19972	0.162713	−2.01012	0.93090	1.27959
3.5	2.59250	0.150857	−1.94631	1.16480	1.63459
4.0	2.98367	0.140292	−1.89878	1.37817	1.97558
4.5	3.37360	0.130929	−1.86198	1.57570	2.30692
5.0	3.76253	0.122627	−1.83263	1.76056	2.63133
6.0	4.53820	0.108656	−1.78875	2.10061	3.26588
8.0	6.08385	0.088240	−1.73408	2.69644	4.50165
10.0	7.62514	0.074166	−1.70136	3.21638	5.71428
12.0	9.16399	0.063919	−1.67957	3.68410	6.91490
14.0	10.7014	0.056140	−1.66402	4.11294	8.10847
16.0	12.2377	0.050040	−1.65236	4.51134	9.29756
18.0	13.7733	0.045130	−1.64329	4.88505	10.4836
20.0	15.3085	0.041095	−1.63604	5.23819	11.6675

TABLE 7.7.

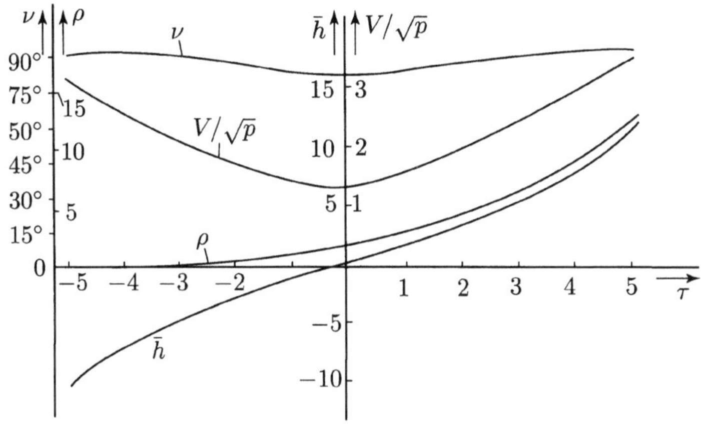

FIGURE 7.5. Time-dependence of the velocity V, energy $\overline{h}$, and distance ρ along the universal spiral

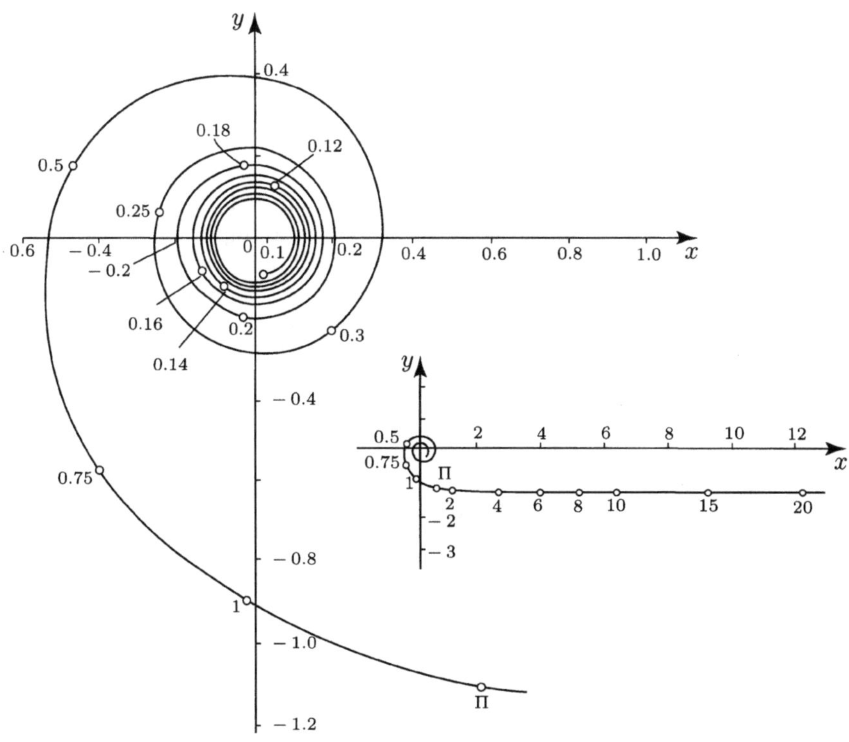

FIGURE 7.6. General shape of the universal spiral

Finally, Figure 7.6 displays the general shape of the described spiral; the values of the parameter p are indicated along the trajectory, and Π indicates the point where escape velocity is attained. At that point[2]

$$V = 1.26846, \qquad \rho = 1.24259.$$

4. Arbitrary trajectories with small eccentricities

Needless to say, the trajectories studied in the preceding section represent only a particular class of solutions of equations (7.3.1). In this class, for a fixed value of the low-thrust acceleration, to a fixed distance from the center of attraction one adds a fully determined unique value of the osculating eccentricity of the orbit. These solutions are very convenient in calculations, for example, of the energy expenditure required for escape (for a set of initial orbits close to one another, with equal initial values of the energy h_*, the energy expenditures for escape are approximately equal to those for the spiral considered above, provided that on that spiral the initial point is taken to be the point corresponding to the energy h_*). However, the trajectory described in the preceding section is "too smooth" – the parameters of the solutions in the corresponding class change in a monotone manner. Rigorously speaking, such a trajectory is not typical with respect to the behavior of its osculating elements.

It is therefore interesting to attempt an analytic study of the behavior of arbitrary escape trajectories. This can be done by exploiting the fact that the reactive acceleration is small. Indeed, the assumption that the "perturbing force" (reactive acceleration) is small allows us to use the asymptotic methods of nonlinear mechanics. Such an investigation was carried out, for example, by Yu. G. Evtushenko in [7.4]. Here we will consider only one of the problems solved in [7.4]; specifically, we will assume that the initial orbit has an arbitrary, yet sufficiently small eccentricity, and we will analyze the behavior of the escape trajectory for as long as the eccentricity remains sufficiently small.

First let us transform equations (7.3.1) to a convenient form, introducing the components of the Laplace vector,

$$\alpha = e \sin \omega, \qquad \beta = e \cos \omega, \tag{7.4.1}$$

as new variables (instead of e and ω). This enables us to eliminate e from the denominators of the right-hand sides of the equations, and hence avoid singularities when $e \to 0$. Further, let us introduce the argument of latitude

$$u = \omega + \nu.$$

[2]Very interesting investigations devoted to the construction of similar, but optimal escape trajectories were carried out by G. B. Efimov [7.5].

Then from (7.3.1) we readily derive the equation

$$\frac{du}{d\tau} = \frac{\sqrt{p}}{\rho^2}.$$

We see that u varies monotonically, and in fact rapidly ($du/d\tau \sim 1$) compared with α, β and p. It is convenient to take u as the independent variable instead of τ. As a result of these transformation equations (7.3.1) become

$$\left.\begin{aligned} \frac{dp}{du} &= 2fp^3F, \\ \frac{d\alpha}{du} &= 2fp^2(\alpha + \sin u)F, \\ \frac{d\beta}{du} &= 2fp^2(\beta + \cos u)F, \\ F &= (1 + \alpha \sin u + \beta \cos u)^{-2} \times \\ &\quad \times \left(1 + \alpha^2 + \beta^2 + 2\alpha \sin u + 2\beta \cos u)^2\right]^{-1/2}, \end{aligned}\right\} \tag{7.4.2}$$

or, retaining only the terms of order one in the eccentricity (i.e., of order one in α and β),

$$\left.\begin{aligned} \frac{dp}{du} &= 2fp^3(1 - 3\alpha \sin u - 3\beta \cos u), \\ \frac{d\alpha}{du} &= fp^2(-\alpha + 2\sin u + 3\alpha \cos 2u - 3\beta \sin 2u), \\ \frac{d\beta}{du} &= fp^2(-\beta + 2\cos u - 3\alpha \sin 2u - 3\beta \cos 2u). \end{aligned}\right\} \tag{7.4.3}$$

To this system we can apply directly the algorithm of the asymptotic method, regarding f as a small parameter and u as a fast variable. Averaging the right-hand sides in (7.3.4) with respect to u we obtain

$$\frac{dp}{du} = 2fp^3, \qquad \frac{d\alpha}{du} = -fp^2\alpha, \qquad \frac{d\beta}{du} = -fp^2\beta. \tag{7.4.4}$$

From these equations it follows that

$$\frac{d(e^2)}{du} \equiv \frac{d}{du}\left(\alpha^2 + \beta^2\right) = -\frac{1}{2} fp^2 \left(\alpha^2 + \beta^2\right) < 0.$$

Therefore, under the action of the reactive acceleration the eccentricity of the orbit decreases monotonically in the mean. This is in annoying disagreement with the properties of the escape trajectories with which we are already familiar. Indeed, at the terminal time of the escape segment the velocity attains the escape value, which corresponds to an eccentricity $e = 1$. Hence, sooner or later the eccentricity must begin to increase, a behavior that is not "grasped" by the solution of the averaged system (7.4.4). This circumstance can be explained as follows: since the solution of the averaged system (7.4.4) corresponds to the solution of the original system (7.4.3) only on a bounded "time" interval, the eccentricity-increase effect begins to manifest itself only after this interval expires.

Two realize this effect we must enlarge the "time interval" on which the solution is sufficiently well approximated. In simpler terms, we must take a higher-order approximation to the solution of system (7.4.3). To this end, following the paper [7.4], we will assume that the quantities f, α, and β are of the same order of smallness (say, ε) and seek the solution of system (7.4.3) in the form

$$p = \overline{p} + \varepsilon p_1, \qquad \alpha = \overline{\alpha} + \varepsilon \alpha_1, \qquad \beta = \overline{\beta} + \varepsilon \beta_1, \tag{7.4.5}$$

where $\overline{p}$, $\overline{\alpha}$, $\overline{\beta}$ is the solution of the averaged second-order approximation system, i.e., of a system of equations of the type

$$\frac{d\overline{p}}{du} = \varepsilon p_1 + \varepsilon^2 p_2, \tag{7.4.6}$$

and p_1, α_1, β_1 must be determined by substituting expressions (7.4.5) in (7.4.3) and equating the like terms in ε. The latter operation yields the equations

$$\frac{dp_1}{du} = 0, \qquad \frac{d\alpha_1}{du} = 2p^2 \sin u, \qquad \frac{d\beta_1}{du} = 2p^2 \cos u. \tag{7.4.7}$$

But a glance at the system (7.4.6) reveals that it is identical with the system (7.4.4). The right-hand sides of the last two equations in (7.4.7) are small of second order because, by assumption, $f\alpha \sim \varepsilon^2$, $f\beta \sim \varepsilon^2$. In the first of equations (7.4.4) the right-hand side is small of first order, the second-order term being identically equal to zero.

The solution of the system (7.4.4) is readily obtained by integrating it, and for the solution of system (7.4.7) we take

$$p_1 = 0, \qquad \alpha_1 = -2p^2 \cos u, \qquad \beta_1 = 2p^2 \sin u.$$

Substituting these expressions in (7.4.5) (considering that $\varepsilon \equiv f$) and taking into account that $\overline{\alpha}$, $\overline{\beta}$, $\overline{p}$ is a solution of the system (7.4.4), we finally obtain

$$\left.\begin{aligned} p &= p_0 \left(1 - 4p_0^2 f u\right)^{-1/2}, \\ \alpha &= \frac{A}{\sqrt{p}} - 2fp^2 \cos u, \\ \beta &= \frac{B}{\sqrt{p}} + 2fp^2 \sin u. \end{aligned}\right\} \tag{7.4.8}$$

The constants A and B are determined from the condition $\alpha = \alpha_0$, $\beta = \beta_0$ for $u = 0$. The solution (7.4.8) approximates the exact solution of the system (7.4.3) with an error $\sim f^2$ on an interval $u \sim f^{-1}$, provided that $e = \sqrt{\alpha^2 + \beta^2} \sim f$.

Let us examine the solution (7.4.8). The focal parameter p increases monotonically, and therefore the orbit increases in size. Squaring the expressions of α and β in (7.4.8), adding the results and averaging with respect to u, we obtain the mean value, per one turn of spiral, of the square $\overline{e^2}$ of the osculating eccentricity:

$$\overline{e^2} = \frac{A^2 + B^2}{p} + 4f^2p^4. \tag{7.4.9}$$

Denoting the initial value of the function $\overline{e^2}$ by $\overline{e_0^2}$, we have

$$A^2 + B^2 = p_0\left(\overline{e_0^2} - 4f^2p_0^4\right).$$

The function (7.4.9) has a minimum when

$$p^5 = p_*^5 \equiv \frac{p_0\left(\overline{e_0^2} - 4f^2p_0^4\right)}{16f^2}.$$

If $p_* > p_0$, then the minimum of $\overline{e^2}$ lies in the domain of real variation of the parameters (indeed, p decreases monotonically from the value p_0). The condition $p_* > p_0$ is equivalent to the following (in a slightly crude approximation, we consider that $\sqrt{\overline{e_0^2}} = e_0$):

$$e_0 > 2\sqrt{3}fp_0^2. \tag{7.4.10}$$

Thus, if the initial eccentricity e_0 and the dimensionless thrust-generated acceleration f satisfy inequality (7.4.10), then the current mean eccentricity of the osculating orbit will first decrease to some minimum value, and then increase monotonically, behaving like $e \sim 2fp^2$ for large p. If inequality (7.4.10) is not satisfied, the mean eccentricity begins to increase right away.

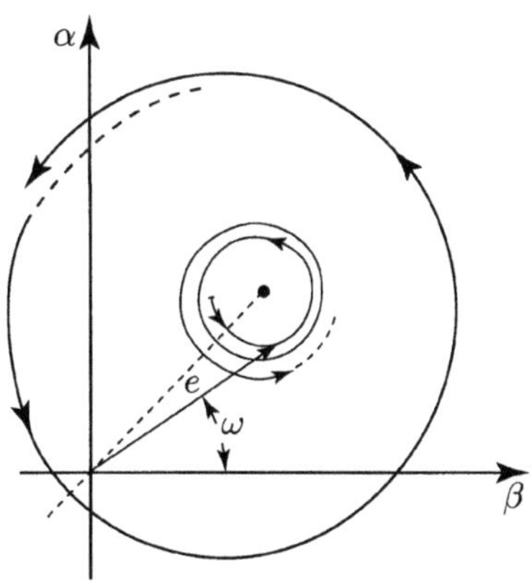

FIGURE 7.7. Hodograph of the osculating Laplace vector of an escape trajectory (schematic representation)

It is readily verified also that the hodograph of the Laplace vector is described by the equation of a circle

$$\left(\alpha - \frac{A}{\sqrt{p}}\right)^2 + \left(\beta - \frac{B}{\sqrt{p}}\right)^2 = 4f^2p^4$$

of variable radius $2fp^2$ and with a variable center of coordinates $(A/\sqrt{p}, B/\sqrt{p})$. As time flows, the radius of the circle increases, while the center tends monotonically (along a line) to the origin. As a result, the hodograph represents an unwinding spiral, each turn of which is traversed over a "time" $\Delta u = 2\pi$ (Figure 7.7). The position vector of the hodograph is equal in magnitude with the eccentricity e of the orbit, and the polar angle is equal to the longitude of the orbit's perigee, ω. Examining the hodograph in Figure 7.7 one can understand the complicated behavior of these two parameters.

Eighth Essay

The Full Force of the Sun Blows in the Sails

In arctic and southern waters,
on the ridges of green waves,
among reefs of basalt and pearl,
rustle the sails of ships.
Swift-winged (ships) are commanded by captains
who discover new lands ...

N. Gumylev, *The Captains*

"Hold your hands out to the Sun ... What do you feel? Heat, of course. But there's pressure as well – though you've never noticed it, because it's so tiny. Over the area of your hands, it comes to only about a millionth of an ounce."

"But out in space, even a pressure as small as that can be important, for it's acting all the time, hour after hour, day after day. Unlike rocket fuel, it's free and unlimited. If we want to, we can use it. We can build sails to catch the radiation blowing from the Sun."

This excerpt is take from Arthur C. Clarke's science fiction story *The Wind from the Sun* [8.1]. Clarke describes poetically and with scientific authenticity a race of solar-sail powered yachts around the Earth. We will repeatedly quote from this story in the present essay. In point of fact, the essay is a sort of scientific commentary on Clarke's story.

The radiation pressure (pressure of light) generated by the incoming flux of solar radiation at positions on the orbit of the Earth is about $p = 4.5 \cdot 10^{-8}$ g/cm^2 – a negligible quantity. However, if we equip a spaceship with a sail that is sufficiently light and at the same time has sufficiently large dimensions, then the total force produced by the radiation pressure on the sail can impart the spaceship an acceleration large enough to allow maneuvering in space. For instance, by controlling the sail intelligently, one could accelerate the spaceship in a spiral trajectory around the Earth and, once escape velocity is attained, fly into open space, to other planets! About this one can read, for example, in the book of G. L. Grodzovskiĭ, Yu. N. Ivanov, and V. V. Tokarev [8.2], which surveys works on the flight dynamics of solar-sail powered spacecraft.

Our analysis here will follow a paper by A. P. Skoptsov [8.3].[1] If the sail is planar and has an ideally reflecting surface, then the resultant radiation-pressure force acting on the sail is given by the formula

$$P = 2p(R_{\text{Sun}}/R)^2 S \cos^2\theta = P_* \cos^2\theta \tag{8.1}$$

and is directed along the normal to the plane of the sail. In (8.1) R and R_{Sun} denote the distance from the Sun to the sail and to the Earth, respectively, S is the area of the sail, and θ is the angle between the direction of the resulting force and that of the solar rays. In what follows we shall consider only orbits of spaceships around the Earth, for which one can take $R_{\text{Sun}}/R = 1$.

However, the sail does not have to be flat (planar). We are interested here in the optimal problem of attaining escape velocity in the shortest time. To achieve this last goal, one has to adequately control the position of the sail during flight, but one can also choose a sail design that allows one to minimize the escape time. One can control in optimal way a flat sail, but one can also control in optimal way a sail of optimal design; obviously, in the second case the flight duration will be shorter than in the first case.

Let $\mathbf{K}_1$ be the total momentum vector of the incoming radiation flux, and $\mathbf{K}_2$ be the momentum vector of the flux reflected by the sail. The resulting radiation-pressure force is $\mathbf{P} \sim \Delta\mathbf{K} = \mathbf{K}_1 - \mathbf{K}_2$. An optimal sail design is specified by the following conditions (see [8.3]):

(1) $|\mathbf{K}_1| = |\mathbf{K}_2|$;

(2) the quantity $|\mathbf{K}_1|$ does not depend on the direction of $\mathbf{K}_2$.

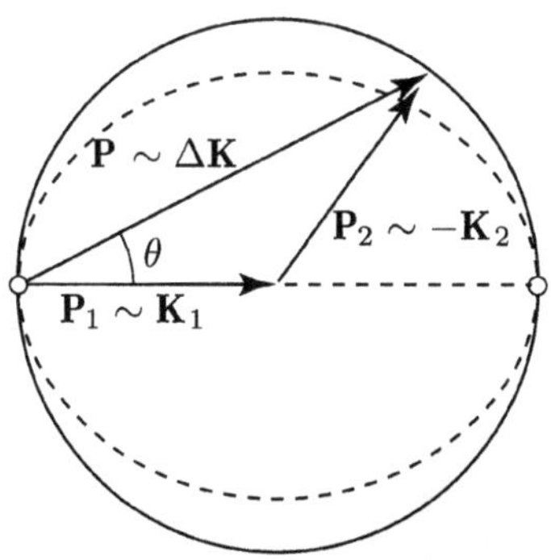

FIGURE 8.1. Conditions for optimal design of a solar sail

If (1) and (2) hold, then (Figure 8.1) $|\Delta\mathbf{K}| = 2|\mathbf{K}_1|\cos\theta$, and then the magnitude of the total radiation-pressure force depends on the angle θ according to the law

$$P = P_* \cos\theta, \qquad -\pi/2 \le \theta \le \pi/2. \tag{8.2}$$

[1] We use this opportunity to thank A. P. Skoptsov for his help in the analysis and computations, which were carried out specially for this essay.

The law (8.2) is associated with a circular diagram of variation of the force with the angle θ (Figure 8.1). The same figure shows (dashed line) the force diagram (8.1) for a flat sail. The two diagrams do not differ by much, though the difference is noticeable. There is no unique design of an "optimal sail." One of the possible choices is described in [8.3].[2] If an optimal sail and a planar sail generate the same maximal thrust P_*, then, for any angle θ, one can associate to the optimal sail a flat sail, placed normally to the thrust $\mathbf{P}$ of the optimal sail. Then the flat sail will yield a thrust in the same direction as the optimal sail (though the thrust of the flat sail will be somewhat smaller).

Let us write the equations of motion in polar coordinates r, φ. We shall take the direction $\varphi = 0$ to be that of the Sun's rays (Figure 8.2). We may assume that the Sun is fixed in space, since our flight will not take too long (just a few days). As usual, let us denote by u and v the radial and transversal components of the velocity of the spaceship. By the foregoing discussion, the thrust acting on the spaceship is the vector sum of two forces, $\mathbf{P}_1$ and $\mathbf{P}_2$. The force $\mathbf{P}_1$ is constant in magnitude and direction (directed along the solar rays); its radial and transversal components are equal to $P_* \cos\varphi/2$ and $-P_* \sin\varphi/2$, respectively. The force $\mathbf{P}_2$ makes a constant angle, denoted here by γ, with the radial direction; its components are $P_* \cos\gamma/2$ and $P_* \sin\gamma/2$, respectively.

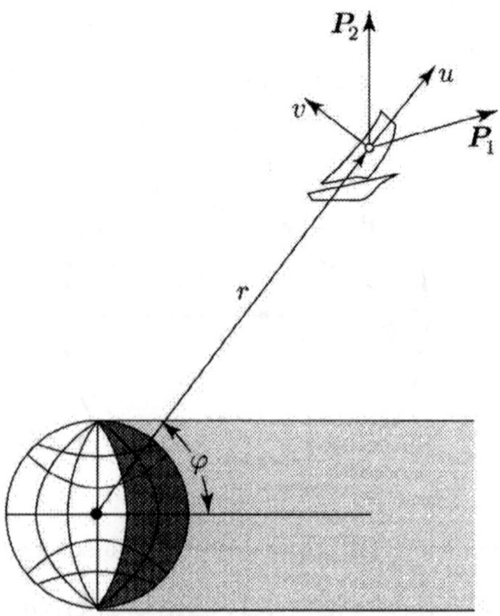

FIGURE 8.2. Coordinate system

[2] It is amusing to remark that this design uses as a component the "hyperboloid of engineer Garin," described in A. Tolstoy's science fiction story with the same title: two confocal hyperboloids with their bowls facing each other, and having very differrent focal distances.

Denote by $a_0 = P_*/m$ the maximal acceleration generated by the optimal sail. Then the equations of motion have the form

$$\left.\begin{aligned}\frac{du}{dt} &= \frac{v^2}{r} - \frac{\mu}{r^2} + \frac{a_0}{2}(\cos\gamma + \cos\varphi) f(r,\varphi),\\ \frac{dv}{dt} &= -\frac{uv}{r} + \frac{a_0}{2}(\sin\gamma - \sin\varphi) f(r,\varphi),\\ \frac{dr}{dt} &= u,\\ \frac{d\varphi}{dt} &= \frac{v}{r}.\end{aligned}\right\} \tag{8.3}$$

Here μ is the Earth's gravitational parameter. Equations (8.3) contain an essential factor – the "shadow function" $f(r,\varphi)$. Indeed, the force due to radiation pressure can act on the sail only outside the Earth's shadow – once the spaceship enters the Earth's shadow it no longer experiences the radiation pressure. To take this into account the components of the radiation-pressure force are multiplied by a function that is equal to 0 in the Earth's shadow and to 1 outside the shadow. The Earth's shadow has a complex structure; its boundary is not sharply defined because of the presence of a penumbra due to the dispersive action of the atmosphere. For this reason it is better to take for f not a "relay" function of the type 0–1, but some continuous function that approximates the real properties of the shadow; a function of this type is shown in Figure 8.3. For example, the following function is used in [8.3]:

$$f(r,\varphi) = \frac{1}{\pi}\arctan\alpha\left(\sqrt{1 - \left(\frac{R_{\text{Sun}}}{r}\right)^2} - \cos\varphi\right) + \frac{1}{2}, \tag{8.4}$$

where α is a parameter, the choice of which determines the rate of growth of the shadow function when one crosses the boundary of the shadow (i.e., the length of the penumbra segment in Figure 8.3).

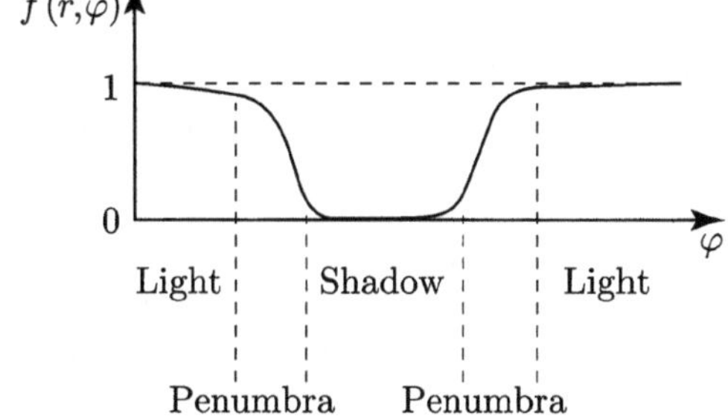

FIGURE 8.3. Shadow function

Equations (8.3) contain a free function $\gamma(t)$, which must be chosen in such a manner that the escape time of our sailship will be minimal. Such an optimal control of a sail is considered in [8.3]. It is rather complicated. However, as shown therein, the result (i.e., the escape time) turns out to be very close to the one obtained if a so-called *locally-optimal control* is used. By that we mean a control $\gamma(t)$ with the property that in each point of the phase plane it gives the maximum increase of the energy $\widetilde{h} = u^2 + v^2 - 2\mu/r$.

Let us calculate the derivative of $\widetilde{h}$:

$$\frac{d\widetilde{h}}{dt} = a_0 f(u\cos\gamma + v\sin\gamma + u\cos\varphi - v\sin\varphi).$$

The maximum of this derivative with respect to the control γ is achieved for the control law

$$\cos\gamma = u/V, \qquad \sin\gamma = v/V, \qquad V = \sqrt{u^2 + v^2}. \tag{8.5}$$

In this case the force $\mathbf{P}_2$ is directed along the velocity vector of the spaceship. Then

$$\frac{d\widetilde{h}}{dt} = a_0 f(V + V_s), \tag{8.6}$$

where V is the magnitude of the total velocity of the spaceship and V_s is the projection of the velocity vector on the direction of the Sun's rays. Since $V + V_s \geq 0$, (8.6) shows that the energy $\widetilde{h}$ increases monotonically, which guarantees that any preassigned value will be attained, in particular, the one necessary to acquire the escape velocity.

We will term the control (8.5) *locally optimal* and from this point on we will consider the equations of motion (8.3), (8.4) with the control law (8.5). We thus arrive at an interesting problem of dynamics, which is a superposition of two problems we are already familiar with: motion under the combined action of an acceleration of constant magnitude and constant direction in space, and of a constant tangential acceleration. In our case the absolute magnitudes of these accelerations coincide.

Fully armed with the equations of motion (8.3)–(8.5), we can now embark upon a dynamical analysis of Clarke's story.

> "The enormous disc of sail strained at its rigging, already filled with the wind that blew between the worlds. In three minutes the race would begin ... the immense sail was taut, its mirror surface sparkling and glittering gloriously in the sun.
>
> "To Merton, floating weightless at the periscope, it seemed to fill the sky. As well it might – for out there were fifty million square feet of sail, linked to his capsule by almost a hundred miles of rigging. All the canvas of all the tea clippers that had once raced like clouds across the China seas, sewn into one gigantic sheet, could not match the single sail that *Diana* had spread beneath the sun. Yet it was little more substantial than a soap bubble;

that two square miles of aluminized plastic was only a few millionth of an inch thick."

Fifty million square feet means 4.65 square kilometers, which corresponds to a circular sail with a diameter of 2.4 kilometers! Only with a solar sail of such huge dimensions one can hope to attain an acceleration that will suffice for maneuvering in space. Of course, the sail must be as light as possible, and hence as thin as possible. Indeed, the radiation-pressure force is proportional to the area of the sail, and the acceleration is obtained by dividing this force by the mass of the whole spaceship (including the sail). It is therefore quite clear how important is to reduce the weight of the sail. The construction of a very thin sail with a very large area is the main technical difficulty in achieving solar-sail powered flights.

Diana's sail, despite its negligible thickness (of the order of ten micrometers = 10^5 m), should weight 700–1,000 kilograms! (The specific weight of an aluminum-enforced plastic foil is about $1.2 \cdot 10^3$ kg/m^3. The *soap bubble* that Clarke is talking about should of course have dimensions close to the dimensions of the sail. If the weight of the remaining part of the spaceship (rigging, cabin, astronaut, and so on) is about 1–1.5 tons, then the acceleration imparted to the ship may reach $\sim 10^{-3}g$. The yachts that participate in Clarke's race are of precisely this "class." There are seven of them, engaged in a quite dramatic competition. But this is not the place to describe the adventures on the race's course and the fate of *Diana*'s courageous captain, John Merton. We are interested in flight mechanics, and for that purpose it suffices to follow the motion of one or two yachts.

> "On the four inhabited worlds, there were scarcely twenty man who could sail a sun yacht; and they were all here, on the starting line or aboard the escort vessels, orbiting twenty-two thousand miles above the equator."

We may assume that the initial orbit in the story is circular. Moreover, it is diurnal: the period of revolution of the satellite on a diurnal orbit is equal is the Earth's period of rotation (24 hours). On an equatorial diurnal orbit the satellite is "suspended" all the time over the same geographic point on Earth. The radius of a diurnal circular orbit is $r_0 \approx 42,190$ km. If we take into account that the radius of the Earth is $R_{\rm E} = 6,370$ km, we conclude that the altitude of the orbit above Earth's surface is $h = 35,820$ km, or 22,260 miles, which is about the 22 thousand miles that Clarke speaks about. A bit farther in the story there is indeed mention of motion in a diurnal orbit. Accordingly, as the initial data for the integration of the equations of motion (8.3)–(8.5) we should take the data corresponding to a diurnal circular orbit:

$$r_0 = 42,188\,\text{km}, \qquad u_0 = 0, \qquad v_0 = \sqrt{\mu/r_0} \tag{8.7}$$

(as is known, the last formula gives the velocity on a circular orbit of radius r_0). The initial angle φ will be chosen below.

> "Seven knife blades sliced through seven thin lines tethering the yachts to the mother ships that had assembled and serviced them. Until this mo-

ment, all had been circling Earth together in a rigidly held formation, but now the yachts would begin to disperse, like dandelion seeds drifting before the breeze. And the winner would be the one that first drifted past the Moon.

"Aboard *Diana*, nothing seemed to be happening. But Merton knew better. Though his body could feel no thrust, the instrument board told him that he was now accelerating as almost one thousandth of a gravity. For a rocket, that figure would have been ludicrous – but this was the first time any solar yacht had ever attained it. *Diana*'s design was sound; the vast sail was living up to his calculations. At this rate, two circuits of the Earth would build up his speed to escape velocity, and then he could head out for the Moon, with the *full force of the Sun* behind him."

Since the target of the race is the Moon, it is natural to consider that the race takes place in the plane of the Moon's orbit (or, approximately, in the ecliptic plane). We will assume that the initial diurnal orbit lies in that plane; this does not contradict the mention in the story of yachts moving above the equator: one can consider that at the time of our description the route of the yachts has already crossed the equator. An example of route of a diurnal ecliptic satellite (i.e., the trace of its radius vector on the surface of the Earth) is shown in Figure 8.4. From the excerpt quoted above it also follows that in equations (8.3) we must put

$$a_0 = 0.001g, \tag{8.8}$$

since this is (roughly) the maximal acceleration imparted to *Diana* by the force due to radiation pressure (of course, the quoted excerpt speaks precisely about this acceleration, and not the total acceleration due to radiation pressure and gravitational forces).

The initial position of the yachts on the diurnal orbit is specified by the angle

$$\varphi_0 = -\frac{\pi}{2}. \tag{8.9}$$

As we will see later, this value of φ_0 follows unequivocally from the analysis of the information given in Clarke's story. The data (8.7)–(8.9) are sufficient for the numerical integration of the equations (8.3)–(8.5). The results of this integration, that is, the trajectory of John Merton's yacht, are displayed in Figure 8.5. On this figure one notes first of all that till the moment it attains escape velocity, *Diana* makes only slightly more than two spiral laps around the Earth, in complete agreement with what is said in the excerpt just quoted. True, in another place in Clarke's story there is an inaccuracy:

"*... it can reach escape velocity in a couple of days*" – and the entire story is constructed upon the assumption that the run-up to escape velocity takes about two days. In reality Merton spends for that purpose about 120 hours, i.e., 5 days (the flight duration is marked along the trajectory in Figure 8.5). The time spent on the first lap of spiral is about 30 hours (more than one day), and the remaining duration is strongly "stretched" along the second lap of the trajectory ...

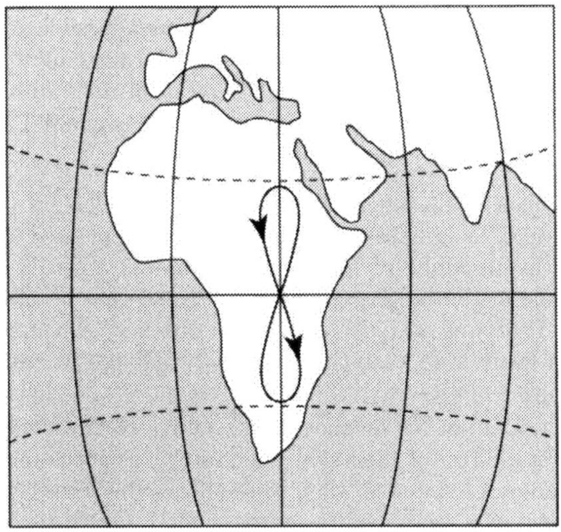

FIGURE 8.4. Route of a diurnal ecliptic satellite

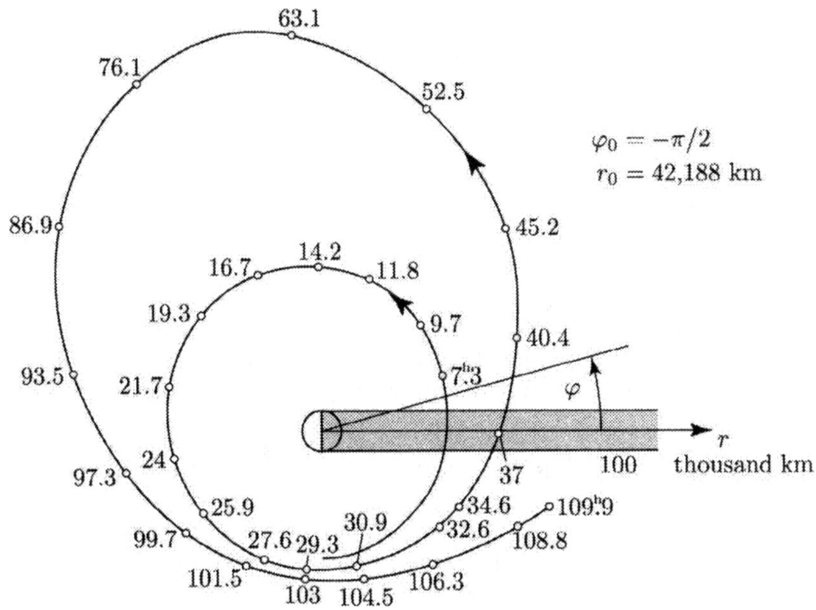

FIGURE 8.5. *Diana*'s trajectory

A considerable reduction of flight duration is not possible since, as we mentioned above, our flight program is nearly optimal, guaranteeing the fastest escape. But let us excuse Clarke for his error: *The Wind from the Sun* is not a dissertation, but a work of fiction, where some literary license is allowed. In return, other details of the story are of brilliant scientific credibility ...

> "*Diana* had made a good start; time to take a look at the opposition... There they were, looking like strange silver flowers planted in the dark fields of space... the Republic of Mars's *Sunbeam* was a flat ring, with a half-mile-wide hole in the center, spinning slowly, so that centrifugal force gave it stiffness. That was an old idea, but no one had ever made it work; and Merton was fairly sure that the colonials would be in trouble when they started to turn.
>
> "That would not be for another six hours, when the yachts had moved along the first quarter of their slow and stately twenty-four-hour orbit. Here at the beginning of the race, they were all heading directly away from the Sun – running, as it were, before the solar wind. One had to make the most of this lap, before the boats swung around to the other side of Earth and then started to head back into the Sun."

Six hours spent on the first quarter of the flight, if the latter is understood as the polar angle varying by 90°, is almost the exact value. Actually, in the absence of a sail this would be the exact value. The changes in the orbit on the initial segment of the flight due to the influence of the sail cannot be very large (they do not manage to accumulate). If the influence of the sail is taken into account, the time spent on the first quarter of a lap is about 6 hours and 5 minutes. As for the assertion that "*one had to make the most of this lap,*" indeed, on the initial segment of the flight the acceleration produced by the radiation pressure is maximal or nearly maximal, since the sail is almost perpendicular to the Sun's rays. Figure 8.6 shows the hodograph of the acceleration due to radiation pressure along *Diana*'s trajectory. The yacht's flat sail must always be perpendicular to the acceleration vector, as displayed in this figure.

From the departure time ($\varphi = -90°$), when the sail is perpendicular to the solar rays and the thrust acceleration is maximal (equal to $a_0 = 0.001g$), till the time the spaceship enters the Earth's shadow ($\varphi = -7.5°$), the sail turned by less than 40°, and the thrust acceleration dropped by only 20% (to $0.8a_0$).

> ... "The little windlasses were continually turning, playing lines in and out as the autopilot kept the sail trimmed at the correct angle to the Sun...
>
> "Now the Earth had almost vanished; it had waned to a narrow, brilliant bow of light that was moving steadily toward the Sun. Dimly outlined within that burning bow was the night side of the planet, with the phosphorescent gleams of great cities showing here and there through gaps in the clouds. The disc of darkness had already blanked out a huge section of the Milky Way. In a few minutes it would start to encroach upon the Sun.

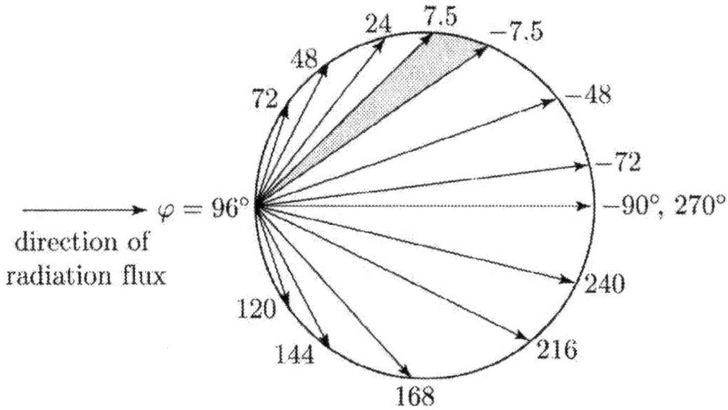

FIGURE 8.6. Hodograph of *Diana*'s perturbing acceleration

> "The light was fading; a purple, twilight hue – the glow of many sunsets, thousands of miles below – was falling across the sail as *Diana* slipped silently into the shadow of Earth. The Sun plummeted below that invisible horizon; within minutes, it was night.
>
> "Merton looked back along the orbit he had traced, now a quarter of the way around the world. One by one he saw the brilliant stars of the other yachts wink out, as they joined him in the brief night. It would be an hour before the Sun emerged from that enormous black shield, and through all that time they would be completely helpless, coasting without power."

Thus, till the moment they enter the Earth's shadow, the yachts make a quarter of a lap, which explains the initial value (8.9): $\varphi = -90°$ that we took for the integration of the equations of motion. More precisely, we should have taken $\varphi = -97.5°$ in order to complete exactly one quarter of a lap till one enters the shadow, because the angular displacements of the yachts when they sail across the entire shadow is about 15°. But we will not attempt here to scrupulously reconstruct the race in exact detail... indeed, the story does not tell us that the yacht sailed exactly a quarter of a lap before entering the shadow; we may assume that it sailed almost a quarter – the exact value is not important (the author became convinced that this is indeed the case after he nevertheless carried out the computation of the trajectory for $\varphi_0 = -97.5°$). Let us mention also that the time spent to cross the shadow is indeed about an hour, as indicated in the excerpt quoted above; after an hour the yachts leave the shadow...

> "From now on, for almost half of his orbit around the Earth, he must keep the whole of this immense area edge-on to the Sun. During the next twelve or fourteen hours, the sail would be a useless encumbrance; for he would be heading *into* the Sun, and its rays could only drive him backward along his orbit."

Let us examine again to the diagram of *Diana*'s accelerations on the first lap (Figure 8.6). We see that indeed, for a major part of the half-lap of the orbit the plane of the sail is almost parallel to the Sun's rays. Thus, from $\varphi = 48°$ to $\varphi = 144°$ the inclination of the sail's plane to the direction of the radiation flux does not exceed 20°. On a large segment of the orbit (from $\varphi = 24°$ to $\varphi = 168°$), on which the motion takes place more or less "against the Sun," the inclination of the sail's plane to the flux of solar rays does not exceed 35–45°. This segment is traversed in 14 hours. The sail is strictly edge-on to the Sun in only one point ($\varphi = 96°$). In this point the thrust generated by the sail is equal to zero.

> "It was a pity that he could not furl the sail completely, until he was ready to use it again..."

Merton's regrets are not justified. Without the sail the energy supply would drop to zero. With the sail on, the energy increases monotonically, although slowly.

Let us remark that the locally-optimal control of the sail is very close in nature to the following control law: rotate the sail continuously in the same direction the spaceship moves, but with an angular velocity equal to half the spaceship's angular velocity. In other words, if θ denotes the angle between the normal to the sail and the direction of the Sun's rays, then the locally-optimal control is close to the law

$$\theta = \frac{\pi}{4} + \frac{\varphi}{2}\,. \tag{8.10}$$

Here it is assumed that both sides of the sail have the same reflectivity characteristics (otherwise, instead of (8.10) one can write an analogous control that is piecewise-linear in φ). Figure 8.7 shows the graph of the dependence $\theta(\varphi)$ for *Diana*'s trajectory. We see that the deviations from the control law (8.10) do not exceed 5°.

> "Within minutes the [number of participants] five had dropped to four. From the beginning, Merton had doubts about the slowly rotating *Sunbeam*; now he saw them justified.
>
> "The Martian ship had failed to tack properly. Her spin had given her too much stability. Her great ring of sail was turning to face the Sun, instead of being edge-on to it. She was being blown back along her course at almost her maximum acceleration."

To simplify the arguments, let us assume that thanks to its extreme stability *Sunbeam*'s sail does not change its position in space at all. Then the acceleration vector generated by the radiation pressure is constant in magnitude and direction. But we have already dealt with such a problem (see [8.3], [8.4]) in the third essay, where we saw that, indeed, there exist trajectories which "turn back" – at least trajectories of the type shown in Figure 3.15. If *Sunbeam*'s crew cannot manage to control their sail, then they cannot increase the ship's energy and attain escape velocity; they will end up "dangling" in the vicinity of Earth on a bounded trajectory, similar to that shown in Figure 3.15.

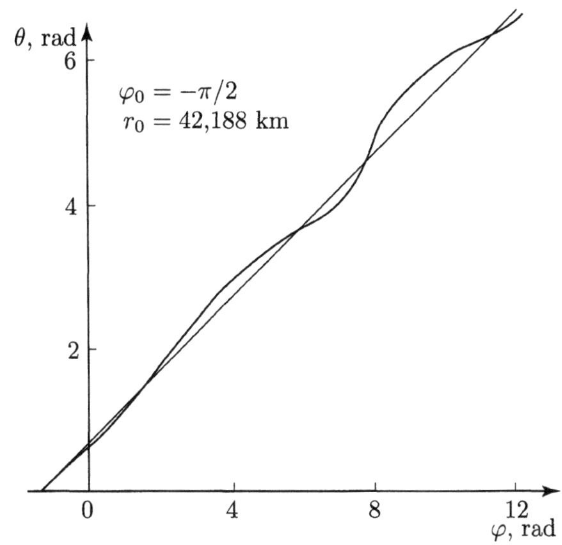

FIGURE 8.7. Control law of *Diana*'s sail along its trajectory

As we recall from the third essay, in such a situation the type of trajectory is determined by the values of the constants $h/\sqrt{f}$ and c (see Figure 3.4). Four our initial data we have

$$c = -\frac{1}{2}f = -\frac{a_0}{2g\left(\dfrac{R_{\text{Sun}}}{r_0}\right)^2} \approx -0.022,$$

$$h = -0.5,$$

$$h/\sqrt{f} = -0.5/\sqrt{0.044} = -2.380.$$

This point in the plane $(h/\sqrt{f}, c)$ falls into the domain of "snake-like" trajectories, so that the radiation pressure obligatorily "blows away" the *Sunbeam*, not allowing it to make even one lap around the Earth! (Here the one-time entrance in the Earth's shadow plays no role.) Figure 8.8 shows a computed trajectory of such a yacht (but with a larger value of the maximal acceleration a_0).

We should point out that even if the sail is controlled intelligently such unpleasant behavior cannot be always avoided. If we make an unfortunate choice of the point of departure on the initial circular orbit, then the motion on the segment where escape velocity is attained may happen to take place with the sail facing the Sun, which will double the time necessary to attain escape velocity! Figure 8.9 shows what would happen to Merton's yacht if the point of departure on the orbit were specified by $\varphi_0 = -135°$. It is interesting that by starting the escape process

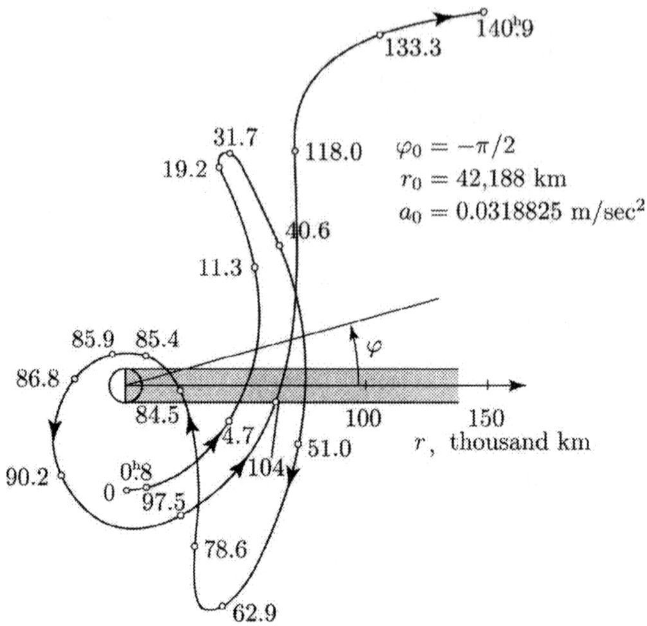

FIGURE 8.8. Trajectory of the yacht *Sunbeam*

in the Earth's shadow ($\varphi_0 = 0$) one can save about 10 hours compared with the escape time for the choice $\varphi = -90°$. The dependence of the escape time (under locally-optimal control) on the point of departure is shown in Figure 8.10.

But let us return to our race.

"So now it was a straight fight between *Diana* and *Lebedev* ... For that matter, it was hard to see what *Lebedev* could do to overtake *Diana*'s lead; but all the way around the second lap, through eclipse again and the long, slow drift around the Sun, Merton felt a growing unease.

"He knew the Russian pilots and designers. They had been trying to win this race for twenty years – and, after all, it was only fair that they should, for had not Pyotr Nikolaevich Lebedev been the first man to detect the pressure of sunlight, back at the very beginning of the twentieth century? But they had never succeeded.

"And they would never stop trying. Dmitri [Markoff, *Lebedev*'s captain] was up to something – and it would be spectacular.

"If all went well, this would be the last circuit, both for him and for the Russians. They had spiraled upward by thousands of miles, gaining energy from the Sun's rays. On this lap, they should escape from Earth completely, and head outward on the long run to the Moon ... Luckily, the most difficult

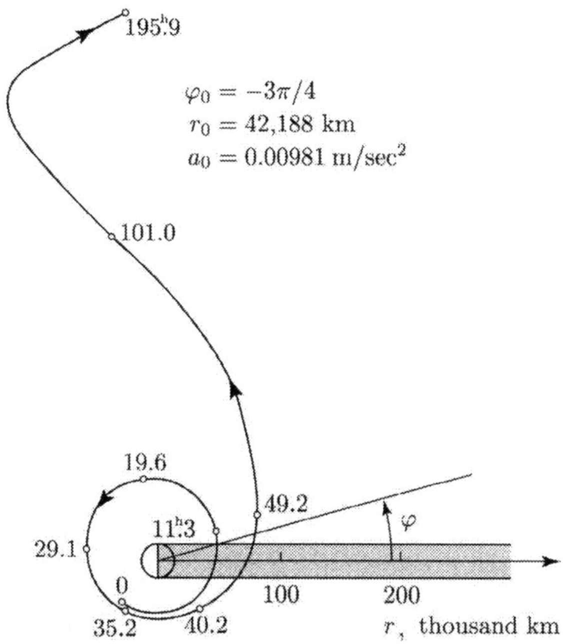

FIGURE 8.9. Escape trajectory against the Sun. To make the picture more intuitive, the Earth's shadow is exaggerated

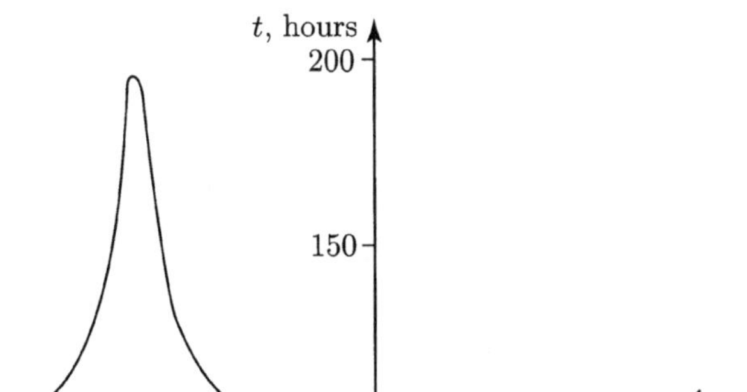

FIGURE 8.10. Dependence of the escape time on the point of departure from the diurnal trajectory

> maneuvers were over; from now on, *Diana* would have the Sun behind her as she sailed straight down the solar wind. And as the old-time sailors had often said, it was easy to handle a boat when the wind was blowing over your shoulder."

Once again we are amazed by the scientific credibility of Clarke's description of the yachts' trajectories. Figure 8.5 shows that at the moment when *Diana* attains escape velocity its velocity vector makes an angle of about 25° with the direction of the Sun's rays, and the angle made by the thrust vector is even smaller (about 15°). As the motion continues further along the escape trajectory these angles can only decrease: "*the winds blows over one's shoulder*." But we know that the segment on which escape velocity is attained is not at all necessarily "with the wind behind" – the opposite may be the case if the point of departure on the initial trajectory is chosen wrongly.

> "And then, in the fiftieth hour of the race, just after the end of the second orbit around Earth, Markoff sprang his little surprise.
>
> "'Hello, John,' he said casually over the ship-to-ship circuit. 'I'd like you to watch this. It should be interesting.'"
>
> "Merton drew himself across to the periscope and turned up the magnification to the limit. There in the field of view, a most improbable sight against the background of the stars, was the glittering Maltese cross of *Lebedev*, very small but very clear. As he watched, the four arms of the cross slowly detached themselves from the central square, and went drifting away, with all their spars and rigging, into space.
>
> "Markoff had jettisoned all unnecessary mass, now that he was coming up to escape velocity and need no longer plod patiently around the Earth, gaining momentum on each circuit. From now on, *Lebedev* would be almost unsteerable – but that did not matter; all the tricky navigation lay behind her. It was as if an old-time yachtsman had deliberately thrown away his rudder and heavy keel, knowing that the rest of the race would be straight downwind over a calm sea."

About the error in measuring time ("*fiftieth hour of the race*") we already spoke above. But here is a question: how effective is Markoff's maneuver of jettisoning part of the sail? (The other, square part remained.) A hint to the answer is provided by the fact that "*old-time yachtsmen*" would throw away the heavy keel, but not the propelling sail. We should add that Markoff did jettison in space not only the excess sail, but also the copilot of his yacht (who had to follow his ship by foot!).

> "Merton ... was too busy doing some hurried calculations, based on what he knew of Lebedev's design. By the time he had finished, he knew that the race was still in doubt."

It would be interesting to follow Merton in his calculations. Will the acceleration of Markoff's yacht increase? The solution to this problem is left in the hands of the reader.[3]

[3]The first serious investigations of the problem of space flight with the assistance of solar radiation pressure were undertaken by F. A. Zander (1887–1933) in 1924–1925. The results of these investigations were published in his book *Problems of Flight of Jet-Driven Spacecrafts: Interplanetary Flights,* 2nd edition, Oborongiz, Moscow, 1961. See also the following books: A. C. Clarke, editor, *Project Solar Sail*, Penguin Books, New York, 1990; L. Friedman, *Starsailing: Solar Sails and Interstellar Travel*, J. Wiley, New York, 1988; C. R. McInnes, *Solar Sailing: Technology, Dynamics and Mission Applications*, Springer Verlag, Berlin, New York, 1999; E. N. Polyakhova, *Space Flight Using a Solar Sail*, Nauka, Moscow, 1986; J. L. Wright, *Space Sailing*, Gordon and Breach, Philadelphia, 1992. The interested reader may also visit the Solar Sail homepage at http://www.kp.dlr.de/SolarSail/ and many similar sites.

Ninth Essay

The Gravity Flyer

... The gauge indicator was showing that the tank was dry, with not even a drop of fuel left ... Krug stood motionless for several minutes, getting used with this thought. No fuel. Then how the hell will he get to the spaceship? And at that moment he realized for the first time that perhaps he will not get there at all ...

... And then he acted in an illogical, yet ordinary way ... doing what any drowning person does: he began swimming. He swam, although that was extremely stupid – to swim in space, where there is nothing to push away from! He threw his hands forward and moved his legs the way a swimmer does. He dutifully gathered in the vacuum with his wide armor-covered palms, stirring it with his heavy legs, and swam, swam ...

B. Mikhaĭlov, *Among Stars*

1. Force of attraction on a body of non-negligible dimensions

Ordinarily there is no need to remember that a satellite flying in an orbit around the Earth is a body, and not a particle (point mass). As a rule, in the calculation of orbits this fact is neglected. But a fact remains a fact: the Earth's force of attraction acting on a body is different from the force acting on a particle with the same mass as the body and placed at the same distance from the center of the Earth as the center of mass of the body. This obvious, but easy to forget fact can have surprising consequences!

For instance, let us consider a dumbbell shaped satellite (Figure 9.1) with two identical balls, each of mass $m/2$; we neglect the mass of the connecting rod. Suppose that the rod is perpendicular to the line connecting the center of the dumbbell and the center of the Earth. Denote by r the distance from the center of mass of the satellite (i.e., the middle of the rod) to the center of the Earth, by l half of the rod's length, and by $R = \sqrt{l^2 + r^2}$ the distance from a ball to the center of the Earth. Each ball is acted upon by the Newtonian force of attraction determined by the force functions $U_1 = U_2 = \mu m/(2R)$, and the total force function is

$$U = \frac{\mu m}{\sqrt{l^2 + r^2}} = \frac{\mu m}{r\sqrt{1+\alpha^2}}, \qquad \alpha = \frac{l}{r}. \tag{9.1.1}$$

If the satellite is not of supergigantic dimensions, the quantity α (and the more so α^2) is very small (compare, say, $l = 7$ m and $r \approx 7,000,000$ m). Hence, the force function U is close to the force function $U_{\text{N}} = \mu m/r$ of the attraction exerted by the Earth on a particle m placed at the center of mass of our dumbbell. Each ball is acted upon by the Newtonian force of magnitude

$$F_i = \frac{\mu m}{2R^2} = \left|\frac{\partial U_i}{\partial R}\right|. \tag{9.1.2}$$

Both forces $\mathbf{F}_i$ are directed toward the center of the Earth. The resultant force $\mathbf{F}$ is obtained from the parallelogram of forces (Figure 9.1) and has the magnitude

$$F = \left|\frac{\partial U}{\partial r}\right| = \frac{\mu m}{r^2}\,\frac{1}{(1+\alpha^2)^{3/2}}. \tag{9.1.3}$$

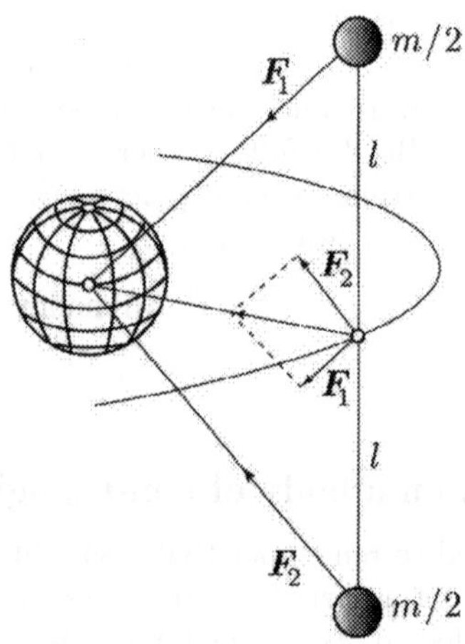

FIGURE 9.1. Newtonian center of attraction acting on a dumbbell

In our scheme the resultant force $\mathbf{F}$ is directed toward the center of the Earth and is almost identical to the "ordinary" Newtonian force $F_{\text{N}} = \mu m/r^2$. Almost, but not completely! In other words, the effect of taking into account that the body is extended is as if an additional repelling radial force is introduced. And although in the case of small satellites this additional force is practically imperceptible, it is nevertheless present! And for large satellites it may become more notable. This fact lies at the foundation of a new and somewhat unexpected method of maneuvering in space, which was proposed by the author and M. E. Giverts [9.1], [9.2]. (A similar idea was independently proposed by J. F. Schaefer in [9.8]).

2. A pulsating spaceship

Our reasoning can be summarized in three assertions:

(1) The force of gravity acting on a body of non-negligible (finite) dimensions differs from the force acting on a particle with the same mass, placed at the center of mass of the body.

(2) Modifying the dimensions and the shape of the body one can modify the magnitude of the gravity force acting on it.

(3) Such modifications of dimensions and shape of the body can be controlled in such a manner that the resulting variations of the gravity force will lead in time to a trajectory that differs substantially from the initial trajectory.

The first of these assertions is obvious, and the second is a consequence of the first. The third assertion will be proved below.

In the general case, the motion of a body (in contrast to that of a particle) in a Newtonian force field cannot be described in closed analytic form and is complicated due to the interaction between the motion of the body's center of mass and the motion around the center of mass. However, as shown in our book [9.3], a symmetric body can move in a Newtonian force field so that its symmetry axis is permanently perpendicular to the orbital plane. Such a situation is indeed depicted in Figure 9.1. The force that acts on the body turns out to be central, depending only on the distance from the body's center of mass to the center of attraction, and is defined by a force function $U(r)$. The motion takes place along a planar orbit and is completely determined by the area and energy integrals:

$$r^2 \frac{d\varphi}{dt} = c, \tag{9.2.1}$$

$$\frac{V^2}{2} - \frac{U(r)}{m} = h. \tag{9.2.2}$$

Let us introduce the osculating focal parameter $p(t)$ and the osculating eccentricity $e(t)$ of the orbit. The radius vector r and the velocity V of the osculating orbit are expressed as

$$r = \frac{p}{1 + e\cos\nu}, \qquad V = \sqrt{\frac{\mu}{p}\left(1 + e^2 + 2e\cos\nu\right)}. \tag{9.2.3}$$

Here ν denotes the true anomaly in the perturbed motion. By (9.2.1), the focal parameter p is constant in the perturbed motion as well – this is a well known property of motion in an arbitrary central force field.

Let us calculate the quantity

$$\frac{V^2}{2} - \frac{\mu}{r} = \frac{\mu}{2p}\left(1 + e^2 + 2e\cos\nu\right) - \frac{\mu}{p}\left(1 + e\cos\nu\right) = \frac{\mu}{2p}\left(e^2 - 1\right). \tag{9.2.4}$$

Using (9.2.4) and the constancy of p, the integrals (9.2.1) and (9.2.2) can be written as

$$p = p_0, \tag{9.2.5}$$

$$e^2 + \frac{2p}{\mu}\left[\frac{\mu}{r} - \frac{1}{m}U(r)\right] = \overline{h}, \tag{9.2.6}$$

where $\overline{h}$ is a new constant. In the case under study $U(r)$ is given by formula (9.1.1) and after some transformations the integral (9.2.6) can be recast as

$$e^2 + \frac{2p}{r}\,\frac{\alpha^2}{1+\alpha^2+\sqrt{1+\alpha^2}} = \overline{h}, \qquad \alpha = \frac{l}{r}. \tag{9.2.7}$$

This shows that e is a function of r and is not constant. During the motion the osculating eccentricity $e(t)$ and the radius vector $r(t)$ vary so that relation (9.2.7) holds. Therefore, we can consider the motion governed by (9.2.7) in the plane with coordinates e^2, r/p. Notice that the actual motion cannot take place in the whole plane e^2, $\overline{r}$ (where $\overline{r} = r/p$), being confined to a certain domain. Indeed, the first of the equalities (9.2.3) yields $e^2\cos^2\nu = (p/r-1)^2$, whence

$$e^2 \geq \left(\frac{p}{r} - 1\right)^2 \equiv e_*^2(\overline{r}). \tag{9.2.8}$$

We see that in the plane e^2, $\overline{r}$ the motion can take place only above the curve $e_*^2(\overline{r})$ defined by the equation $e_*^2 = (\overline{r}^{-1} - 1)^2$. Clearly, $e_*^2(\overline{r}) \to \infty$ when $\overline{r} \to 0$, $e_*^2(1) = 0$, and $e_*^2(\overline{r}) \to 1$ when $\overline{r} \to \infty$. Moreover, $e_*^2(1/2) = 1$.

Let us construct in the region of real motions (9.2.8) the curve (9.2.7) for fixed p and $\overline{h}$ (Figure 9.2). When r increases from 0 to ∞ the eccentricity increases monotonically from $e^2 = -\infty$ to $e^2 = \overline{h}$. The real motion is given by the arc of the curve (9.2.7) which lies in the domain defined by inequality (9.2.8). The intersection of the curve (9.2.7) with the boundary curve $e_*^2 = e_*^2(\overline{r})$, defined by the relation that follows from (9.2.8), gives the extremum points of the trajectory: $\overline{r} = \overline{r}_{\min}$ (and then $e = e_{\min}$) and $\overline{r} = \overline{r}_{\max}$ (and then $e = e_{\max}$). The motion of the body is periodic with respect to e and $\overline{r}$, so that in the plane e^2, $\overline{r}$ it takes place, for example, along the arc 1–2 (Figure 9.2), from the point 1 ($e = e_{\min}$, $\overline{r} = \overline{r}_{\min}$) to the point 2 ($e = e_{\max}$, $\overline{r} = \overline{r}_{\max}$), then back from the point 2 to the point 1, and so on. If instead of a body we would have a particle of the same mass (i.e., if we would set $l = 0$), then, as follows from (9.2.7), the particle would move in an unperturbed Keplerian elliptic orbit $e =$const (for example, to and fro along the segment 2–3 in Figure 9.2).

Thus, it turns out that in the motion of a dumbbell normal to its orbital plane the eccentricity oscillates periodically between its maximum and minimum values. When the dumbbell is in its closest position to the Earth the eccentricity reaches its minimum; in the point of the orbit farthest from the Earth the eccentricity reaches its maximum. These oscillations of the eccentricity are not large, and as

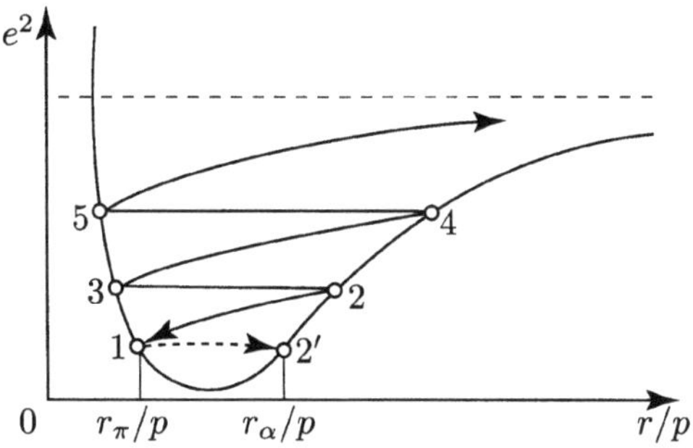

FIGURE 9.2. Diagram of the motion of a pulsating dumbbell spaceship

a whole the orbit of the dumbbell resembles a Keplerian ellipse; the dumbbell satellite will move inside a bounded neighborhood of Earth, without distancing itself from it.

Suppose one is able to modify the eccentricity of the orbit in a noticeable manner. Then by means of successive modifications of the eccentricity the original elliptic ($e < 1$) orbit could be transformed into a hyperbolic orbit ($e > 1$), and as a result the satellite will fly away to remote space. But for a satellite regarded as a particle the eccentricity of the orbit does not change; and in the case of the dumbbell satellite that we have just analysed, although the eccentricity changes, the change is periodic and small. There is no systematic variation of the eccentricity. So, if the satellite has bounded dimensions no substantial change of the eccentricity can be achieved.

But let us make the following assumption (and here we reach the most important point of our exposition): our dumbbell is capable of pulsating; at requisite moments of time it can shrink to a point or instantaneously extend to full length. Incidentally, the feasibility of instantaneous pulsations is not essential and is assumed here only for the sake of simplicity; one can also consider slow variations of the length of the dumbbell.

So, let our dumbbell start its motion from the closest-to-Earth point and traverse a half-turn of orbit to the farthest-to-Earth point (along the segment 1–2 in Figure 9.2). At that point (2) the eccentricity – as we observed above – reaches its maximum value ($e = e^{(1)}_{\text{max}}$, $\overline{r} = \overline{r}^{(1)}_{\text{max}}$). There we bring together the two sections of the dumbbell, instantaneously reducing its length to "zero" (i.e, the length of the "folded" dumbbell satellite is negligibly small compared with the length of the "extended" satellite). From this moment on the satellite will move as a particle, that is, along an arc of Keplerian ellipse (along the segment 2–3 in Figure 9.2), preserving the newly-attained value of the eccentricity $e = e^{(1)}_{\text{max}}$.

Returning to the closest-to-Earth point (which, of course, will differ from the first "closest-to-Earth" point), the dumbbell "carries with it" the previously acquired value of the eccentricity. At that point we instantaneously extend the

dumbbbell satellite to its full length! Now on the next half-turn of orbit, because the satellite is extended, the value of the eccentricity will again increase (along the segment 3–4 in Figure 9.2). When we reach the farthest-to-Earth point for the second time, the eccentricity $e = e_{\max}^{(2)}$ will be larger than for the first farthest-to-Earth point ($e = e_{\max}^{(2)} > e = e_{\max}^{(1)}$). Next, we again collapse the satellite and retain the acquired value of the eccentricity until we again extend the dumbbell, and so on. Eventually, after multiple pulsations of the dumbbell, one can achieve a substantial modification of the eccentricity of the orbit, and possibly even attain the hyperbolic value of the eccentricity ($e > 1$) and escape the Earth's gravitational field.

As a result of the process described above, the energy spent in order to "switch on" and "switch off" the dumbbell is pumped into energy of orbital motion of the satellite and allows us to accelerate the satellite to the point that it leaves the Earth's gravitational field. The corresponding orbit of the satellite is a spiral with the number of turns equal to the number of pulsations of the dumbbell. A spaceship whose orbit changes as a result of the variation of the gravitational forces acting on it will be called here a *gravity flyer*.

The internal forces used to "switch on" and "switch off" the dumbbell produce a notable external effect because (and only because) there exists an external force field with which they interact. In order for this mechanism to work a certain resonant tuning must be achieved between the external and internal forces. Isn't this what the baron Münchhausen had in mind in his famous story?

3. Left behind by your spaceship? Swim breaststroke!

Isn't this a truly captivating image: instead of a jet engine, use an electric motor to extend or shrink, at the right moment, the rod that connects the balls of the dumbbell, or even ... assign this task to a "sailor"?

The actual estimates will cool down somewhat the burning enthusiasm generated by this image. Over n cycles (of switching on and off the dumbbell), the eccentricity will change its value to roughly $e_n^2 \approx e_0^2 + n(l/p)^2(1+e_0^2)$ (from its original value e_0). Hence, to attain a value $e \sim 1$ it is necessary to perform $n \sim (p/l)^2$ cycles (orbital revolutions). If for the initial orbit $p = 10,000\,\mathrm{km}$ and $l = 1\,\mathrm{km}$, then $n \sim 10^8$ revolutions. Since even at the level of the Earth's surface one revolution of a satellite takes an hour and a half, and since the farther one is from the Earth the larger the period of revolution, it follows that the flight duration required for performing that many revolutions is $T > 1.5 \cdot 10^8$ hours, or about 20,000 years! No "sailor" would enlist for such a boring job. Note that the ratio of the perturbing acceleration to the background acceleration of gravitation for the initial orbit is of order $f \sim (l/p)^2$.

However, if for the same orbits one consider a dumbbell of length $l = 10\,\mathrm{km}$, or even $l = 100\,\mathrm{km}$, then we have $f \sim 10^{-6}$ or $f \sim 10^{-4}$, respectively. The acceleration generated for such values of l is already of the same order of magnitude as that generated by a ion or plasma low-thrust engine. The price paid is of course that

the spacecraft must be tens of even hundreds of kilometers long! For $l = 100\,\text{km}$ the number of orbital laps needed in order to attain escape velocity is cut down to $n \sim 10^4$, and the lower estimate of the escape time, which is now reduced to about two years, does at least no longer sound ridiculous; but one has to keep in mind the problems attached with extending and contracting a rod hundreds of kilometers long. If we let our imagination run wild and conceive a 1,000 km long spacecraft (there is no lack space in cosmos!), then such a spacecraft would attain escape velocity after as little as 100 orbital laps, which should take about a week of flight!

Let us mention yet another circumstance. The efficiency of a spacecraft is the larger the greater the mass of the attracting central body is and the closer the spacecraft can be to the center of that body. The Sun's mass is large, but so is its size. One can fly closer to the center of Earth, but the mass of Earth is relatively small. However, in the infinite domains of the cosmos there are many stars that seem to have been deliberately designed for maneuvering a gravity flyer in their vicinity. They are the "white dwarfs," whose dimensions are comparable with those of a planet, but whose mass is comparable with that of the Sun. In the vicinity of a white dwarf the escape time from the star's sphere of attraction is tens or hundreds of times shorter than the escape time in the vicinity of Earth.

Table 9.1 shows the escape characteristics of a 140 km long spacecraft (i.e., the half-length of the gravity flyer is $l = 70\,\text{km}$) in the vicinity of the Earth, the Sun, and a white dwarf. Perhaps the inhabitants of planetary systems of white dwarfs use gravity flyers and not rockets for their space travel?

Let us add that it is not obligatory to interpret "pulsations" as changes in the shape of the spacecraft. The same results can probably be achieved by letting a mass of fluid pulsate inside a spacecraft of fixed shape [9.4].

It is probably possible to manage altogether without pulsations, provided that at requisite moments of time one makes the spacecraft spin in some way or another (using fly-wheels, for instance). Indeed, the force of attraction depends also on how the spacecraft is positioned with respect to the Earth. Then the energy spent to spin the spacecraft is transformed into energy that modifies its orbit; the spacecraft itself does not change shape, which of course is more convenient than using a pulsating gravity flyer. But the principle remains the same – the orbit of the gravity flyer changes as a result of varying the force of gravity that acts on it.

Celestial body	d_0	$f \sim (l/p)^2$	$\#_{\text{E}}$	T_{E}
Earth	$7 \cdot 10^3$	10^{-4}	10^4	at least 2 years
Sun	$7 \cdot 10^5$	10^{-8}	10^8	at least 80 years
White dwarf (Sirius B)	$2 \cdot 10^4$	10^{-5}	10^5	a few hors (no less than 1.5 hours)

TABLE 9.1.

In Table 9.1 d_0, $\sharp_{\mathrm{E}}$, and T_{E} denote the initial distance of the spacecraft to the center of the celestial body (in kilometers), the number of orbit turns till escape, and the escape time, respectively.

Our goal here, however, is not the engineering design of such a spacecraft, but proving that in principle it is possible to modify the orbit of a pulsating body. Again in principle, an astronaut that was left behind by his spaceship and lost his individual rocket engine (a situation much loved by science fiction writers) may catch up with the spaceship by modifying his orbit using purposeful pulsations. Let him swim breaststroke! But don't ask me how much time this will take!

4. The gravity flyer and the readers

The publication of the papers [9.1] and [9.2] on the gravity flyer provoked a quite animated reaction from part of the readers. The engineer I. Troitskiĭ from Tula wrote to the to the editors of the magazine *Tekhnika–Molodezhi* the following comment in connection with the paper of V. Beletsky and M. Giverts, published in the journal *Kosmicheskie Issledovaniya*: "This communication produced a stunning impression on me. Already from our school years we are used with the idea that redistributing the mass inside a 'closed system' (for example, a satellite), has no influence on the trajectory of its motion. As it turns out, this is not at all the case... A very real possibility arises – by varying the shape of a spaceship, or by pumping a fluid from one place to another inside it, or by special maneuvers one can modify the flight direction and embark upon a long trip. *Kosmicheskie Issledovaniya* is a specialized scientific journal of limited circulation... It would not be a bad idea to ask one of the authors to write a popular-science article addressed to the general reader." In response, *Tekhnika–Molodezhi* published the paper [9.2]. And immediately the editors received the following letter from an angry Moscow engineer.

"Dear comrades! In the 3rd issue of the current year you published the paper of professor V. Beletsky '*Gravity flyer*' ... The arguments given in that paper are based simply on ignorance and witness to the wrong practice of awarding scientific titles to people who have no representation whatsoever about the most elementary matters... It is about time to disrobe a few doctors who refuse to accept obvious, experimentally established laws, among them the author of the '*Gravity flyer*'."

How about that for a letter!

A reader from Dzhezkazgan also found impossible to "keep silent about the completely false interpretations of V. Beletskiĭ:" "The internal forces in a spacecraft can have no influence on the Newtonian force of attraction exerted on the spacecraft by the Celestial Body. Hence, professor Beletskiĭ's 'dumbbell' must occupy, both before and after its unfolding, a position relative to the Celestial Body such that its mass is always at a constant distance from the center of the latter, because this Newtonian force keeps the mass of the spacecraft 'tied' to the Celestial Body. For this reason the cunning fabrications of professor and doctor in physico-mathematical sciences V. Beletskiĭ appear to be the result of his complete lack of understanding of the laws of celestial mechanics, in the realm of which he probably 'swims breaststroke'."

The author has all the respect for the opinions of his readers and understands that "*Nature cares nothing about our academic titles; in front of Nature man is like a poor naked soul.*"[1] But both the fervent and the indignant readers were victims of two misunderstandings:

The roots of the first misunderstanding are in the thesis "internal forces are not capable of changing the motion of the center of mass of a system." This is of course true, but only in the *total absence* of external forces. However, in real conditions there always are external forces acting on a given system. And in many cases the motion is determined by precisely the interaction between internal and external forces. One example is the railway steam engine (friction between the wheels and the rail!). Another example is the gravity flyer.

As for the second misunderstanding, its origin lies in the fuzzy understanding of the fact that Newtonian gravity acts on *extended* bodies (i.e., not on point particles) and for that reason the motion of those bodies, strictly speaking, is not Keplerian.

Let us give a vivid example of the "gravity flyer" effect [9.3].

The equations of motion of a dumbbell-shaped body in a central Newtonian force field admit the following exact particular solution: the center of mass of the dumbbell moves in a planar circular orbit of arbitrary constant radius R_0, the center of the orbit coinciding with the center of attraction; during this motion the dumbbell itself is all the time aligned along the current radius vector of the orbit. The velocity V of the center of mass of the dumbbell is given by the formula

$$V^2 = \frac{1+\alpha^2}{(1-\alpha^2)^2} \cdot \frac{\mu}{R_0},$$

where $\alpha = l/R_0$ and l is half of the length of the dumbbell. This circular orbit is *not* Keplerian (for a Keplerian orbit the velocity is given by the formula $V^2 = \mu/R_0$.) When $\alpha \geq 0.47$, it turns out that $V^2 \geq 2\mu/R_0$, i.e., V is larger than the escape velocity. Therefore, if a dumbbell with such a value of α is "contracted" to a point mass, then it will tear itself away from the circular orbit and fly to infinity. For this to happen, however, the dumbbell's size must be of the same order of magnitude as the size of the orbit! For a spacecraft of moderate size one can, as we saw above, use the same principle, replacing the one-time contraction of the dumbbell by intelligently programmed multiple "switch on" and "switch off" operations.

It is important to emphasize here that the proof of the feasibility of gravity-flyer type maneuvering was established not by verbal reasoning, but by solving the rigorous equations of motion in closed form.

---

[1] This aphorism belongs to the well-known Siberian geologist and academician V. P. Solonenko (see the newspaper *Vostochno-Sibirksaya Pravda*, December 18th, 1966).

5. The gravity flyer as a resonance phenomenon

The gravity flyer offers an example of the amazing laws governing resonance phenomena. The control of the gravity flyer is essentially resonant, which is precisely the meaning of the expression "intelligent control" used above. Indeed, the entire cycle of pulsations of the gravity flyer coincides with the duration of a complete orbital revolution. In other words, we are dealing with an $1:1$ resonance between two characteristic frequencies of a system. For other resonance relations between characteristic frequencies the evolution of the orbit may possibly take place at a slower rate. If the absence of a resonance, no significant evolution of the orbit could be achieved, in much the same way as by dangling our legs in a disordered manner one cannot make a swing actually swing.

The resonant nature of the gravity-flyer concept was successfully demonstrated by A. B. Mitnitskiĭ, whose results are discussed below.[2]

Let us confine our discussion to planar motion and a planar design of a gravity flyer that is dynamically-symmetric with respect to the axis normal to the orbital plane.

Then in the force function (6.1.6), which describes with high accuracy the action of the Newtonian force of attraction on the gravity flyer, we must put $C = A + B = J$, $\gamma'' = 1$, which yields

$$U = \frac{\mu m}{r} - \frac{1}{2}\frac{\mu}{r^3}J. \tag{9.5.1}$$

The equations of motion in polar coordinates r, φ have the form

$$\ddot{r} = -\frac{\mu}{r^2} + \frac{c^2}{r^3} - \frac{3\mu J}{2mr^4} \tag{9.5.2}$$

$$r\dot{\varphi} = c. \tag{9.5.3}$$

Further, for the angle ψ between two fixed directions, one in space and one in the gravity flyer, we obtain the equation

$$\dot{\psi} = \frac{c_1}{2J} - \frac{c}{r^2}. \tag{9.5.4}$$

In equations (9.5.2)–(9.5.4) c and c_1 are constants.

The moment of inertia J is used as the control parameter – it varies due to the pulsations of the gravity flyer. We require that $0 < J_{\min} \leq J \leq J_{\max}$ and we will assume that $J = J(\varphi) = J_0 \cdot f(\varphi)$ is an explicit function of the polar angle φ.

Let us perform the so-called *Binet transformation*, introducing the new variable $u = p/r$, where $p = c^2/\mu$, and transform equation (9.5.2), choosing φ as the new independent variable, with (9.5.3) taken into account. This yields the equation

$$\frac{d^2u}{d\varphi^2} + u = 1 + \varepsilon f(\varphi)u^2, \qquad \varepsilon = \frac{3J_0}{2mp^2}. \tag{9.5.5}$$

[2]The author thanks A. B. Mitniskiĭ for allowing him to use this material.

Now let us choose a concrete form for the function $f(\varphi)$. By the setting of the problem, $f(\varphi) > 0$. To account for the "pulsations" of the gravity flyer we must require that $f(\varphi)$ be a periodic function of its arguments. Let us choose the simplest rule that satisfies all these requirements:

$$f(\varphi) = 1 + \delta \sin(\alpha\varphi - \beta). \tag{9.5.6}$$

Here the constant δ obeys the condition $0 < \delta < 1$, and the constant α and β are arbitrary for the moment.

Let us examine equation (9.5.5). For $\varepsilon = 0$ ($J_0 = 0$) we obtain the case of the motion of a particle in the Newtonian force field and, naturally, the solution of equation (9.5.5) corresponding to Keplerian orbits. This solution reads

$$u = 1 + e\cos(\varphi - \omega), \tag{9.5.7}$$

where e and ω are interpreted as the eccentricity of the orbit and the longitude of the perigee, respectively. Recalling that $r = p/u$, we conclude that p is the focal parameter of the orbit. The parameter ε is very small. When $\varepsilon \neq 0$ it is natural to seek the solution of equation (9.5.5) in the same form (9.5.7), where now e and ω are no longer constants. The problem is then reduced to finding the functions $e(\varphi)$ and $\omega(\varphi)$. One can express this by saying that (9.5.7) is a *generating solution* for the sought one.

But we prefer to improve somewhat the generating solution based on the following considerations. Our motion is equivalent to the motion of a particle in some central, but not Newtonian force field (9.5.1). An unavoidable consequence of the fact that the field is not Newtonian is that the orbit "opens," in the sense that its perigee undergoes a systematic drift. It is advisable to already take this effect into account, at least partially, in the generating solution. To that end we put $u = 1 + q$ and substitute this expression in (9.5.5), using also (9.5.6). This yields the following equation for q:

$$\left.\begin{aligned} &\frac{d^2q}{d\varphi} + \lambda q^2 = \varepsilon F(q,\varphi), \qquad \lambda^2 = 1 - 2\varepsilon, \\ &\quad F(q,\varphi) = 1 + q^2 + \delta\sin(\alpha\varphi - \beta)\left[1 + 2q + q^2\right]. \end{aligned}\right\} \tag{9.5.8}$$

As the generating solution for equation (9.5.8) we may again take the solution of the same equation with $F \equiv 0$:

$$q = e\cos(\lambda\varphi - \omega), \qquad q' = -e\lambda\sin(\lambda\varphi - \omega). \tag{9.5.9}$$

The corresponding trajectory in the original polar variables is given by

$$r = \frac{p}{1 + e\cos(\lambda\varphi - \omega)}.$$

Since $\lambda \neq 1$, this trajectory is obviously a nonclosed rosette of the kind shown in Figure 1.12 (a). Next, let us seek the solution of equation (9.5.8) in the form (9.5.9), where e and ω are no longer constant. We are now in an already familiar situation, leading to equations of the type (1.6.4):

$$\left.\begin{aligned} \frac{de}{d\varphi} &= -\frac{\varepsilon}{\lambda} F(e\cos(\lambda\varphi - \omega), \varphi)\sin(\lambda\varphi - \omega), \\ \frac{d\omega}{d\varphi} &= \frac{\varepsilon}{\lambda e} F(e\cos(\lambda\varphi - \omega), \varphi)\cos(\lambda\varphi - \omega). \end{aligned}\right\} \tag{9.5.10}$$

In view of the structure of the function $F(q, \varphi)$ (see (9.5.8)), only the following resonance relations between the frequency α of pulsations of the gravity flyer and the frequency λ of its orbital motion are possible:

$$\alpha = k\lambda, \qquad k = 1, 2, 3. \tag{9.5.11}$$

Applying the standard averaging procedure to equations (9.5.10) we reach right away the following conclusion: if α is not close to λ, 2λ, or 3λ, then the orbit does not evolve – the mean values of the right-hands sides of equations (9.5.10) are equal to zero and $e \approx e_0$, $\omega \approx \omega_0$. Conversely, if one of relations (9.5.11) holds then the orbit undergoes a substantial evolution that in the first approximation of the averaging method is described by the equations

$$\left.\begin{aligned} \frac{de}{d\varphi} &= -\frac{\varepsilon\delta}{2\lambda} J_e^k(e)\cos(\omega - \beta), \\ \frac{d\omega}{d\varphi} &= \frac{\varepsilon\delta}{2\lambda} J_\omega^k(\omega)\sin(\omega - \beta), \end{aligned}\right\} \tag{9.5.12}$$

where the expressions $J_e^k(e)$ and $J_\omega^k(\omega)$ take the following form for the resonances $k = 1, 2, 3$:

$$\left.\begin{aligned} J_e^1 &= \left(1 + \frac{1}{4}e^2\right), & J_\omega^1 &= \frac{1}{e}\left(1 + \frac{3}{4}e^2\right), \\ J_e^2 &= e, & J_\omega^2 &= 1, \\ J_e^3 &= \frac{1}{4}e^2, & J_\omega^3 &= \frac{1}{4}e. \end{aligned}\right\} \tag{9.5.13}$$

From (9.5.13) and (9.5.12) it follows that the highest rate of evolution of the orbit with respect to the eccentricity corresponds to the resonance 1 : 1. In this case $de/d\varphi \sim \varepsilon$, whereas for the resonances 1 : 2 and 1 : 3 we have $de/d\varphi \sim \varepsilon e$ and $de/d\varphi \sim \varepsilon e^2$, respectively.

The system (9.5.12) can be integrated by quadratures. However, as this point this is not so interesting: our aim here was to convince the reader of the resonance nature of the gravity-flyer effect.

6. The gravity flyer and writers

Not only was the concept of the gravity flyer developed further in the scientific literature (see [9.5]–[9.8]), but it also found a place in the popular science literature, and even in works of fiction. The journal *Kvant* responded by publishing in its 2nd 1974 issue the paper *An extraordinary trip* by I. I. Vorob′ev. Vorob′ev reformulated the gravity flyer problem in such a manner that it became possible to treat it in an elementary mathematical language: the trajectory of the gravity flyer is obtained by gluing together segments of purely Keplerian trajectories. (An inessential error found its way in Vorob′ev's paper: instead of $\cos\frac{\alpha}{2}$ should be $\cos^2\frac{\alpha}{2}$.) In his popular-science book *The New Practical Astronomy* (Nauka, Moscow, 1972) V. N. Komarov devoted a special section to the concept of gravity flyer. But perhaps the most vivid representation of this concept appeared in the pages of the 2nd 1972 issue of the journal *Zemlya i Vselennaya*, in Komarov's science-fiction story *Gambit*. Here is an excerpt from that story:

> The transport starship *Omicron* was on a routine trip to Megos, carrying a crew of 12 and 360 passengers. Captain Meng and the navigator Gascondi silently looked at the monitor, both reaching the unavoidable conclusion that the situation was desperate ... The error had occured the moment the ship left the hyperspace. Something in the ship's complex automatic control system had failed. A minute error in the program, a random fluctuation, which nevertheless was enough to make the starflyer find itself five parsecs from the computed target point ... And here waited a white dwarf – a tiny star with huge density and powerful gravity.
>
> All engines were turned to full power. This managed to save *Omicron* from falling into the abyss, but not to break the chains of the dwarf's gravitational attraction. Presently the ship moved on a closed orbit at a mean distance of twenty thousand kilometers from the center of the star, and the full force of the engines was not enough to escape. On top of all this, the energy supply was nearly exhausted. Time was running out.
>
> ... The dwarf was too close. Although the ship was protected by its shield, Meng almost physically felt the hot breath of the star. For the moment they were safe ... But in six and a half hours the energy supply will run dry, and then ...
>
> During his long service in the space fleet captain Meng had found himself in a critical situation more than once. But those were situations in which there was a way out, in which everything depended on the experience and inventiveness of the person in command, since an optimal course of action had to be found in a few seconds. And until today Meng had always succeeded in finding one.
>
> But now there was no way out, a fact silently demonstrated by a simple computation that any student could have carried out.
>
> ... No, one nevertheless had to fight. One shouldn't give up under any circumstances. Even if is hopeless ....

"... We need to check all possible variants again." "But this is an elementary case!" burst out Gascondi. "What variants can there be?"

Meng understood that as well as his navigator. This was a classical situation that had been thoroughly studied at the dawn of space flights, a situation in which everyone had lost interest many years ago. New navigation tools had freed astronauts from such horrors. At least in the last fifty years no spaceship had fallen into a gravity trap. Only *Omicron* found itself out of luck ...

But perhaps precisely the fact that no one did theoretical work on this problem anymore offered a unique chance? Science does not step in place. And with a fresh look at their hopeless situation, armed with contemporary knowledge, perhaps they could find a variant overlooked by the classical art of navigation.

At any rate, one had to search. But how to convince Gascondi? Meng stood up and walked to the navigator's chair:

"Let's think together. And what if ..."

He did not notice Vel entering the deck cabin, and saw him only as he stood near the main control panel watching the monitor, and narrowing his eyes to fight his near-sightedness.

As a rule, passengers were strictly forbidden to enter the commanding deck. But Vel was not a simple passenger. At the foundation of *Omicron*'s design lay a scientific theory he had created. Vel was the author of countless original ideas that had exerted a notable influence on the development of physics and astrophysics. On Megos he was scheduled to deliver a course on the theory of hyperspace at the local university.

Nonetheless, on *Omicron* Vel was a passenger, and Meng anxiously understood that their desperate position was no longer a secret.

"An interesting situation, isn't it?"

These words sounded quite strange, not only because of the mounting danger of their situation, but also because they were uttered with a nuance of both sarcasm and hard to understand satisfaction.

Gascondi answered with a shrug.

"Not enough power," asked Vel, finally tearing himself away from the control panel.

"As you can see," barked out Gascondi, rather brusquely.

"And the radiation shield will be down in a few hours?"

"In six and a half," answered Meng absentmindedly.

"So," vaguely muttered the theoretician. "Hmm, so ..."

Sparks of excitement lit in his deeply set eyes, and at that instance he reminded Meng of a hunter who unexpectedly faces his game. Apparently Vel was not at all interested in the fact that, in the current situation, he was the game ...

With a habitual surge of determination, Vel ignored everything but the data of the unusual problem posed by the absurd concurrence of circumstances. But all his life Vel had solved precisely this kind of problems.

"May I use your computer?" he asked, leaving his reverie for a moment.

...Meng and Gascondi waited silently. Finally, Vel tore himself away from the keyboard and took a deep breath showing neither relief nor disappointment; but the serene sparks were again flickering in his gray eyes.

"Do you ever play chess?" he inquired in a dull voice.

"Yes," said Meng.

"Do you know what a gambit is? The position is decidedly lost, but there is a path that seems to lead the most rapidly to defeat. And it is exactly this strange path that in fact brings victory..."

At that moment Meng new for sure that Vel had found a way out.

"Well, what of that?" Meng asked, no longer capable of keeping his impatience in check.

"...we must turn the propulsion on," said Vel. He rapidly scribbled a few numbers on a piece of paper and handed it to Meng.

"But this will achieve nothing," mumbled Gascondi, confused. "It will only make the orbit more oblate."

"Exactly, exactly," said Vel.

...But deep in his heart Meng trusted Vel. Without hesitating, he reached for the main instrument board and successively turned four colored control levers a few notches.

Gascondi turned pale. The air filled with the characteristic hum of the engines and the clicks of the anti-overload protection relays.

"Perhaps its time for you to explain?" asked Meng.

"If I am not mistaken," began Vel slowly, "*Omicron* consists of two individual sections?"

"Yes," confirmed Meng. "One holds the command complex and the engines, and the other the cabins and the auxiliary premises."

"And these sections can be separated and moved a considerable distance away from one another?"

"Yes, this ability was envisaged as desirable in case of an accident or a need to repair the power plants. The two units can be brought closer or farther apart by means of a special 'pulsator'."

"And what is the maximum separation that can be attained?"

"One hundred and fifty kilometers."

"One hundred and forty will do," muttered Vel.

"You want to jettison the passenger compartment?" Gascondi finally began to speak. "But the thrust will still be insufficient."

"No," retorted Vel energetically. "That would be too simple. The dwarf is not going to let us go so easily... The idea here is completely different."

"We are losing time," intervened Meng. "Perhaps..."

"Oh! We have plenty of time," said Vel unperturbed. So here it is ... you are of course familiar with the idea of a pulsating spacecraft?"

Gascondi and Meng exchanged puzzled glances.

"Yes," consented Vel. "It is a very old and long forgotten idea ..."

"I seem to remember something vaguely," slowly articulated Meng. "Something I saw in old textbooks ...[3] "If I am not wrong, the issue is that a spacecraft is not a point, and its mass is distributed over a certain volume."

"That's it!" said Vel, coming to life. "If one divides our starship into two sections, then the resultant of the gravity forces that act upon them will be smaller than the force that acts on *Omicron* now."

He spoke distinctly and with clarity, as if lecturing a group of students.

"And that means," picked up Meng, "that the stretched starship will be acted upon by a repelling force?"

"And if at the apocenter we bring the two sections together, and at the pericenter we separate them, then *Omicron* will leave the Keplerian orbit and begin to move along an untwining spiral."

"Right," continued Meng.

"I recall now, too," said the agitated Gascondi unexpectedly . "Wonderful, magnificent, brilliant! ... " he burst out in a nervous laughter. "But, as far as I recall, to overcome even the force of gravitational attraction of the Earth in this manner a spaceship needs a few years. But the attraction of a white dwarf? ..."

"But this exactly is the trick," continued Vel unperturbed. "In our case the attraction works for us. The more massive a star or a planet, the faster one reaches escape velocity. That's the paradox!"

"How many hours will we need?" asked Meng.

"One and a half hours, I think, not longer."[4]

"You are a genius," smiled the captain and took his place at the control desk.

"We need only choose the optimal moments for bringing the sections far and close to one another," cautioned Vel.

"I understand," echoed Meng, punching the keyboard of the computer. "I will start the procedure in six minutes ..."

It was an unprecedented spectacle. The gigantic starship seemed to break into two parts, which alternately moved apart, distancing from one another, and then back together, joining to form a single unit. An in the course of this fantastic 'cosmic dance' the death orbit on which *Omicron* moved began to gradually unravel.

---

[3]By chance, the journal *Zemlya Vselenaya* (*Earth and Universe*) placed a review of one of these "old textbooks" (the first edition of the author's *Essays*) right after Komarov's story from which this excerpt is taken.

[4]Alás, no less than one and a half hours. The escape time may differ from this *lower* estimate by several orders of magnitude.

The powerful force of gravity, surrendered to the power of the human intelligence, was steadily carrying away the starship farther and farther from the menacing star.

Tenth Essay

Interplanetary Flights: Low Thrusts for High Goals

We were but prisoners on a humble planet
And how many times, in the countless succession of years,
Earth's steady gaze into the dark expanse
Followed with longing the motion of the spheres!

V. Bryusov, *Son of Earth*

We must carry to other planets
The gospel of our little Earth!

V. Bryusov, *Child Dreams*

1. Prelude

The achievements of the human race in the conquest of cosmic space have been enormous. During an interval of 12 years, between 1957 and 1969, we went from the launching of the first artificial satellite to the landing of the first mission to the Moon. But our greatest dream – missions to other planets – still remains to be fulfilled. Carrying out such missions is a difficult task, which can be accomplished by various means. Perhaps spaceships will be assembled and their engines fueled in an orbit around the Earth. Or perhaps spaceships will depart from such an orbit by using not chemically-fueled jet engines, but low-thrust engines – for instance, ion or plasma engines that are being designed in many laboratories and were already mentioned in the 7th essay. The maximal acceleration that a spaceship equipped with such engines may attain is of only several mm/sec^2, but acting unceasingly over the entire duration of the flight (for several months!), even such a low thrust gives ample capabilities for maneuvering the spaceship in cosmic space. One can picture how such a spaceship, moving along an unwinding spiral trajectory around the Earth, attains escape velocity and, free to move in space, embarks upon a course to Mars. In the vicinity of Mars the low-thrust engines gradually slow the spaceship to parabolic (relative to Mars) velocity; then the spaceship engages in a spiral trajectory which leads to the final orbit around Mars and parks on it until the expedition lands on the planet's surface using a landing spacecraft and then returns.

Low-thrust spiral trajectories in the vicinity of a planet were already considered in one of the preceding essays of this book. Here we will consider low-thrust *interplanetary* trajectories. This flight segment starts after parabolic velocity with respect to the first planet is attained and ends when the parabolic velocity with respect to the second planet is reached. Hence, on this segment the attraction of the two planets is extremely small and can be ignored.

The motion of the spaceship is determined by the attraction of the Sun and the thrust of the jet engines. This is precisely the setting in which we will consider the problem. The exposition will follow mainly the papers [10.1]–[10.3] of the author and his colleagues V. V. Golubkov, V. A. Egorov, and V. G. Ershov. The reader interested in the dynamics of low-thrust flights can access a large volume of information in the monograph of G. L. Grodzovskiĭ, Yu. N. Ivanov, and V. V. Tokarev [10.4].

2. Larger payloads, less fuel

That is the usual requirement for flights in space (and not only in space – and of course not necessarily for flights only). In other words, the flight should be controlled in such a manner that its target is attained with the smallest possible mass (fuel) expenditure. Here *controlling* the flight means finding the requisite law of variation of the reactive acceleration with time, $\mathbf{f}(t)$. We known (by the definition of the reactive force) that

$$f = -\frac{V_{\mathrm{r}}}{m}\frac{dm}{dt}, \tag{10.2.1}$$

where V_{r} is the relative (with respect to the rocket) outflow rate of combustion products (the speed of particles in the reactive jet), $m(t)$ is the variable mass of the rocket, and dm/dt is the instantaneous mass expenditure. For a wide class of rockets (with chemical fuel, for example) one can take $V_{\mathrm{r}} = \mathrm{const}$; then (10.2.1) yields

$$\frac{m_{\mathrm{t}}}{m_0} = \exp\left(-\frac{1}{V_{\mathrm{r}}}\int_0^T f\,dt\right), \tag{10.2.2}$$

where T is the flight duration. The ratio of the terminal mass m_{t} (payload) to the initial mass m_0 is therefore the larger the smaller the integral

$$S = \int_0^T f\,dt. \tag{10.2.3}$$

The control of the acceleration $\mathbf{f}(t)$ must be chosen so that the desired flight will be carried out with the smallest possible value of the integral (10.2.3).

However, for many low-thrust engines the quantity V_{r} is not constant. What is constant is the jet power N, i.e., that part of the power of the on-board engine that is transformed into kinetic energy of the reactive jet. By definition,

$$N = -\frac{dm}{dt}\frac{V_{\mathrm{r}}^2}{2}. \tag{10.2.4}$$

Then, by (10.2.4) and (10.2.1), $(-1/m^2)dm/dt = -f^2/(2N)$, whence

$$\frac{m_t}{m_0} = \frac{1}{1 + \dfrac{m_0}{2N} \displaystyle\int_0^T f^2\,dt}. \tag{10.2.5}$$

We see that the mass ratio is the larger the smaller the integral

$$I = \int_0^T f^2\,dt. \tag{10.2.6}$$

Thus, to carry out a desired flight using a low-thrust engine one needs to arrange that the reactive acceleration $\mathbf{f}(t)$ will vary so as to minimize the value of the integral I given by (10.2.6).

Problems of this type are known as *optimal control problems*.

This essay will deal mainly with optimal flights that minimize the functional (10.2.6). However, at the end of the essay we will consider one important problem on the minimization of the functional S given by (10.2.3).

3. The Pontryagin maximum principle

The optimal control problem that arises in the situation we are interested in can be formulated as follows. Assume that the equations of motion are written in the vector form

$$\dot{\mathbf{x}} = \mathbf{f}(\mathbf{x}, \mathbf{u}), \tag{10.3.1}$$

where $\mathbf{x}$, $\mathbf{f}$, $\mathbf{u}$ are vectors with components $x_1, \dots, x_n$, $f_1, \dots, f_n$, $u_1, \dots, u_r$, respectively. The vector $\mathbf{u}$ is called the *control* vector; at each moment of time $\mathbf{u}$ is constrained to take values in a closed set U: $\mathbf{u} \in U$.

In practical problems the control is almost always subject to some constraints: the reactive thrust cannot exceed its maximal value, rudders cannot turn more than a maximal angle, and so on. This explains why the control is confined to a closed set, which actually means that the control is *bounded*: the point $\mathbf{u} = (u_1, \dots, u_r)$ may lie only in the interior of the set U or on its boundary. However, in particular situations one can also consider values of $\mathbf{u}$ that are not constrained in any way. We call *admissible control* any piecewise-continuous vector-functions $\mathbf{u}(t)$ such that $\mathbf{u}(t) \in U$.

Let some quality indicator of the controlled motion process (10.3.1) be characterized by the functional

$$I = \int_{t_0}^{t_1} f_0(\mathbf{x}, \mathbf{u})dt, \tag{10.3.2}$$

where t_0 and t_1 are the initial and terminal moments of time of the motion and $f_0(\mathbf{x}, \mathbf{u}) \geq 0$ (which incidentally is not essential). The optimization problem of the motion governed by (10.3.1) is then stated as follows: *among all admissible controls* $\mathbf{u} = \mathbf{u}(t)$ *under the action of which the phase point* $\mathbf{x}_0 = \mathbf{x}_0(t)$ *is carried*

by the solution of the equation of motion (10.3.1) *into the phase point* $\mathbf{x}_1 = \mathbf{x}_1(t)$, *find the controls that minimize the value of the functional* (10.3.2).

The most widely applicable method of solving optimal control problems is the *Pontryagin maximum principle*. Rigorous statements and proofs can be found, e.g., in the monographs [10.5] and [10.6]. Here we are interested in the formal apparatus for solving optimal control problems.

Let us introduce the auxiliary vector

$$\boldsymbol{\psi} = (\psi_0, \psi_1, \dots, \psi_n), \tag{10.3.3}$$

where ψ_0 is a constant quantity, $\psi_0 \leq 0$, and $\psi_1, \dots, \psi_n$ are some functions of time: $\psi_i = \psi_i(t)$.

Using the right-hand sides of equation (10.3.1), the function f_0, and the vector $\boldsymbol{\psi}$, let us construct the function

$$\mathcal{H} = \sum_{i=1}^{n} \psi_i f_i(\mathbf{x}, \mathbf{u}). \tag{10.3.4}$$

Note that equations (10.3.1) can be then written in the form

$$\dot{x}_i = \frac{\partial \mathcal{H}}{\partial \psi_i}, \qquad i = 1, \dots, n. \tag{10.3.5}$$

Now let us determine the functions $\psi_i(t)$, $i = 1, \dots, n$, from the differential equations

$$\frac{d\psi_i}{dt} = -\frac{\partial \mathcal{H}}{\partial x_i}, \qquad i = 1, \dots, n. \tag{10.3.6}$$

It turns out (see, e.g., [10.5], [10.6]) that for the optimality (in the sense of the minimum of the functional (10.3.2)) of the process $\mathbf{x}(t)$, $\mathbf{u}(t)$ that steers the phase point $\mathbf{x}_0$ to the phase point $\mathbf{x}_1$ it is necessary that there exist a nontrivial solution of the system (10.3.6) and a constant $\psi_0 \leq 0$ such that, for any time τ with $t_0 \leq \tau \leq t_1$, the following maximum condition is satisfied:

$$\mathcal{H}(\boldsymbol{\psi}(\tau), \mathbf{x}(\tau), \mathbf{u}(\tau)) = \max_{\mathbf{v} \in U} \mathcal{H}(\boldsymbol{\psi}(\tau), \mathbf{x}(\tau), \mathbf{v}). \tag{10.3.7}$$

Thus, an optimal control $\mathbf{u}$ is found from the condition (10.3.7) – practically in terms of $\boldsymbol{\psi}$ and $\mathbf{x}$. This then allows us, by jointly integrating equations (10.3.6) and (10.3.5) (or, equivalently, (10.3.1)), to construct an optimal trajectory. Usually the most difficult part of the problem is finding the initial data for the integration of the system (10.3.6).

Note that the system (10.3.5)–(10.3.6) is of order $2n$; consequently, its solution depends on $2n$ arbitrary constants, which indeed allows us to satisfy $2n$ boundary conditions

$$\mathbf{x}_i(t_0) = \mathbf{x}_{i0}, \qquad \mathbf{x}_i(t_1) = \mathbf{x}_{i1}, \qquad i = 1, \dots, n.$$

The Pontryagin maximum principle gives only *necessary* conditions for optimality: it necessarily holds for optimal processes. But it may happen that the maximum principle holds for some process, but the process in question is *not* optimal. Hence, generally speaking, for problems solved by means of the maximum principle one still needs to verify that the optimality conditions obtained are sufficient. Often (for example – in linear problems), the maximum principle gives conditions that are also sufficient. But we need to remember that this is not always the case.

4. The equation of optimal flight

Let us consider the motion of a rocket of mass m in a force field with potential $\overline{U}$. Denote

$$\left.\begin{aligned} &\frac{\overline{U}}{m} = U(x,y,z), \qquad \operatorname{grad} U = \left(\frac{\partial U}{\partial x}, \frac{\partial U}{\partial y}, \frac{\partial U}{\partial z}\right), \\ &\mathbf{r} = (x,y,z), \qquad \mathbf{V} = (u,v,w), \qquad \mathbf{f} = (f_x, f_y, f_z). \end{aligned}\right\} \tag{10.4.1}$$

Here x, y, z are the coordinates of the rocket, u, v, w are the components of its velocity, and f_x, f_y, f_z are the components of the reactive force. Then the equations of motions in vector form read

$$\dot{\mathbf{V}} - \operatorname{grad} U = \mathbf{f}, \qquad \dot{\mathbf{r}} = \mathbf{V}. \tag{10.4.2}$$

Let us assume that the power of the reactive jet generated by the rocket's engine is constant. Then we must find trajectories for which the functional (10.2.6) attains its minimum. Let us also subject the reactive acceleration to the following constraints:

$$-a \le f_x \le a, \qquad -a \le f_y \le a, \qquad -a \le f_z \le a. \tag{10.4.3}$$

Following the recipe (10.3.4), we form the function

$$\begin{aligned} \mathcal{H} = &- (f_x^2 + f_y^2 + f_z^2) + \psi_u \left(f_x + \frac{\partial U}{\partial x}\right) + \psi_x u + \\ &+ \psi_v \left(f_y + \frac{\partial U}{\partial y}\right) + \psi_y v + \psi_w \left(f_z + \frac{\partial U}{\partial z}\right) + \psi_z w. \end{aligned} \tag{10.4.4}$$

In (10.4.4) we took $\psi_0 = -1$, but one can use any $\psi_0 < 0$ provided that the quantities $\psi_x, \psi_y, \ldots, \psi_w$ are suitably redefined.

According to the maximum principle (10.3.7), for the controls f_x, f_y, f_z to be optimal they must maximize the quantity $\mathcal{H}$ given by formula (10.4.4). Since these controls appear symmetrically in the expression of $\mathcal{H}$, it suffices to consider only one of them, say, f_x. The dependence of $\mathcal{H}$ on f_x is shown in Figure 10.1

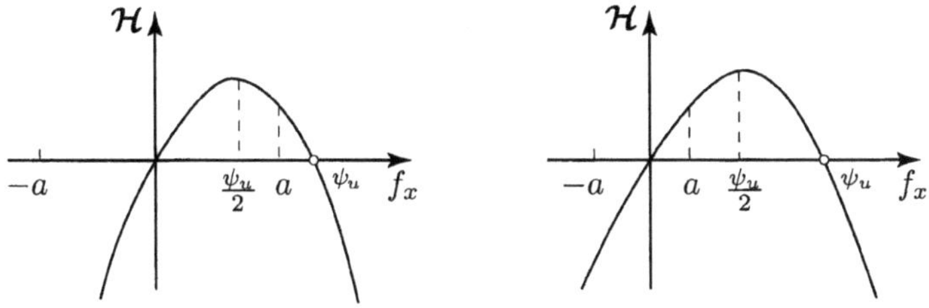

FIGURE 10.1. Regarding the construction of an optimal control

(for $\psi_u > 0$; for $\psi_u < 0$, the picture is reflected symmetrically). We see that the maximum of $\mathcal{H}$ is attained for $f_x = \psi_u/2$, if $|\psi_u/2| \le a$, and for $f_x = a$, if $|\psi_u/2| > a$. The situation is similar for f_y and f_z. We conclude that the optimal control is given by the formula

$$f_x = \begin{cases} a, & \text{if } \psi_u/2 > a, \\ \psi_u/2, & \text{if } |\psi_u/2| \le a, \\ -a, & \text{if } \psi_u/2 < -a. \end{cases} \tag{10.4.4'}$$

Analogous formulas hold for f_y and f_z. The dependence, say, of f_x on ψ_u is shown in Figure 10.2. Moreover, according to (10.3.6) and (10.4.4′), the quantities ψ must satisfy the system of equations

$$\left.\begin{aligned} &\frac{d\psi_x}{dt} = -\left(\psi_u \frac{\partial^2 U}{\partial x^2} + \psi_v \frac{\partial^2 U}{\partial y \partial x} + \psi_w \frac{\partial^2 U}{\partial z \partial x}\right), \\ &\cdots\cdots\cdots\cdots\cdots\cdots\cdots\cdots \\ &\frac{d\psi_u}{dt} = -\psi_x, \\ &\cdots\cdots\cdots\cdots\cdots\cdots\cdots\cdots \end{aligned}\right\} \tag{10.4.5}$$

Here the dots stand for the analogous equations for the other variables ψ. In vector form equations (10.4.5) can be written as

$$\dot{\boldsymbol{\psi}}_{\mathbf{r}} + A\boldsymbol{\psi}_{\mathbf{V}} = 0, \qquad \dot{\boldsymbol{\psi}}_{\mathbf{V}} = -\boldsymbol{\psi}_{\mathbf{r}}, \tag{10.4.6}$$

where

$$A = \begin{pmatrix} \frac{\partial^2 U}{\partial x^2} & \frac{\partial^2 U}{\partial x \partial y} & \frac{\partial^2 U}{\partial x \partial z} \\ \frac{\partial^2 U}{\partial y \partial x} & \frac{\partial^2 U}{\partial y^2} & \frac{\partial^2 U}{\partial y \partial z} \\ \frac{\partial^2 U}{\partial z \partial x} & \frac{\partial^2 U}{\partial z \partial y} & \frac{\partial^2 U}{\partial z^2} \end{pmatrix},$$

$\boldsymbol{\psi}_{\mathbf{r}}$ is the vector with components ψ_x, ψ_y, ψ_z, and $\boldsymbol{\psi}_{\mathbf{V}}$ is the vector with components ψ_u, ψ_v, ψ_w. Equations (10.4.2), (10.4.4′), and (10.4.6) form a closed system of

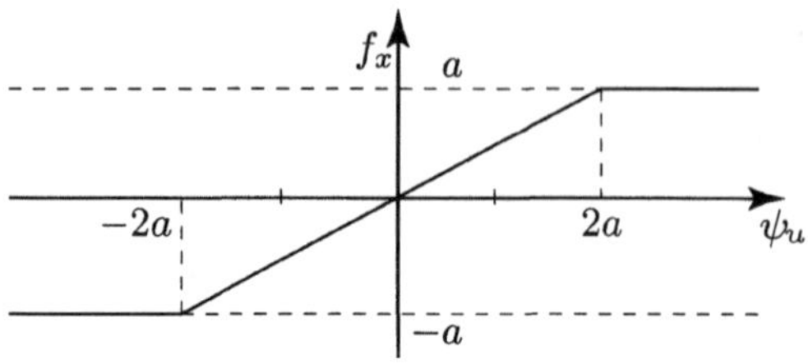

FIGURE 10.2. Optimal control

equations of optimal (in the sense of the minimum of the integral (10.2.6)) flight of the rocket with the constraints (10.4.3) on the control accounted for.

A major advantage of the Pontryagin maximum principle it that in the calculation of optimal motions it allows us to take into account, in a simple and unified manner, the various constraints imposed on controls. This is very important for modern problems of mechanics. We have demonstrated the effectiveness of the maximum principle on the example of equations (10.4.2), (10.4.4′), (10.4.6). However, to simplify the exposition, in what follows we shall consider the case when the controls f_x, f_y, f_z are subject to no constraints ($a = \infty$). This is equivalent to considering flight trajectories for the optimal controls of which it is not required that the components of the reactive acceleration attain a limit value $a \neq \infty$.

5. No constraints

If we remove the constraints, then, as it readily follows from (10.4.4), the optimal control is given by the formula $\mathbf{f} = \psi_{\mathbf{v}}/2$, and instead of equations (10.4.6) we can write directly the differential equations of the optimal control! Together with (10.4.2), they give the closed system of equations of optimal flight, which can be written as follows:

$$\left.\begin{aligned} \ddot{\mathbf{r}} - \operatorname{grad} U &= \mathbf{f}, \\ \ddot{\mathbf{f}} - A\mathbf{f} &= 0. \end{aligned}\right\} \tag{10.5.1}$$

Here the matrix A is the same as in (10.4.6).

We have agreed earlier to neglect the attraction of planets on the flight segment between them. Then the boundary conditions for problem (10.5.1) are approximately given by the condition of departure from a (nonattracting!) planet with some initial relative (to that planet) velocity and arrival to another planet with some relative (to the target planet) velocity. The relative initial and terminal velocities may be considered equal to zero (since they are small compared with the velocity of translational motion of the planets themselves). Thus, as boundary conditions we will henceforth take the coordinates and components of the velocity of the planet of origin (say, Earth) and of the target planet (say, Mars).

The 12 integration constants of the system (10.5.1) must be defined so that the following 12 boundary conditions will be satisfied:

$$\left.\begin{aligned} \mathbf{r} &= \mathbf{r}^0, & \dot{\mathbf{r}} &= \dot{\mathbf{r}}^0 & \text{for} \quad t &= t_0, \\ \mathbf{r} &= \mathbf{r}^{\mathrm{t}}, & \dot{\mathbf{r}} &= \dot{\mathbf{r}}^{\mathrm{t}} & \text{for} \quad t &= t_1. \end{aligned}\right\} \tag{10.5.2}$$

Naturally, the system (10.5.1) admits the first integral

$$\mathcal{H} = \mathcal{H}_0, \tag{10.5.3}$$

where $\mathcal{H}$ is given by the expression (10.4.4), in which we replace ψ_u, ψ_v, ψ_w by $2f_x$, $2f_y$, $2f_z$, respectively, and ψ_x, ψ_y, ψ_z by $-2\dot{f}_x$, $-2\dot{f}_y$, $-2\dot{f}_z$, respectively. If we also assume that the force function U is the force function of a central force field (which is the case for our problem), then equations (10.5.1) will admit three additional first integrals, equivalent to a single vector integral:

$$\mathbf{r} \times \dot{\mathbf{f}} + \mathbf{f} \times \dot{\mathbf{r}} = \mathbf{k}. \tag{10.5.4}$$

However, the integrals (10.5.3) and (10.5.4) are not sufficient for the integration of the system (10.5.1); apparently, this system is *not* integrable.[1] It is therefore necessary to seek approximate methods of solving the boundary value problem (10.5.1)–(10.5.2).

6. The method of carrier trajectories

The boundary value problem (10.5.1)–(10.5.2) can be solved by numerical methods, using computers. To this end we must know some approximate solution of the problem and use it as the first approximation in an iteration process for solving the boundary value problem. Such an iterative method (method of successive approximations) is almost always unavoidable in studying nonlinear boundary value problems. Iteration requires considerable machine time, which makes it hard to obtain transparent results.

Numerical methods are always at our disposal. However, we can try to manage without them, if not for all boundary value problems (10.5.1)–(10.5.2), then at least for a sufficiently wide class of them. Incidentally, if we are successful in obtaining an approximate solution of the problem, then we can always use it as a first approximation in a numerical iteration method.

T. M. Eneev proposed to use in the dynamics of low-thrust flight a method that he called "*the method of carrier trajectories*."

[1] All the attempts of the author to break this wall have resulted only in a particular case of optimal control that satisfies the system (10.5.1): $\mathbf{f} = \frac{2\beta}{2+\beta t}\mathbf{V}$, where β is an arbitrary constant. Of course, using this control one cannot solve every optimal flight problem, but only a problem that depends on a total of seven boundary conditions.

Suppose that we want to accomplish an interplanetary flight, departing at time t_0 from the planet of origin and arriving at time t_1 at the target planet. First, let us find the ordinary Keplerian trajectory that solves this problem. This can be done because, as we know, the equations of Keplerian motion are integrable. If in this situation the boundary value problem is solvable, we obtain an elliptic trajectory that passes through the point of departure at time t_0 and through the terminal point at time t_1. The low-thrust trajectory in which we are interested must not differ too significantly from the Keplerian trajectory just found. Indeed, if the flight duration is not too long, a small reactive acceleration simply does not manage to drive the flight trajectory too far from the properly chosen Keplerian trajectory, as if the Keplerian trajectory draws the low-thrust trajectory after it. Let us call such a Keplerian trajectory a "*carrier trajectory*." A coordinate system that moves forward along a carrier trajectory is called a "*carrier coordinate system*."

The idea is then to consider the motion of a low-thrust rocket in a carrier coordinate system. As is readily understood, the trajectory of such a relative motion does not drift too far from the origin of coordinates, which allows us to linearize the boundary value problem and solve it to the very end. This is shown in the author's joint paper with V. A. Egorov [10.1]. Following that paper, we assume that the trajectory of the motion is written as

$$\mathbf{r} = \mathbf{r}_0 + \boldsymbol{\rho}, \tag{10.6.1}$$

where $\mathbf{r}_0$ is the known carrier trajectory and the quantity $|\boldsymbol{\rho}|$ is small compared with $|\mathbf{r}_0|$. The vector-function $\mathbf{r}_0(t)$ satisfies the equation of unperturbed motion

$$\ddot{\mathbf{r}}_0 - \operatorname{grad} U_0 = 0, \tag{10.6.2}$$

where U_0 is the value of the force function U along the carrier trajectory $\mathbf{r}_0(t)$. Let us substitute (10.6.1) in (10.5.1), expand $U(x, y, z)$ in a series, and neglect the small terms of order two and higher. Here we take into account that, by assumption, not only the quantities $\boldsymbol{\rho}$ and $\dot{\boldsymbol{\rho}}$ are small, but also $\mathbf{f}$, which means that we also neglect the terms involving products of components of $\boldsymbol{\rho}$ and $\mathbf{f}$ and similar ones. Then instead of the system (10.5.1) we obtain the linearized system

$$\ddot{\boldsymbol{\rho}} - A_0\boldsymbol{\rho} = \mathbf{f}, \tag{10.6.3}$$

$$\ddot{\mathbf{f}} - A_0\mathbf{f} = 0. \tag{10.6.4}$$

Here A_0 is the value of the matrix-function A along the carrier trajectory $\mathbf{r}_0(t)$; the elements of $A_0(t)$ are known functions of time.

A linear system of arbitrary order is integrable when the coefficients of its homogeneous part are constant, which is manifestly not the case for our system. Generally speaking, linear systems with variable coefficients are not integrable. However, it turns out that our system (10.6.3)–(10.6.4) is! (This is precisely the "secret" reason for introducing the carrier coordinate system.) Indeed, let us assume for a moment that in the system (10.6.3)–(10.6.4) $\mathbf{f} \equiv 0$. Then, by the idea of our method, the remaining (left-hand) part of the vector equation (10.6.3) has as solution a variation of the Keplerian motion! The Keplerian motion is described by quadratures, and hence so is its variation. Further, the vector equation (10.6.4) has exactly the form of equation (10.6.3) with null right-hand side. Therefore, equation (10.6.4) is also integrable! Finding its solution $\mathbf{f}(t)$ and substituting it in the right-hand side of equation (10.6.3), we obtain a nonhomogeneous vector equation whose right-hand side is a known function of time, and whose corresponding homogeneous equation is integrable. As is known, the solution of such an equation is found by quadratures (for instance, by the variation-of-constants method). Thus, the system (10.6.3)–(10.6.4) is integrable by quadratures.

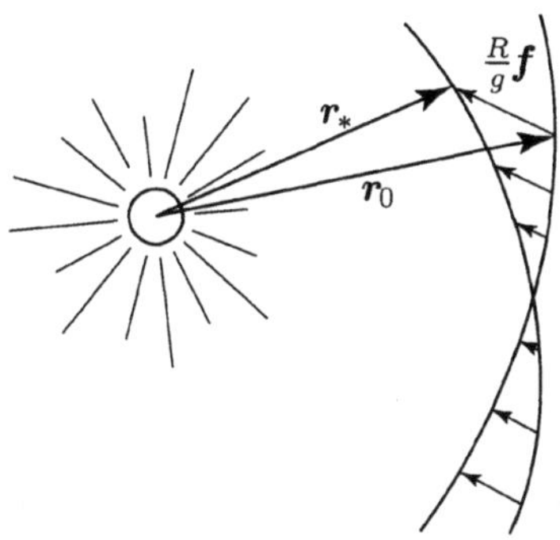

FIGURE 10.3. Optimal control in the linear approximation

Incidentally, from these considerations (and dimensional considerations) it follows that in the approximation adopted by us the optimal control of the reactive acceleration vector is given by the formula

$$\mathbf{f} = \alpha \frac{g}{R} (\mathbf{r}_* - \mathbf{r}_0), \tag{10.5.6}$$

where $r_0(t)$ is the carrier Keplerian elliptic trajectory, $r_*(t)$ is the Keplerian elliptic trajectory whose parameters are determined by the boundary conditions of the problem, g is the acceleration of gravity at distance R from the center of attraction, and $\alpha \sim 1$ is a dimensionless scalar constant. Figure 10.3 shows the control scheme corresponding to (10.6.5).

The actual integration of the system (10.6.3)–(10.6.4) will be carried out a bit later.

7. The scheme for solving the boundary value problem

The linear equations (10.6.3)–(10.6.4) are nice not only because they are integrable, but also because they allow us to solve the boundary value problem "in one strike," without resorting to successive approximations. For nonlinear problems this, generally speaking, is not possible.

Thanks to (10.6.1), the boundary conditions (10.5.2) take now the form

$$\left.\begin{aligned} \boldsymbol{\rho} &= \boldsymbol{\rho}_0, & \dot{\boldsymbol{\rho}} &= \dot{\boldsymbol{\rho}}_0 & \text{for} \quad t &= t_0, \\ \boldsymbol{\rho} &= \boldsymbol{\rho}_{\mathrm{t}}, & \dot{\boldsymbol{\rho}} &= \dot{\boldsymbol{\rho}}_{\mathrm{t}} & \text{for} \quad t &= t_1. \end{aligned}\right\} \tag{10.7.1}$$

Here

$$\left.\begin{aligned} \boldsymbol{\rho}_0 &= \mathbf{r}^0 - \mathbf{r}_0(t_0), & \dot{\boldsymbol{\rho}}_0 &= \dot{\mathbf{r}}^0 - \dot{\mathbf{r}}_0(t_0), \\ \boldsymbol{\rho}_{\mathrm{t}} &= \mathbf{r}^{\mathrm{t}} - \mathbf{r}_0(t_1), & \dot{\boldsymbol{\rho}}_{\mathrm{t}} &= \dot{\mathbf{r}}^{\mathrm{t}} - \dot{\mathbf{r}}_0(t_1). \end{aligned}\right\} \tag{10.7.2}$$

Since the carrier trajectory $\mathbf{r}(t)$, $\dot{\mathbf{r}}(t)$ is known at any time t, the boundary conditions for the relative motion (10.7.1) are readily calculated in terms of the boundary conditions for the absolute motion via formulas (10.7.2).

The scheme for integrating the system of equations (10.6.3)–(10.6.4) and solving the boundary value problem (10.7.1) can be described as follows. Redefine $\boldsymbol{\rho}$ to denote the six-dimensional vector formed by the coordinates *and* by the components of the velocity. Then we can seek the solution of system (10.6.3) in the form

$$\boldsymbol{\rho} = \boldsymbol{\rho}_* + \boldsymbol{\rho}_{**}, \tag{10.7.3}$$

where $\boldsymbol{\rho}_*$ is the solution of the homogeneous system (10.6.3) and $\boldsymbol{\rho}_{**}$ is a particular solution of the nonhomogeneous system (10.6.3).

Let $\boldsymbol{\rho}_i$ be particular linearly independent solutions of the homogeneous system (10.6.3). They are known functions, since the variation of the Keplerian motion is known. Then

$$\boldsymbol{\rho}_* = \sum_{i=1}^{6} c_i \boldsymbol{\rho}_i, \tag{10.7.4}$$

where c_i are integration constants. From the foregoing discussion it follows that the solution of system (10.6.4) is another linear combination of the same functions $\boldsymbol{\rho}_i$:

$$\mathbf{f} = \sum_{i=1}^{6} a_i \boldsymbol{\rho}_i, \tag{10.7.5}$$

where a_i are new integration constants. Of course, in formula (10.7.5) $\mathbf{f}$ denotes the six-dimensional vector formed by the components of the acceleration and their time derivatives.

Now substituting (10.7.5) in (10.6.3) and applying the variation-of-constants method we obtain the particular solution

$$\boldsymbol{\rho}_{**} = \sum_{i=1}^{6} a_i \boldsymbol{\varphi}_i, \tag{10.7.6}$$

where we choose $\boldsymbol{\rho}_{**}$ so that $\boldsymbol{\rho}_{**}(t_0) = 0$, i.e.,

$$\boldsymbol{\varphi}_i(t_0) = 0. \tag{10.7.7}$$

Finally, using (10.7.4) and (10.7.3) we obtain

$$\boldsymbol{\rho} = \sum_{i=1}^{6} (c_i \boldsymbol{\rho}_i + a_i \boldsymbol{\varphi}_i). \tag{10.7.8}$$

The arbitrary constants c_i and a_i allow us to satisfy the 12 boundary conditions (10.7.1). More precisely, c_i and a_i are determined from the following system of linear algebraic equations (recall that $\boldsymbol{\rho}$, $\boldsymbol{\rho}_i$, and $\boldsymbol{\varphi}_i$ are six-dimensional vectors, with $\boldsymbol{\varphi}_i$ satisfying (10.7.7)):

$$\boldsymbol{\rho}_0 = \sum_{i=1}^{6} c_i \boldsymbol{\rho}_i(t_0), \tag{10.7.9}$$

$$\boldsymbol{\rho}_{\mathrm{t}} - \sum_{i=1}^{6} c_i \boldsymbol{\rho}_i(t_1) = \sum_{i=1}^{6} a_i \boldsymbol{\varphi}_i(t_1). \tag{10.7.10}$$

Solving the system (10.7.9) one finds the constants c_i, and then solving the system (10.7.10) gives the constants a_i. The algorithm described solves completely the problem (10.6.3), (10.6.4), (10.7.1).

8. Integration

Let us choose an absolute coordinate system XYZ with the origin S at the center of attraction such that the plane of the carrier trajectory coincides with the XY-plane and the X-axis is directed toward the pericenter of the carrier trajectory. The axes of the carrier coordinate system $Oxyz$ are parallel to the corresponding axes of the XYZ-system (Figure 10.4). We will assume that the equations of motion (10.6.3) and the controls (10.6.4) are written in dimensionless variables. If R, V, T, and f_{d} are quantities whose dimensions are distance, velocity, time, and acceleration, respectively, then the corresponding dimensionless quantities r, v, t, and f are defined as follows

$$r = \frac{R}{R_0}, \qquad v = \frac{V}{\sqrt{gR_0}}, \qquad t = T\sqrt{\frac{g}{R_0}}, \qquad f = \frac{f_{\mathrm{d}}}{g}, \tag{10.8.1}$$

where R_0 is the initial distance to the center of attraction and $g = \mu/R_0^2$ is the acceleration of gravity at that distance.

For such a choice of a coordinate system and dimensionless variables the elements of the matrix A_0 in the equations (10.6.3)–(10.6.4) have the following

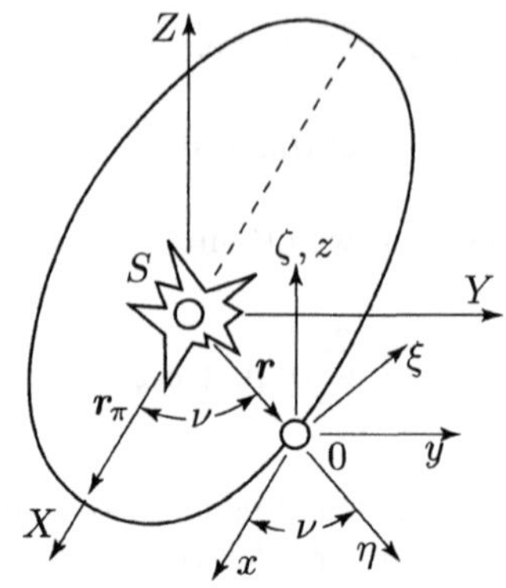

FIGURE 10.4. Coordinate systems

expressions:

$$\left.\begin{array}{lll} U^0_{xx} = \dfrac{1}{r_0^3}(3\cos^2\nu - 1), & U^0_{yy} = \dfrac{1}{r_0^3}(3\sin^2\nu - 1), & \\ U^0_{zz} = -\dfrac{1}{r_0^3}, \quad U^0_{xy} = \dfrac{1}{r_0^3}3\cos\nu\sin\nu, & & U^0_{xz} = U^0_{yz} = 0. \end{array}\right\} \tag{10.8.2}$$

In these formulas $r_0(t)$ is the current radius of the carrier trajectory and ν is the true anomaly along it, which are connected by the relations of the Keplerian motion

$$r_0 = \frac{p}{1 + e\cos\nu}, \qquad \frac{d\nu}{dt} = \frac{\sqrt{p}}{r_0^2}; \tag{10.8.3}$$

here e denotes the eccentricity of the carrier trajectory and p is its dimensionless focal parameter.

To integrate the system (10.6.3)–(10.6.4) it is convenient to pass to a rotating coordinate system $O\eta\xi\zeta$ with the origin at the point O of the carrier trajectory (Figure 10.4), and with the η-axis directed along the radius vector, the ξ-axis directed along the transversal to the carrier trajectory (in the direction of motion), and the ζ-axis parallel to the z-axis. Then

$$\left.\begin{array}{l} x = \eta\cos\nu - \xi\sin\nu, \\ y = \eta\sin\nu + \xi\cos\nu, \\ z = \zeta, \end{array}\right\} \tag{10.8.4}$$

and after some transformations equations (10.6.3) take on the form

$$\left.\begin{array}{r} \ddot\eta - 2\omega\dot\xi - \omega^2\eta - \dot\omega\xi - \dfrac{2}{r_0^3}\eta = f_\eta, \\ \ddot\xi + 2\omega\dot\eta - \omega^2\xi + \dot\omega\eta + \dfrac{1}{r_0^3}\xi = f_\xi, \\ \ddot\zeta + \dfrac{1}{r_0^3}\zeta = f_\zeta. \end{array}\right\} \tag{10.8.5}$$

Here $\omega = \dot{\nu}$ is the angular velocity of the motion along the carrier trajectory and f_η, f_ξ, f_ζ are the projections of the reactive acceleration vector on the axes η, ξ, ζ. Equations (10.8.5) could have been derived straight from the general equations of relative motion.

Let us remark that the coefficients in (10.8.5) depend on time only through the true anomaly ν. Hence, it is quite natural to use relations (10.8.3) and replace the time t in (10.8.5) by the new independent variable ν. This transformation yields the system

$$\left.\begin{aligned} \eta'' - \frac{2\rho'}{\rho}\eta' - 2\xi' - (1+2\rho)\eta + \frac{2\rho'}{\rho}\xi &= p^3\rho^4 f_\eta, \\ \xi'' - \frac{2\rho'}{\rho}\xi' + 2\eta' - (1-\rho)\xi - \frac{2\rho'}{\rho}\eta &= p^3\rho^4 f_\xi, \\ \zeta'' - \frac{2\rho'}{\rho}\xi' + \rho\zeta &= p^3\rho^4 f_\zeta. \end{aligned}\right\} \qquad (10.8.6)$$

Here and in what follows we use the notations

$$\rho = (1 + e\cos\nu)^{-1}, \qquad \rho' = e\sin\nu(1+e\cos\nu)^{-2}, \qquad (10.8.7)$$

and the primes stand for differentiation with respect to ν.

Let us ignore for the moment the fact that in our problem the components f_η, f_ξ, f_ζ satisfy the differential equation (10.6.4) and consider them arbitrary, but known functions of the true anomaly ν.

Thus, we have produced a closed system (10.8.6), with explicitly written coefficients that depend periodically on the independent variable ν. At first glance this system seems impregnable. But, armed with our confidence in its integrability, we proceed courageously to its attack.

Since the third equation of the system (10.8.6) is independent of the first two, we will begin with it. First, let us try to get rid of the term containing the first derivative ζ'. As is known from the theory of linear differential equations, this can be always achieved by means of the substitution

$$\zeta = \varkappa \exp\left(-\frac{1}{2}\int p_1 d\nu\right), \qquad (10.8.8)$$

where p_1 is the coefficient of ζ' and $\varkappa$ is a new variable replacing ζ. In our case $p_1 = -2\rho'/\rho$ and the substitution (10.8.8) has the simple form

$$\zeta = \varkappa\rho. \qquad (10.8.9)$$

In the variable $\varkappa$ the third equation of system (10.8.6) reads

$$\varkappa'' + \left[\rho - 2\left(\frac{\rho'}{\rho}\right)^2 + \frac{\rho''}{\rho}\right]\varkappa = p^3\rho^3 f_\zeta. \qquad (10.8.10)$$

But, by (10.8.7),

$$\rho - 2\left(\frac{\rho'}{\rho}\right)^2 + \frac{\rho''}{\rho} = 1,$$

so (10.8.10) takes a simple form, which allows a direct integration:

$$\varkappa'' + \varkappa = p^3 \rho^3 f_\zeta. \tag{10.8.11}$$

The homogeneous equation corresponding to (10.8.11) has the solution

$$\left.\begin{aligned} \varkappa &= c_5 \sin\nu + c_6 \cos\nu, \\ \varkappa' &= c_5 \cos\nu - c_6 \sin\nu, \end{aligned}\right\} \tag{10.8.12}$$

where c_5 and c_6 are integration constants.

Following the variation-of-constants method, we seek the solution of the non-homogeneous equation (10.8.11) in the same form (10.8.12) for $\varkappa$ and $\varkappa'$, where now the coefficients c_5 and c_6 depend on ν. Then, as is known, we obtain the following linear algebraic system for the derivatives of the functions $c_5(\nu)$ and $c_6(\nu)$:

$$\begin{aligned} \frac{dc_5}{d\nu} \sin\nu + \frac{dc_6}{d\nu} \cos\nu &= 0, \\ \frac{dc_5}{d\nu} \cos\nu - \frac{dc_6}{d\nu} \sin\nu &= p^3 \rho^3 f_\zeta. \end{aligned}$$

This yields

$$\frac{dc_5}{d\nu} = p^3 \rho^3 f_\zeta \cos\nu, \qquad \frac{dc_6}{d\nu} = -p^3 \rho^3 f_\zeta \sin\nu.$$

Integrating these expressions, substituting the results in (10.8.12), and returning to the original variable ζ by means of (10.8.9), we finally obtain

$$\begin{aligned} \zeta &= c_5 \rho \sin\nu + c_6 \rho \cos\nu + \\ &\quad + p^3 \rho \left[\sin\nu \int f_\zeta \rho^3 \cos\nu \, d\nu - \cos\nu \int f_\zeta \rho^3 \sin\nu \, d\nu \right]. \end{aligned} \tag{10.8.13}$$

This completes the integration of the third equation of system (10.8.6). Differentiating (10.8.13) one readily obtains the expression for ζ'.

Here, however, we are interested in the technique for integrating the system (10.8.6) more than in deriving a complete list of explicit formulas usable in computations (such a list is available in the primary source [10.1] that we are following here). For this reason, referring to [10.1] for the explicit expression of $\zeta'(\nu)$, we turn our attention to the first two equations in (10.8.6). The integration of the third equation in (10.8.6) was considerably facilitated by the substitution (10.8.6), so it is only natural to try a similar substitution for the first two equations:

$$\xi = \rho\alpha, \qquad \eta = \rho\beta. \tag{10.8.14}$$

This yields

$$\beta'' - 3\rho\beta - 2\alpha' = p^3\rho^3 f_\eta, \qquad \alpha'' + 2\beta' = p^3\rho^3 f_\xi. \tag{10.8.15}$$

The substitution (10.8.14) proved fruitful: the second equation in (10.8.15) can be integrated once right away:

$$\alpha' = c_1 - 2\beta + I_\xi, \qquad I_\xi(\nu) = p^3 \int \rho^3 f_\xi \, d\nu, \tag{10.8.16}$$

where c_1 is a new integration constant, and $I_\xi(\nu_0) = 0$, by definition. Substituting this expression of α' in the first of equations (10.8.15), we obtain

$$\beta'' + (4 - 3\rho)\beta = \Phi(\nu), \tag{10.8.17}$$

where

$$\Phi(\nu) = 2c_1 + 2I_\xi + p^3\rho^3 f_\xi. \tag{10.8.18}$$

What to do now with equation (10.8.17) is not very clear, but if we return to the original variable η (the substitution (10.8.14) already did its job), we obtain in place of (10.8.17) the following equation:

$$(1 + e\cos\nu)\eta'' - 2e\sin\nu\,\eta' + (1 + 3e\cos\nu)\eta = \Phi. \tag{10.8.19}$$

A quick examination of this equations reveals that the corresponding homogeneous equation admits the simple particular solution

$$\eta_2 = \sin\nu. \tag{10.8.20}$$

But from the theory of linear differential equations one knows that an equation of the form

$$\eta'' + a_1(\nu)\eta' + a_2(\nu)\eta = 0$$

that admits the particular solution $\eta = \eta_2$ also admits a second, linearly independent particular solution, given by the formula

$$\eta_3 = \eta_2 \int \eta_2^{-2} \exp\left(-\int a_1 \, d\nu\right) d\nu.$$

In our case

$$a_1 = -\frac{2e\sin\nu}{1 + e\cos\nu},$$

and η_2 is given by formula (10.8.20). Consequently, the second particular solution of the homogeneous equation in question is

$$\eta_3 = I^* \sin\nu, \qquad I^* = \int \frac{\rho^2}{\sin^2\nu} \, d\nu. \tag{10.8.21}$$

The integral I^* can be calculated explicitly [10.1], but the resulting expression of $I^*(\nu)$ is rather complicated. To avoid not seeing the forest for the trees, we will retain the compact expression (10.8.21) for $I^*(\nu)$. We merely note that, for a circular orbit ($e = 0$, $\rho = 1$), $I^*(\nu) = -\cot\nu$, and also that up to second order in the eccentricity

$$I^*(\nu)\sin\nu \approx -\cos\nu + 2e. \tag{10.8.22}$$

Thus, we have found two particular solutions of the homogeneous equation. This already allows us to apply the variation-of-constants method and obtain the general solution of equation (10.8.19) in the form

$$\left.\begin{aligned}
\eta &= \frac{c_1}{e}(I^*\sin\nu + \cos\nu) + c_2\sin\nu + c_3 I^*\sin\nu + \\
&\qquad\qquad + F_2\sin\nu + I^* F_3\sin\nu,\\
F_2 &= -p^3\int\frac{I^*\sin\nu}{\rho}\left(\frac{2I_\xi}{p^3} + \rho^3 f_\eta\right)d\nu,\\
F_3 &= \ \ p^3\int\frac{\sin\nu}{\rho}\left(\frac{2I_\xi}{p^3} + \rho^3 f_\eta\right)d\nu.
\end{aligned}\right\} \tag{10.8.23}$$

It remains to determine $\xi(\nu)$. To this end, let us return to equation (10.8.16). Since $\eta(\nu)$, and hence $\beta(\nu) = \eta/\rho$ are now known, a single quadrature yields

$$\alpha = c_1\nu - 2\int\frac{\eta}{\rho}\,d\nu + \int I_\xi\,d\nu + c_4. \tag{10.8.24}$$

Using (10.8.14) and (10.8.23), we obtain

$$\left.\begin{aligned}
\xi\ &= c_1\xi_1 + c_2\xi_2 + c_3\xi_3 + c_4\xi_4 + \rho\Psi,\\
\xi_1 &= \rho\left(\nu - \frac{2}{e}\int\frac{I^*\sin\nu + \cos\nu}{\rho}\,d\nu\right),\\
\xi_2 &= \rho(2 + e\cos\nu)\cos\nu,\\
\xi_3 &= -2\int\frac{I^*\sin\nu}{\rho}\,d\nu,\\
\xi_4 &= \rho,\\
\Psi &= \int\left[I_\xi - \frac{2}{\rho}(F_2\sin\nu + I^* F_3\sin\nu)\right]d\nu.
\end{aligned}\right\} \tag{10.8.25}$$

The expressions for η' and ξ' are obtained from (10.8.23) and (10.8.25) by differentiation. This completes the integration by quadratures of the system (10.8.6).

The quadratures figuring in ξ_1 and ξ_3 can be calculated in closed form; these explicit expressions are given in [10.1], with a complete list of computation formulas. Finding closed-form expressions for the quadratures I_ξ, F_2, F_3 and Ψ, is generally speaking, problematic – the answer depends on the specific form of the controls f_η, f_ξ, f_ζ. In practical large-scale calculations of trajectories it is apparently more convenient to find the above quadratures by numerical methods; however, there are totally realistic situations in which these quadratures can be written in explicit form.

The difficulties encountered in the integration of the system (10.8.6) carried out above are connected with the fact that it has variable coefficients, or, in other words, with the elliptic character of the carrier trajectory. This forced us to solve (10.8.6) for arbitrary values of the eccentricity $e < 1$. In return, the formulas obtained can be used (and were indeed used in [10.1]–[10.4]) in the calculation of trajectories of controlled interplanetary flights with an arbitrary reactive acceleration (in particular, with an optimal one).[2]

9. Some problems of relative motion

We must say here that the formulas obtained above have a considerable wider range of applicability and can be used to solve a whole class of problems of relative motion of celestial bodies that are close to one another as well as other problems. Let us list a few of them.

1. Variation of a Keplerian motion and isochronous derivatives. No trajectories computed beforehand can be realized exactly. That task is hindered by numerous (though usually small) errors in implementing the trajectory, in the assumptions made in its calculation, and so on. The deviation of an actual Keplerian interplanetary-flight trajectory from the computed trajectory (or, in the customary terminology, from the *nominal* trajectory) can be found from the formulas obtained above by setting $|\mathbf{f}| \equiv 0$. Here the nominal trajectory serves as carrier trajectory. By determining the integration constants in terms of the initial data and collecting the terms involving the initial coordinates and velocities (η_0, η_0', etc), we obtain expressions for the so-called *isochronous derivatives* of the Keplerian motion. Isochronous derivatives play an important role in the theory of space flight. They allow one to calculate the deviation of each coordinate (or component of velocity) at any moment of time. If the deviations are too large, one can command a trajectory correction and calculate the impulse necessary to implement it.

2. The problem of the motion of a body near a satellite. One can consider, for instance, the motion of a astronaut in the vicinity of a space station both in the case when the astronaut carries a "backpack-type" reactive device ($f \neq 0$) and when he does not carry one. One can single out "unsafe" regimes (or initial positions relative to the space station) for which the astronaut is in danger of getting too far from the station.

[2] Al the qudratures of optimal flight problems have been integrated explicitly in the recently published paper of A. A. Sukhanov, *Optimization of low-thrust trajectories*, Kosmicheskie Issled., **37** (1999), no. 2, 181–191 (Russian). To that end another, nonrotating carrier coordinate system was used.

3. The approach problem of two celestial objects. Assuming that in a previous stage a celestial object was already brought in the neighborhood of another such object, one can consider the final stage, that of controlling the reactive acceleration f so that the objects will meet with the smallest possible relative velocity (docking a rocket to a space station, assembling a space station, etc.), or with arbitrary velocity. In this last problem one may decide not to use a reactive acceleration (i.e., keep $f \equiv 0$), but seek an impulse (i.e., initial values of the components of the velocity) that guarantees the rendezvous.

4. Motion of a cloud of particles, ejected from a satellite, relative to the satellite.

5. Achieving a prescribed orbit by means of an additional impulse or reactive acceleration, i.e., the problem of small modification of the original orbit of a celestial body.

6. Short-duration action of perturbing factors on an orbit.

For many of the problems listed above the carrier trajectory can be chosen to be weakly elliptic (small e), and sometimes even circular. This simplifies formulas considerably, since one can retain only terms of first order in the eccentricity or even take $e = 0$. In this last case all the coefficients in the left-hand sides of the original equations (10.8.6) are constant ($\rho' = 0$, $\rho = 1$), which in principle simplifies their integration. Of course, the solution can also be obtained by letting $e \to 0$ in the general formulas (10.8.13), (10.8.23), (10.8.25). To deal with the indeterminacy in the first term of expression (10.8.23) for η and, correspondingly, for ξ, we need to use the approximation (10.8.22). For the rest, the passage to the limit $e \to 0$ is carried out with no difficulty: we set $I^* = \cot\nu$, $\rho = 1$, and also $p = 1$. This yields the following solution of the system of equations (10.8.6) for the carrier trajectory:

$$\begin{aligned}\eta = 2c_1 + c_2 \sin\nu - c_3\cos\nu + \sin\nu \int \cos\nu \left(2\int f_\xi\, d\nu + f_\eta\right) d\nu +& \\ + \cos\nu \int \sin\nu \left(2\int f_\xi\, d\nu + f_\eta\right) d\nu,&\end{aligned} \tag{10.9.1}$$

$$\begin{aligned}\xi = -3c_1\nu + 2c_2\cos\nu + 2c_3\sin\nu + c_4 +& \\ + \int \Big\{\int f_\xi\, d\nu - 2\Big[\sin\nu \int \cos\nu \left(2\int f_\xi\, d\nu + f_\eta\right) d\nu +& \\ + \cos\nu \int \sin\nu \left(2\int f_\xi\, d\nu + f_\eta\right)\Big]\Big\}\, d\nu,&\end{aligned} \tag{10.9.2}$$

$$\zeta = c_5\sin\nu + c_6\cos\nu + \sin\nu \int f_\zeta \cos\nu\, d\nu + \cos\nu \int f_\zeta \sin\nu\, d\nu. \tag{10.9.3}$$

The quadratures figuring in these formulas are already easier to calculate than in the case of an elliptic carrier trajectory. In a number of cases (for instance, when the components f_η, f_ξ, f_ζ are constant) the quadratures become very simple. In formulas (10.9.1)–(10.9.3), as in the original general formulas (10.8.13), (10.8.23), (10.8.25), the controls f_η, f_ξ, f_ζ were considered to be arbitrary functions of the true anomaly ν. However, we are mainly interested in the case when these components satisfy the optimal control equations (10.6.4). Then, as follows from the arguments of Section 7 (see formula (10.7.5)), in the general case of an elliptic carrier trajectory the solution of the optimal control equations has the form

$$\left.\begin{aligned} f_\eta &= \frac{a_1}{e}(I^* \sin\nu + \cos\nu) + a_2 \sin\nu + a_3 I^* \sin\nu, \\ f_\xi &= a_1\xi_1 + a_2\xi_2 + a_3\xi_3 + a_4\xi_4, \\ f_\zeta &= a_5\rho\sin\nu + a_6\rho\cos\nu. \end{aligned}\right\} \qquad (10.9.4)$$

Here $a_1, \dots, a_6$ are new arbitrary constants and $\xi_1, \dots, \xi_4$ are given by formulas (10.8.25).

Substituting the expressions (10.9.4) in formulas (10.8.23), (10.8.25), and (10.8.13), we obtain the complete solution in quadratures of the optimal flight problem (10.6.3)–(10.6.4). These formulas were used for a whole series of computations of flight trajectories to Mars, Venus, and Jupiter [10.1]–[10.3]. Some of the results of these computations will be discussed in the next section.

10. Computational results for optimal interplanetary trajectories

In the mechanics of space flight the usefulness of a solved problem is often determined by the possibility of taking the solution all the way "to numbers." As a rule, achieving the latter is not easy. Thus, in the method for calculating optimal trajectories considered above we need to perform at least the following computational steps: determine the position and velocity of the planets at the initial and terminal times, t_0 and t_1; calculate the carrier trajectory; calculate the boundary conditions in the forward-moving carrier coordinate system $Oxyz$ and then recalculate these conditions in the rotating carrier coordinate system $O\eta\xi\zeta$; calculate the quadratures (10.8.13), (10.8.23), (10.8.25) figuring in the solution of our problem; solve the boundary value problem following the scheme described in Section 7 and calculate the trajectory and the control in the carrier coordinate system $O\eta\xi\zeta$; finally, recalculate them in the system $Oxyz$.

To estimate the accuracy of the method of carrier trajectories in concrete cases one has to calculate (by means of an even more complex algorithm!) the trajectory corresponding to the exact equations (10.5.1) and compare it with the trajectory calculated by using our approximation. All these "kitchen chores" are not covered in our exposition, since they do not fit into the plan and style of our book. However, here is the place to remind our student-reader that serious work in modern mechanics of space flight is often unavoidably connected with such computational "chores," which of course are unthinkable without computers.

So, leaving aside the description of these computational chores, let us described the results that they yield.

Figures 10.5–10.8 show some characteristics of a mission to Mars. To get an idea about the accuracy of the method of carrier trajectories, these figures display exact as well as approximate characteristics. Figures 10.5 and 10.6 show how the reactive acceleration changes in magnitude and direction. Figure 10.7 shows trajectories of flight to Mars: exact, approximate, and carrier. The flight duration is 212 days. Finally, Figure 10.8 shows the flight trajectory (calculated by means of the approximate formulas) in the carrier coordinate system. We see that the maximal deviation from the origin of coordinates does not exceed 7 million kilometers, while the flight itself has a length of hundreds of million of kilometers. The graphs show how small the required magnitudes of the reactive acceleration are (maximum 1.5 mm/sec^2).

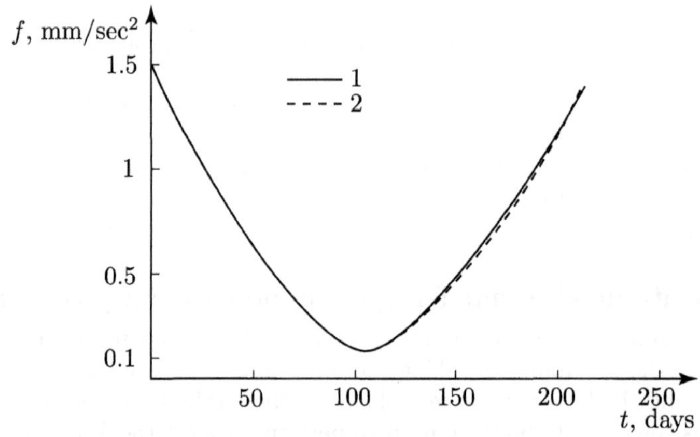

FIGURE 10.5. Variation of the magnitude of the reactive acceleration with time: 1–exact calculation; 2–approximate calculation

An important characteristic of the flight is its *angular range* Φ, that is, the angle swept by the radius vector of the spaceship from the origin of coordinates to the end of the flight. For the example considered in Figures 10.5–10.8, $\Phi \sim 150°$. The method of carrier trajectories proves to be sufficiently accurate precisely for such moderate angular ranges. Indeed, in such cases the flight trajectory does not differ very much from an arc of the properly chosen ellipse. If the angular ranges are large (360°, 720°, and so on), so that the flight represents, for example, a segment of a spiral connecting the initial and target orbits, then no Keplerian trajectory can be sufficiently close to the actual trajectory of a low-thrust flight, and the method of carrier trajectories fails.

One has to keep in mind, however, that the flights of highest practical interest have a small angular range, because this ensures that the flight duration is short. Hence, the method of carrier trajectories is useful precisely because it allows a fast and highly accurate computation of the trajectories that are of most practical

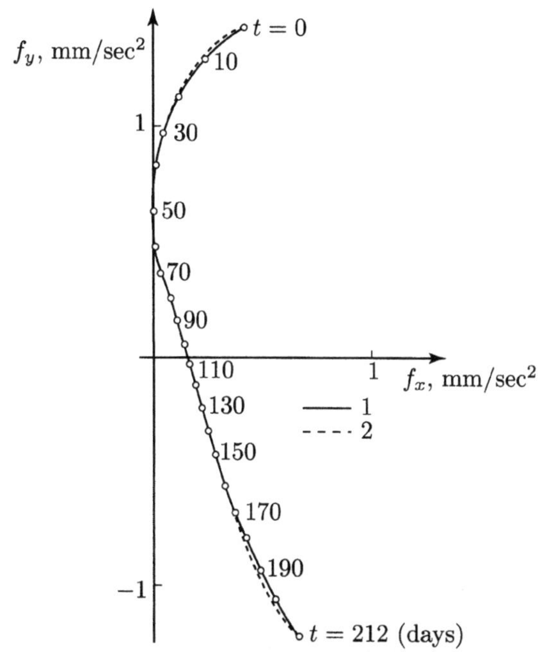

FIGURE 10.6. Variation of the magnitude and direction of the reactive acceleration: 1–exact calculation; 2–approximate calculation

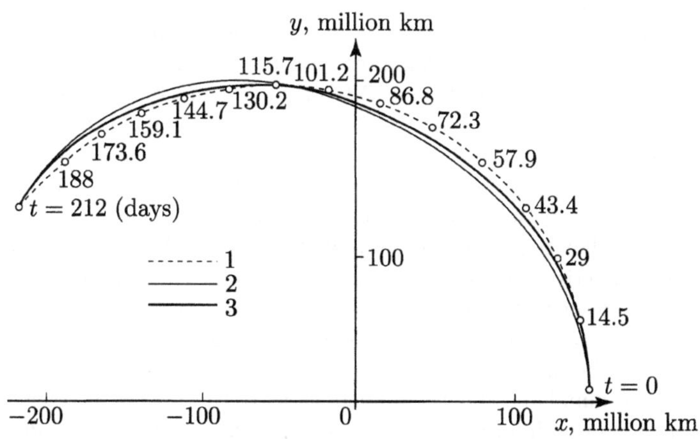

FIGURE 10.7. Flight trajectory to Mars: 1–carrier ellipse; 2–approximate trajectory; 3–exact trajectory

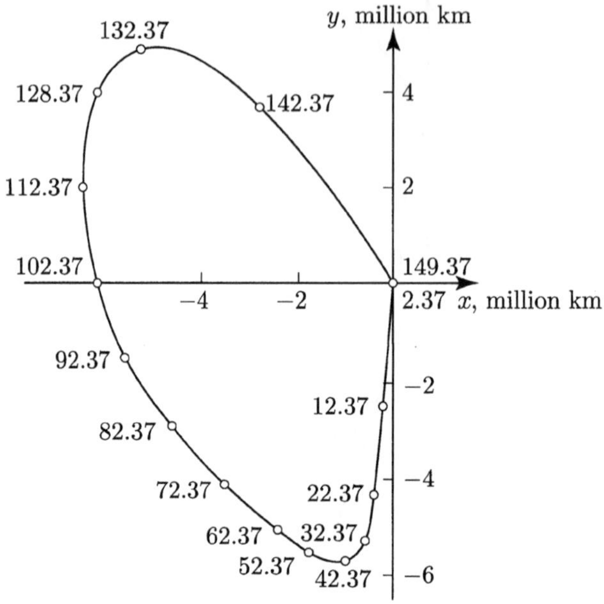

FIGURE 10.8. Trajectory in the carrier coordinate system. Values of the true anomaly on the carrier ellipse are shown along the trajectory

interest. Figure 10.9 gives a representation of the error made in the calculation of a basic flight characteristic, the integral I, when the method of carrier trajectories is used. The graph shows the dependence on Φ of the ratio $\Delta I/I$ (percentage), where I is the exact value of the integral and ΔI is the error made in its computation by the method of carrier trajectories. For flights with an angular range of 225–235°, the error is of only 1–5%, but it grows sharply when Φ increases above 235°. For $\Phi \sim 180°$ the error is negligible – only tenth or even hundredths of one percent. Let us mention also that flights with very small angular range ($\Phi < 50°$) are not very interesting, since they require rather large magnitudes of the reactive acceleration.

Some characteristics of flights to Mars and Venus, calculated by the approximation method described above, and their comparisons with the exact characteristics are given in Table 10.1, where T denotes the total flight duration, Φ is the angular range of the flight between the spheres of influence of the two planets, f_t and f_0 are the terminal and the initial magnitudes of the reactive accelerations; the last column shows the difference between the approximate value I_A and the exact value I_E in percents of I_E. The columns for I, f_t, and t_0 show the approximate (upper) and exact (lower) values of those quantities.

Let us emphasize here that, as one can see, the optimal control is very difficult to achieve. The reactive acceleration changes quite significantly in magnitude and

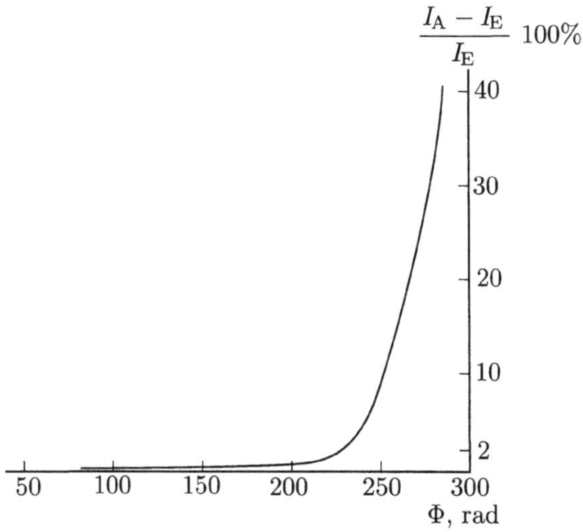

FIGURE 10.9. Dependence of the error made in the approximate calculation of the quantity I on the angular range Φ for an Earth-to-Mars flight

	Departure and arrival dates	I m^2/sec^2	T days	Φ	f_t mm/sec^2	f_0 mm/sec^2	$\frac{I_A - I_E}{I_E}$ 100%
to Mars	09.28.1960	11.53	212	147°.38	1.44	1.51	0.09
	04.28.1961	11.54			1.43	1.52	
	11.15.1964	3.23	330	222°.65	0.405	0.89	0.94
	10.11.1965	3.20			0.42	0.88	
to Venus	03.22.1964	26.44	121	136°.35	4.02	1.85	0.04
	07.21.1964	26.43			4.08	1.845	
	12.11.1963	11.975	220	234°.79	2.25	1.26	5.18
	07.18.1964	11.385			2.24	1.15	

TABLE 10.1. Some characteristics of flights to Mars and Venus

direction. And one of the engineering requirements is always that the control should be as simple as possible. A gain in the mass of the cargo delivered at the target planet translates into a loss of simplicity, and hence into a loss of reliability of the flight. Therefore, the following legitimate question arises: can we approximate the optimal control by one that is simpler to implement? Suppose, for example, that at the beginning of the flight the acceleration is constant in magnitude and direction, that in the middle flight segment the engines are turned off altogether (passive segment), and that at the end of the flight the acceleration is again constant in magnitude and direction. One can hope that such a "step-wise" control will to some degree approximate the optimal one.

V. V. Golubkov proposed two very simple step-wise control laws [10.3].

Let

$$\mathbf{f} = \begin{cases} \mathbf{f}, & \text{if } 0 \leq t \leq \tau, \\ 0, & \text{if } \tau \leq t \leq \tau + \Delta\tau, \\ \mathbf{g}, & \text{if } \tau + \Delta\tau \leq t \leq T. \end{cases} \tag{10.10.1}$$

The two cases of control law considered by Golubkov are as follows: 1) $f = g$, and 2) $f \neq g$; here the constant vectors **f** and **g** are the reactive accelerations in the first and respectively the third segment of the flight, t is the current time, T is the flight duration, τ is the duration the engines operate in the first segment of the flight, $\Delta\tau$ is the duration of the passive segment of the flight, $f = |\mathbf{f}|$, $g = |\mathbf{g}|$. Thus, generally speaking the trajectory has two active segments and one passive segment. The parameters of the control law are adjusted to satisfy the boundary conditions and the minimization condition of I.

Numerical calculations of Earth-to-Mars flights have been carried out for some fixed departure date by means of the method of carrier trajectories and in the exact formulation (in the latter the equations of motion were integrated numerically). Some results of these computations are shown below in tables 10.2 and 10.3, where θ [resp., $\Delta\theta$] denotes the ratio (τ, in fractions of T)/(τ, in days) [resp., ($\Delta\tau$, in fractions of T)/($\Delta\tau$, in days)].

An examination of these tables reveals that the accuracy of the calculation of the step-wise control is identical to that of the calculation of the optimal control (Figure 10.9). This justifies a comparison of the magnitude of the integral I in the step-wise and optimal controls in the approximate calculation of the corresponding trajectories.

Such a comparison shows what we loose when we replace the optimal control by the step-wise one. The corresponding data for Earth-to-Mars flights are given below in Table 10.4, in which δ stands for the ratio $(|I_1 - I_{\text{opt}}^{(1)}|)/I_1\, 100\%$, where $I_{\text{opt}}^{(1)}$ is the value of the integral I for the optimal control in the first approximation and I_1 is the value of I for the step-wise control in the first approximation, T is the flight duration, and Φ is the angular range.

Furthermore, the comparison shows that from an energy-balance point of view (in the sense of the quantity I) the optimal control law differs noticeably from the step-wise control. As seen from Table 10.4, this difference depends on the flight duration. For a flight duration of 150 days, the energy gain in using the optimal control law compared to the step-wise one is 9.2% when $f \neq g$ and 15.5% when $f = g$. For a flight duration of 350 days this gain is 39.4% and 51.5%, respectively. The step-wise control with $f \neq g$ is clearly more suitable than that with $f = g$.

An analysis of the exact computation of Earth-to-Mars flights with step-wise control (see tables 10.2 and 10.30) shows that I has a minimum with respect to the flight duration T for a fixed departure date. This minimum has a sufficiently gentle slope, which allows one to vary the flight duration within rather wide limits (of the order of 200 days) without affecting significantly the value of I. Furthermore, as tables 10.2 and 10.3 demonstrate, this minimum is achieved for a sufficiently long flight duration T (530 days when $f = g$ and 500 days when $f \neq g$).

We note also that when $f = g$ the minimum of I may be attained on trajectories that have no passive segment (see Table 10.2).

In the case of optimal trajectories the situation is different. Exact computations of optimal trajectories show that there are no flight durations T for which I attains its minimum. The larger the duration T is, the smaller I becomes (and hence the larger the number of laps the spiral flight trajectory makes around the Sun becomes). When $T \to \infty$, $I \to 0$ and we could save a lot of fuel if only a human's life would be not so short.

		Approximate problem					Exact problem					
T days	Φ degrees	I $\frac{\text{m}^2}{\text{sec}^2}$	f $\frac{\text{mm}}{\text{sec}^2}$	g $\frac{\text{mm}}{\text{sec}^2}$	θ	$\Delta\theta$	I $\frac{\text{m}^2}{\text{sec}^2}$	f $\frac{\text{mm}}{\text{sec}^2}$	g $\frac{\text{mm}}{\text{sec}^2}$	θ	$\Delta\theta$	$\frac{\lvert I_A - I_E \rvert}{I_E}$ 100%
150	128	93.77	2.46	3.943	$\frac{0.311}{47}$	$\frac{0.343}{52}$	93.00	2.451	3.896	$\frac{0.311}{47}$	$\frac{0.399}{51}$	0.83
250	176	8.690	0.784	0.849	$\frac{0.271}{68}$	$\frac{0.402}{101}$	8.639	0.792	0.837	$\frac{0.267}{67}$	$\frac{0.402}{100}$	0.57
350	232	4.251	0.694	0.255	$\frac{0.288}{80}$	$\frac{0.302}{106}$	4.151	0.695	0.263	$\frac{0.218}{76}$	$\frac{0.318}{111}$	2.41
400	263	4.623	0.612	0.238	$\frac{0.276}{110}$	$\frac{0.192}{77}$	3.870	0.644	0.198	$\frac{0.213}{85}$	$\frac{0.182}{73}$	19.7
500	326						3.510	0.467	0.181	$\frac{0.261}{130}$	$\frac{0.000}{0}$	
600	381						3.865	0.340	0.218	$\frac{0.391}{235}$	$\frac{0.000}{0}$	
700	428						4.464	0.344	0.211	$\frac{0.398}{279}$	$\frac{0.000}{0}$	

TABLE 10.2. $f \neq g$

To conclude this section let us mention that by using the approximation method of carrier trajectories one can reduce the computational time 7–10 times on the average, compared with the time spent for the computations with the exact equations. This proves to be very significant in large-scale computations of whole series of trajectories.

		Approximate problem				Exact problem				
T days	Φ degrees	I $\frac{\text{m}^2}{\text{sec}^2}$	f $\frac{\text{mm}}{\text{sec}^2}$	θ	$\Delta\theta$	I $\frac{\text{m}^2}{\text{sec}^2}$	f $\frac{\text{mm}}{\text{sec}^2}$	θ	$\Delta\theta$	$\frac{\lvert I_A - I_E \rvert}{I_E}$ 100%
150	128	98.173	3.493	$\frac{0.255}{34}$	$\frac{0.379}{57}$	97.098	3.456	$\frac{0.228}{34}$	$\frac{0.372}{56}$	1.11
250	176	8.710	0.824	$\frac{0.250}{63}$	$\frac{0.406}{102}$	8.649	0.819	$\frac{0.253}{63}$	$\frac{0.403}{101}$	0.70
350	232	5.070	0.520	$\frac{0.362}{127}$	$\frac{0.379}{133}$	5.027	0.518	$\frac{0.361}{126}$	$\frac{0.381}{133}$	0.85
400	263	5.442	0.461	$\frac{0.390}{156}$	$\frac{0.278}{111}$	4.909	0.441	$\frac{0.399}{160}$	$\frac{0.268}{107}$	10.9
500	326					4.570	0.342	$\frac{0.451}{225}$	$\frac{0.095}{48}$	
600	381					4.325	0.306	$\frac{0.476}{286}$	$\frac{0.025}{15}$	
700	428					7.287	0.419	$\frac{0.331}{232}$	$\frac{0.312}{219}$	

TABLE 10.3. $f = g$

Optimal control law			$f = g$		$f \neq g$	
Φ degrees	T days	$I_{\text{opt}}^{(1)}$ m^2/sec^2	I_1 m^2/sec^2	δ	I_1 m^2/sec^2	δ
128	150	85.0	98.2	15.5	93.8	9.2
152	200	21.4	25.4	18.7	24.1	12.6
176	250	7.5	8.71	16.1	8.69	15.9
203	300	4.2	5.4	28.6	5.0	19.0
232	350	3.3	5.0	51.5	4.3	39.4

TABLE 10.4.

11. Presentation of results of the computation of series of trajectories

When one computes a sufficiently large number of trajectories it is important to present the results of the computations in a transparent manner, in order to be able to choose a trajectory with desired characteristics.

As the basic determining parameters we take the initial and terminal moments of the flight between the spheres of influence of the two planets, t_1 and t_2. In all cases, in the computations it was required that the velocity of the spacecraft relative to the planet at the moment of exit from its sphere of influence be zero.

To present the results we will use the (t_1, t_2)-plane (Figure 10.10). In this plane it is convenient to consider various families of isolines, in particular, curves of equal values of the reactive acceleration or of the fuel expenditure. To each "to" flight there corresponds in this plane a point (t_1, t_2) above the bisector of the first quadrant, because $t_2 > t_1$.

The magnitude of the reactive acceleration (or of the quantity I) necessary for a given flight is determined by the isoline that passes through this point. The bisector corresponds to flight with infinite velocity. If for return flights one marks on the axis of ordinates the departure time t'_1 and on the axis of abscissae the arrival time t'_2, then to return flights there correspond points lying under the bisector of this plane, because $t'_2 > t'_1$.

In order for a pair of points (t_1, t_2) and (t'_1, t'_2) to be suitable for representing a round-trip flight it is necessary that $t'_1 \geq t_2$, in which case the difference $t'_1 - t_2$ is the time spent by the spacecraft near the target planet. The difference $T_\Sigma = t'_2 - t_1$, equal to the distance along the horizontal between the points (t_1, t_2) and (t'_1, t'_2), represents the total duration of the mission.

We represent the results of the computation of the integral I by curves I =const in the (t_1, t_2)-plane. In addition, we trace the isolines of maximal magnitudes of the reactive acceleration.

An analysis of the (t_1, t_2)-plane allows us to determine the characteristics of families of trajectories of flight to other planets, and also of trajectories with return to Earth.

To give an example, Figure 10.10 shows the isolines $f_{\max}$=const, I =const for missions to Mars carried out in the years 1964–1966 (the reader should not be surprised by these dates – they are only an example). Analogous isolines in the (t_1, t_2)-plane were calculated in [10.2], [10.3] for flights to Venus and Jupiter.

Tables 10.5–10.7 give characteristics of the most interesting trajectories with departure during 1964–1965. These characteristics were successfully obtained precisely by analyzing the isolines $f_{\max}$=const and I =const in the (t_1, t_2)-plane. The tables display characteristics of round-trip flights to Mars, Venus, and Jupiter, respectively. The first columns show the values of the maximal reactive acceleration $f_{\max}$ in the flight under consideration; the second column shows the departure date t_{dep}; the third column shows the total flight duration T_Σ (in days), which includes the time required for braking in the sphere of influence of the planet (to an orbit at an altitude h of 300 km for the flight to Mars and Venus, and the altitude of the fifth satellite of Jupiter, Amalthea, for flights to Jupiter) and the duration of the return escape; the fourth column shows the required value $I_{\text{T,R}}$ of the integral I for the flight between the spheres of influence of the planets ("to" and "return"); finally, the fifth column shows the required total value I_Σ of the integral I, including the expenditure for braking and return escape. The expenditures for braking and return escape are readily calculated if one knows the characteristics of the braking and escape segments (see, e.g., the seventh essay of this book).

Examining tables 10.5–10.7 we see that a maximal acceleration of 2 to 3 mm/sec^2 ensures round-trip flights from Earth to Mars or Venus [resp., Jupiter] with an approximate duration of one and a half years [resp., three years]. A one and a half year long flight to Mars is possible even for $f_{\max} = 1$ mm/sec^2.

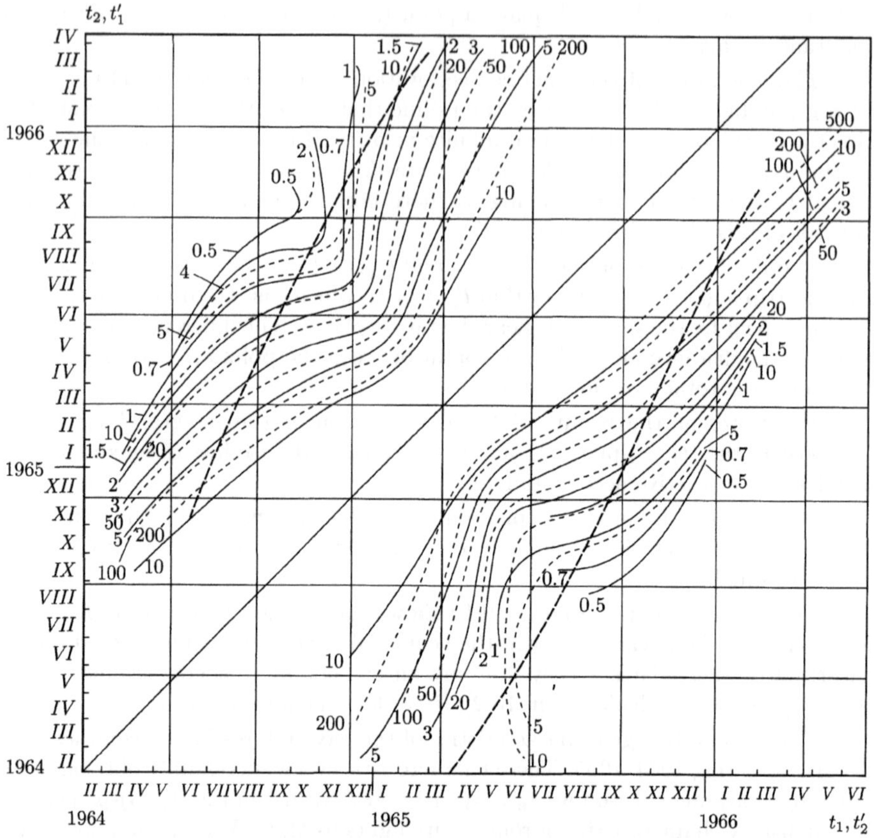

FIGURE 10.10. Isolines of characteristics of flights to Mars. The continuous lines are isolines of maximal accelerations (in mm/sec^2); the dashed lines are isolines of values of I (in m^2/sec^3)

Examining the (t_1, t_2)-plane one notes the following property of optimal trajectories: existence of a departure date that ensure a sharp maximum for the values of I and $f_{\max}$ for a given flight duration. Indeed, the geometric locus of the points of equal flight durations T_0 is a line parallel to the first quadrant's bisector in the (t_1, t_2)-plane. This line intersects some isolines in two points, which means that a flight of duration T_0 is possible (with a given value of I) for two different departure dates. A unique isoline $I = I_0$ is tangent to the line $t_2 - t_1 = T_0$; I_0 is precisely the minimal value of I for the flight of duration T_0, and the corresponding departure date is optimal. The optimal departure dates happened to fall in October 1964–January 1965, as seen in Figure 10.10.

I	II	III	IV	V
f_{max} mm/sec^2	t_{dep}	T_Σ days	$I_{T,R}$ m^2/sec^3	I_Σ m^2/sec^3
1	11.20.1964	564	13.5	18.7
2	12.20.1964	470	37	47.4
5	01.03.1965	330	170	196
10	02.06.1965	236	500	550

TABLE 10.5. Some characteristics of round-trip flights to Mars with minimal total flight duration

I	II	III	IV	V
f_{max} mm/sec^2	t_{dep}	T_Σ days	$I_{T,R}$ m^2/sec^3	I_Σ m^2/sec^3
3	05.04.1964	490	45	84
5	09.01.1964	340	125	185
10	01.09.1965	200	350	550

TABLE 10.6. Some characteristics of round-trip flights to Venus with minimal flight duration

I	II	III	IV	V
f_{max} mm/sec^2	t_{dep}	T_Σ days	$I_{T,R}$ m^2/sec^3	I_Σ m^2/sec^3
2	07.02.1965	1090	85	178
3	08.21.1965	1000	135	275
5	10.10.1965	890	320	562

TABLE 10.7. Some characteristics of round-trip flights to Jupiter

For $I < I_0$ a flight of duration T_0 is impossible (the corresponding isolines fail to intersect the line $t_2 - t_1 = T_0$). The dependence of the integral I on the departure date for a fixed duration of an Earth-to-Mars flight is shown in Figure 10.11, borrowed from the paper of W. G. Melbourne and C. G. Sauer [10.7]. They solved numerically the boundary value problem for the exact equations (10.5.1) of optimal motion. Figure 10.11 is interesting first of all because of the wide range of departure dates (four years) for the computed trajectories and for singling-out in this range two families of trajectories. We see, for example, that the optimal departure dates were in January–April 1969 and March–July 1971.

The intersection points (in the period September–October 1970) of the curves of these two families, shown in Figure 10.11, correspond to very interesting trajectories, which yield a nonunique solution of the boundary value problem. Indeed,

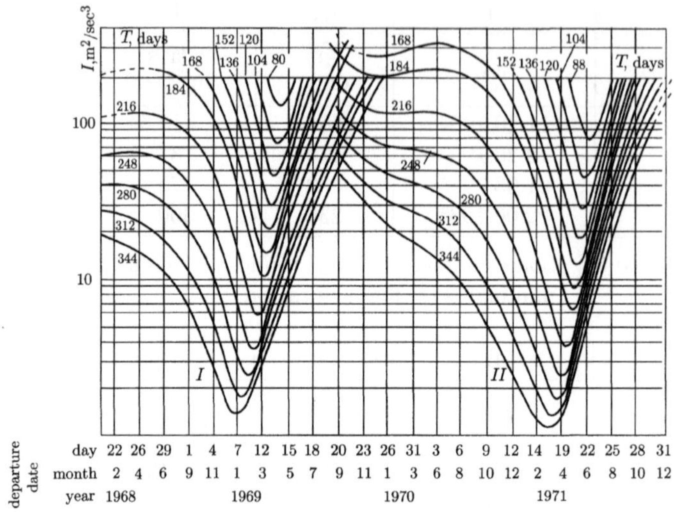

FIGURE 10.11. Dependence of I on the departure date and flight duration. Earth–to–Mars flight

in these points everything is the same: departure date, flight duration (and hence the initial and terminal values of the coordinates and velocities, too), the value of the integral I. Different are only the controls, and hence the trajectories. An example of such a nonunique solution of the boundary value problem is shown in Figure 10.12. The flight duration is 5 months, but one of the trajectories loops around the Sun (its angular range is $\Phi = \Phi_0 + 360° > 360°$), whereas the other does not (its angular range Φ_0 is not large – it is smaller than 90°).

The conditions under which the solution of the boundary value problem is not unique repeat themselves periodically with a period of about 2 years and 3 months. The optimal departure dates repeat themselves with the same period. This period is characteristic for any flights to Mars (and not only for low-thrust flights): the mutual positions of Earth and Mars relative to the Sun repeat themselves after about that time interval (from 2 years and 34 days to 2 years and 80 days).

If the value of the integral I necessary to carry out the desired mission is known, we can use formula (10.2.5) to calculate the ratio of the terminal mass to the initial one. To do this we must know the initial mass m_0 and the power of the engine N. Since N is bounded and depends on the weight of the engine, which in its turn is part of m_0, such a mass-balance analysis of a flight is not an easy task. In this regard the reader is referred to the already mentioned monograph of Grodzovskiĭ, Ivanov, and Tokarev [10.4].

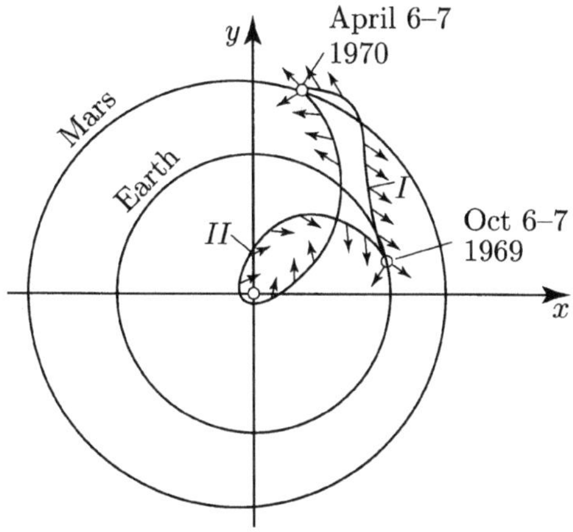

FIGURE 10.12. Example of two distinct Earth-to-Mars flight trajectories with identical departure and arrival dates and fuel expenditures

12. Correction of interplanetary trajectories

In recent years we have witnessed with our eyes a whole series of flights of unmanned spacecraft to Venus and Mars. The flight trajectories of these spacecraft were close to Keplerian, because the flights were carried out with the engines turned off. However, at launching some errors in placing a spacecraft in the interplanetary flight trajectory are unavoidable, and consequently perturbations that are poorly known may accumulate during flight. As a result, an interplanetary spacecraft flies, as a rule, not exactly along the desired trajectory, and if one does not "correct" the trajectory, then it can strongly deviate and miss the target planet. For this reason in real-life flights one resorts to *trajectory corrections*. Such corrections are carried out by turning on the reactive engine for a certain period of time. This prompts an avalanche of questions: When and for how long should we turn on the engine? Should we turn it on once, or several times? How should we direct the turned-on reactive acceleration in space? And, most important, how can we achieve the required correction in the most economic way? Indeed, the fuel supply carried by an interplanetary spacecraft is rather limited and must be used wisely.

We also need to take into account that the answers to the above questions depends on the real trajectory, whose parameters are measured with some error and hence are not known exactly, and that the process of turning on the engine is also affected by errors. Then it becomes clear that the theory of corrections of interplanetary flights cannot be simple. This theory was constructed and continues to be developed by a large number of scientists.

Here we will describe only a simplified correction problem, considered in a more general formulation by A. K. Platonov in [10.8]. It is natural to describe the motion along the true trajectory in terms of the deviation from the nominal trajectory, by using the already familiar linearized equation of motion in vector form (10.6.3). We will simplify this equation further by ignoring the gravitational perturbations of the trajectory, that is, the term $A_0\rho_0$. Investigations have shown that this does not introduce a large error in computations, and in return it allows us to understand without difficulty the qualitative features of the correcting maneuvers. Thus, our equations take the form

$$\dot{\mathbf{v}} = \mathbf{f}, \qquad \dot{\mathbf{r}} = \mathbf{v}, \tag{10.12.1}$$

and the correction problem can be stated as follows: *at the terminal time $t = T$ of the flight, the radius vector $\mathbf{r}$ and the velocity $\mathbf{v}$ must have the nominal values $\mathbf{r}_t$ and $\mathbf{v}_t$, respectively. At the same time, the control $\mathbf{f}$ must ensure that the functional S given by* (10.2.3) *takes its minimum value.* We shall assume that $\mathbf{f}$ is bounded in magnitude, i.e.,

$$-a \le |\mathbf{f}| \le a. \tag{10.12.2}$$

Let us form the Pontryagin function for the present case:

$$\mathcal{H} = -|\mathbf{f}| + (\mathbf{\Psi}_{\mathbf{v}}, \mathbf{f}) + (\mathbf{\Psi}_{\mathbf{r}}, \mathbf{v}), \tag{10.12.3}$$

where $(\cdot,\cdot)$ denotes the scalar product of two vectors, and write the equations in the canonically conjugate variables $\mathbf{\Psi}_{\mathbf{r}}$ and $\mathbf{\Psi}_{\mathbf{v}}$:

$$\dot{\mathbf{\Psi}}_{\mathbf{r}} = 0, \qquad \dot{\mathbf{\Psi}}_{\mathbf{v}} = -\mathbf{\Psi}_{\mathbf{r}}. \tag{10.12.4}$$

The optimal control law is the law under which the value of the function $\mathcal{H}$ is maximal with respect to the control $\mathbf{f}$. This function attains its maximum for

$$\left.\begin{array}{ll} |\mathbf{f}| = 0, & \text{if } |\mathbf{\Psi}_{\mathbf{v}}| < 1, \\ |\mathbf{f}| = a, \ \mathbf{f} = \alpha \mathbf{\Psi}_{\mathbf{v}}, & \text{if } |\mathbf{\Psi}_{\mathbf{v}}| > 1. \end{array}\right\} \tag{10.12.5}$$

This formula (in which α is an arbitrary constant scalar) gives the optimal correction law. The engine must be turned on only when the function $|\mathbf{\Psi}_{\mathbf{v}}| > 1$; when one does so, the value of the acceleration must be maximal and its direction must coincide with that of the vector $\mathbf{\Psi}_{\mathbf{v}}$. Note that equations (10.12.4) are readily integrated and yield the explicit dependence

$$\mathbf{\Psi}_{\mathbf{v}}(t) = -\mathbf{\Psi}_{\mathbf{r}}^0 t + \mathbf{\Psi}_{\mathbf{v}}^0, \tag{10.12.6}$$

whence

$$\begin{gathered} |\mathbf{\Psi}_{\mathbf{v}}(t)| = \sqrt{m^2t^2 + 2nt + k^2}\,, \\ m^2 = |\mathbf{\Psi}_{\mathbf{r}}^0|^2, \qquad n = -(\mathbf{\Psi}_{\mathbf{r}}^0, \mathbf{\Psi}_{\mathbf{v}}^0), \qquad k^2 = |\mathbf{\Psi}_{\mathbf{v}}^0|^2. \end{gathered} \tag{10.12.7}$$

From the character of the time-dependence (10.12.7) it is immediately clear that on the closed time interval $[t_0, T]$ of flight the quantity $|\mathbf{\Psi}_\mathbf{v}(t)|$ can exceed the value 1 ($|\mathbf{\Psi}_\mathbf{v}(t)| > 1$) on at most two segments, and then one of these segments is located at the beginning of the flight interval, and the other at its end. Thus, we have already reached an instructive conclusion: in our formulation of the problem, *an optimal correction contains at most two active segments, one starting at the initial moment of the flight, t_0, and the other ending at the terminal moment of the flight, T.*

Under favorable conditions one of these segments may become unnecessary, while at the other extreme, in the case of unfavorable conditions (very large initial deviations from the nominal trajectory), the two segments merge and the entire flight is carried out with the engines turned on.

Let us mention also that, by (10.12.5) and (10.12.6), under the optimal control the components of the reactive acceleration generated by the turned-on engine change linearly with time. The concrete control law of the reactive acceleration is determined by the expressions of the vectors $\mathbf{\Psi}_\mathbf{v}^0$ and $\mathbf{\Psi}_\mathbf{v}^0$ in terms of the initial conditions $\mathbf{r}_0$, $\mathbf{r}_\mathrm{t}$, $\mathbf{v}_0$, $\mathbf{v}_\mathrm{t}$ of our problem.

A survey of studies on trajectory correction can be found in [10.9].

From the character of the time-dependence (10.12.7) it is immediately clear that on the constant time interval $[t_0, T]$ of flight the error $\delta y(\Psi, T)$ can exceed the value [illegible] at most two segments, and that one of these segments is located at the beginning of the interval and the other at its end. [illegible]

[illegible]

Eleventh Essay

Relative Motion of Orbiting Bodies

Yes he died, but no, he couldn't fall,
Having entered the orbits of planetary motion.
A bottomless mouth gaped below,
But the forces of gravity were too weak.
  The firmament was pierced with rays,
  Divinely cold rays.
  He forward flew, free from decay,
  Gazing with dead eyes upon the stars.

N. Gumilev, *The Eagle*

1. In orbit – two satellites

Each of them moves around the Earth. And we know how. But how do the satellites move relative to one another? This motion must be known, for example, in order to solve the problem of docking two satellites.

Suppose the two satellites are connected by a cable. How does this influence their relative motion?

Or suppose an astronaut separates himself from his spaceship, to which he is connected by a cable (tether). The tether is used not merely because one is "overcautious" – rather, it is an absolutely necessary security measure. Otherwise, according to the laws of celestial mechanics, the astronaut will unavoidably be left behind the spaceship. Or he will outrun it, with the same serious consequences (see the above epigraph). How does an astronaut attached by a tether move relative to his spaceship?

The above questions (and many similar ones) fall into the class of problems of relative motion of cosmic objects – satellites, astronauts, or sections of orbital stations that are flexibly connected to one another.

We begin our analysis of relative motion with the case of the motion of two particles relative to one another. If convenient, one can think of one of these particles as being a satellite, and the other an astronaut. As long as the tether connecting the astronaut to the satellite is not stretched to the maximum, the motion of the astronaut can be regarded as free. When the tether becomes taut, a constraint reaction acts on the satellite, and an equal force of opposite direction acts on the astronaut.

First we will consider the problem of the orbital motion of two particles connected by a cable, and then we will carry out the analysis of the free motion of the particles. The motion of a two-body link (a system of two particles connected by a weightless, flexible, inextensible string) was analyzed in the papers of the author and E. T. Novikova [11.1], [11.2].

Following the paper [11.1], we denote the two masses by m_α and m, and the corresponding geocentric radius vectors by $\mathbf{r}_\alpha$ and $\mathbf{r}$. Suppose that the two particles are connected by a weightless, flexible, inextensible string of length l (Figure 11.1). Then during the motion the following constraint condition must be satisfied:

$$|\mathbf{r}-\mathbf{r}_\alpha| \leq l. \tag{11.1.1}$$

In other words, the distance between the two points cannot exceed the length of the string. The vector equations of motion can be written as Lagrange equations of the first kind.

Let us explain what we mean by this. Suppose that a system of n particles with coordinates x_i, y_i, z_i are subject to the ideal unilateral constraints

$$f_j(x_1, y_1, z_1, \dots, x_n, y_n, z_n) \leq 0, \qquad j = 1, \dots, m \tag{a}$$

The equations of motion

$$\left.\begin{aligned} m_\nu \frac{d^2 x_\nu}{dt^2} &= X_\nu + \sum_{j=1}^{m} \lambda_j \frac{\partial f_j}{\partial x_\nu}, \\ m_\nu \frac{d^2 y_\nu}{dt^2} &= Y_\nu + \sum_{j=1}^{m} \lambda_j \frac{\partial f_j}{\partial y_\nu}, \\ m_\nu \frac{d^2 z_\nu}{dt^2} &= Z_\nu + \sum_{j=1}^{m} \lambda_j \frac{\partial f_j}{\partial z_\nu}, \\ \nu &= 1, \dots, n, \end{aligned}\right\} \tag{b}$$

are called the *Lagrange equations of the first kind* and describe the motion of the system of particles of masses m_ν subject to the action of forces with components X_ν, Y_ν, Z_ν and the constraints (a). Here the Lagrange multiplier λ_j is different from zero only in the case where the constraint f_j is "tight" (i.e., equality holds in (a)); then necessarily $\lambda_j < 0$. If the constraint is not "tight" (strict inequality in (a)), then the corresponding Lagrange multiplier $\lambda_j = 0$. The $3n$ equations of motion together with the m constraint relations determine $3n + m$ unknown quantities: the $3n$ coordinates x_ν, y_ν, z_ν and the m Lagrange multipliers λ_j.

For the problem considered here the Lagrange equations of first kind (in vector form) read

$$\left.\begin{aligned} m\ddot{\mathbf{r}} + \frac{m\mu\mathbf{r}}{r^3} &= -2\lambda(\mathbf{r}-\mathbf{r}_\alpha), \\ m_\alpha\ddot{\mathbf{r}}_\alpha + \frac{m_\alpha\mu\mathbf{r}_\alpha}{r_\alpha^3} &= 2\lambda(\mathbf{r}-\mathbf{r}_\alpha). \end{aligned}\right\} \tag{11.1.2}$$

Equations (11.1.2) take into account the Newtonian forces by which each particle is attracted to the center of the Earth. The right-hand sides of these equations represent the reaction of the constraints. The multiplier λ needs to be determined.

If the motion is free, then in (11.1.1) one has equality and in (11.1.2) one puts $\lambda \equiv 0$. On the contrary, if the motion is constrained, then in (11.1.1) one has strict inequality and in (11.1.2) $\lambda \neq 0$.

We are interested in the relative motion of the particles m and m_α, which will be considered relative to their common center of mass. Naturally, in this situation we must know how the center of mass moves. Generally speaking, it will not move in a Keplerian orbit. This can be understood by considering the following simple "mental experiment."

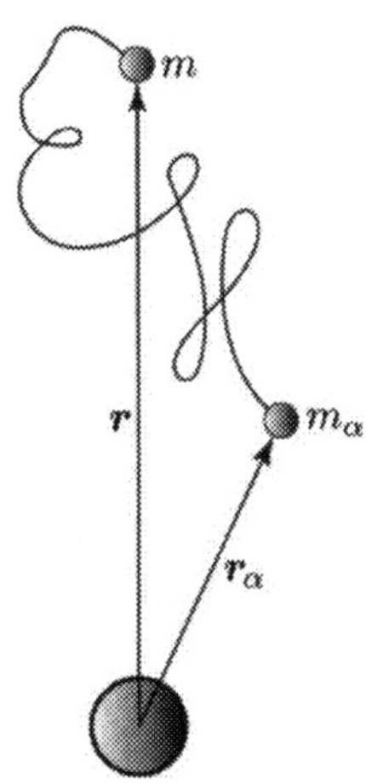

FIGURE 11.1. Link of two bodies in orbit

Suppose that two non-connected particles of equal mass start moving from one point in space, one in a circular orbit and the other in a very close contiguous elliptic orbit. Thanks to the very small difference between the two periods of revolution, after some sufficiently large number of revolutions the particles will lie at almost the same distance from the center of attraction, but the angular distance between their radius vectors will be close to 180°. Clearly, at that moment the center of mass of the system will almost coincide with the center of attraction, whereas at the initial moment of time it was at the point of tangency of the two orbits considered. Not only did the radius vector of the center of mass of the system change substantially in magnitude, but also, as is readily understood, it made a large number of circuits around the center of mass. Hence, over the time interval considered the trajectory of the center of mass has a spiral character (with, of course, local pulsations along the spiral). Over an infinite time interval one obtains a trajectory of pulsating-spiral type, so that the length of the radius vector of the center of mass will decrease to a very small value, and then increase to a value close to the initial one.

In our problem, however, the analysis is simplified considerably if we argue as follows. Let us make the natural assumption that l, the maximal possible distance between the two particles, is always small compared with the distances r and r_α from the particles to the center of attraction, i.e., $l/r \ll 1$ and $l/r_\alpha \ll 1$. Then with high accuracy we may assume that the center of mass of our system moves in a Keplerian orbit. To show this, let us denote the radius vector of the center of mass by $\mathbf{R}$. Then

$$\mathbf{R} = \frac{m\mathbf{r} + m_\alpha \mathbf{r}_\alpha}{m + m_\alpha}. \tag{11.1.3}$$

If we denote $m + m_\alpha = M$, the system (11.1.2) yields

$$\left. \begin{aligned} &M\ddot{\mathbf{R}} + \mu \left(\frac{m\mathbf{r}}{r^3} + \frac{m_\alpha \mathbf{r}_\alpha}{r_\alpha^3} \right) = 0, \\ &\mathbf{r} = \mathbf{R} + \boldsymbol{\rho}, \quad \mathbf{r}_\alpha = \mathbf{R} + \boldsymbol{\rho}_\alpha . \end{aligned} \right\} \tag{11.1.4}$$

Here $\boldsymbol{\rho}$ and $\boldsymbol{\rho}_\alpha$ are the radius vectors of the particles m and m_α relative to their common center of mass. Since $\rho \ll l$ and $\rho_\alpha \ll l$, we have $\rho \ll r$ and $\rho_\alpha \ll r_\alpha$, whence $r \approx r_\alpha \approx R$ and

$$\frac{\rho}{R} \ll 1, \qquad \frac{\rho_\alpha}{R} \ll 1. \tag{11.1.5}$$

Let us expand the expression in parentheses in equation (11.1.4) in powers of the small quantities (11.1.5) and retain only the terms of order less than or equal to two. This yields the equation

$$M\ddot{\mathbf{R}} + \frac{\mu M}{R^3}\mathbf{R} = \mathbf{F}, \tag{11.1.6}$$

where (below $(\cdot,\cdot)$ denotes the scalar product of vectors)

$$\left. \begin{aligned} \mathbf{F} &= \mathbf{F}_1 + \mathbf{F}_2, \\ \mathbf{F}_1 &= -\frac{3\mu}{R^5}(\mathbf{R}, m\boldsymbol{\rho} + m_\alpha \boldsymbol{\rho}_\alpha)\mathbf{R} + \frac{3\mu}{R^3}(m\boldsymbol{\rho} + m_\alpha \boldsymbol{\rho}_\alpha), \\ \mathbf{F}_2 &= \frac{3}{2}\frac{\mu}{R^3}\left\{ m\left[\left(\frac{\rho}{R}\right)^2 - 5\left(\frac{\mathbf{R}}{R}, \frac{\boldsymbol{\rho}}{R}\right)^2 \right] + m_\alpha \left[\left(\frac{\rho_\alpha}{R}\right)^2 - 5\left(\frac{\mathbf{R}}{R}, \frac{\boldsymbol{\rho}_\alpha}{R}\right)^2 \right] \right\}\mathbf{R} + \\ &\qquad + \frac{3\mu m}{R^5}(\mathbf{R}, \boldsymbol{\rho})\boldsymbol{\rho} + \frac{3\mu m_\alpha}{R^5}(\mathbf{R}, \boldsymbol{\rho}_\alpha)\boldsymbol{\rho}_\alpha. \end{aligned} \right\} \tag{11.1.7}$$

According to the notations (11.1.3) and (11.1.4),

$$\left. \begin{aligned} \boldsymbol{\rho}_\alpha &= \frac{m}{m + m_\alpha}(\mathbf{r}_\alpha - \mathbf{r}), \\ \boldsymbol{\rho} &= \frac{m_\alpha}{m + m_\alpha}(\mathbf{r} - \mathbf{r}_\alpha), \end{aligned} \right\} \tag{11.1.8}$$

and

$$m_\alpha \boldsymbol{\rho}_\alpha + m\boldsymbol{\rho} = 0. \tag{11.1.9}$$

If in equation (11.1.6) the perturbing force $\mathbf{F}$ were absent, then its solution $\mathbf{R}(t)$ would simply be a Keplerian motion. The presence of $\mathbf{F}$ "spoils" the Keplerian orbit. However, thanks to relation (11.1.9), the first-order terms in the perturbing force $\mathbf{F}$ (11.1.7) are identically equal to zero: $\mathbf{F}_1 \equiv 0$. Hence, the principal term in $\mathbf{F}$ is small of order two in the quantities (11.1.5) (it is given by formula (11.1.7) for $\mathbf{F}_2$). Consequently, in equation (11.1.6) $|\mathbf{F}|$ is very small compared to the principal term $\left|\frac{\mu M}{R^3}\mathbf{R}\right| = \frac{\mu M}{R^2}$. And with a high degree of accuracy (up to second-order effects) we may consider that

$$\ddot{\mathbf{R}} + \frac{\mu}{R^3}\mathbf{R} = 0, \tag{11.1.10}$$

that is, the center of mass moves in a Keplerian orbit, as claimed above.

2. The equations of relative motion

Thus, we arrived at the usual restricted formulation of the problem: study the motion of the system relative to its center of mass, which moves along a given Keplerian orbit.

By (11.1.8),

$$\boldsymbol{\rho} = -\frac{m_\alpha}{m}\boldsymbol{\rho}_\alpha. \tag{11.2.1}$$

Therefore, it suffices to consider the motion of a single particle, say, m_α (the astronaut); the motion of the second particle (the satellite) is then determined using (11.2.1). Note that if $m_\alpha/m \ll 1$ (the mass of the astronaut is considerably smaller than the mass of the satellite), then (in the limit $m_\alpha \to 0$) we have $|\boldsymbol{\rho}| \sim 0$, and so the motion of the particle m_α relative to the center of mass of the system (m_α, m) is almost identical to the motion of the particle m_α relative to the point m.

Since the motion must obey the condition (11.1.5) throughout the entire infinite time interval due to the constraint (11.1.1), in the study of the relative motion it is natural to confine ourselves to the equation governing small deviations from the center of mass of the system.

To this end let us linearize equation (11.1.2) with respect to $\boldsymbol{\rho}_\alpha$ and $\boldsymbol{\rho}$. This is conveniently carried out as follows. Divide the first equation (11.1.2) by m, the second by m_α, and subtract the results from one another. This yields

$$\left.\begin{aligned} \Delta\ddot{\mathbf{r}} + \mu\left(\frac{\mathbf{r}}{r^3} - \frac{\mathbf{r}_\alpha}{r_\alpha^3}\right) + 2\lambda\left(\frac{1}{m} + \frac{1}{m_\alpha}\right)\Delta\mathbf{r} = 0, \\ \Delta\mathbf{r} = \mathbf{r} - \mathbf{r}_\alpha. \end{aligned}\right\} \tag{11.2.2}$$

Now linearize this equation in $\Delta\mathbf{r}$ using the relation $\Delta\mathbf{r} = -((m + m_\alpha)/m)\boldsymbol{\rho}_\alpha$, which is another way of writing (11.1.8). We obtain

$$\ddot{\boldsymbol{\rho}}_\alpha + \frac{\mu}{R^3}\boldsymbol{\rho}_\alpha - \frac{3\mu}{R^5}(\mathbf{R}, \boldsymbol{\rho}_\alpha)\mathbf{R} = \widetilde{\lambda}_\alpha \boldsymbol{\rho}_\alpha, \tag{11.2.3}$$

where $\widetilde{\lambda}_\alpha = -2\lambda(m + m_\alpha)/(mm_\alpha)$ is a new multiplier, subject to determination. Here instead of (11.1.1) the constraint is given by the inequality

$$\rho_\alpha \leq \frac{m}{m + m_\alpha} l$$

and in the constrained motion governed by (11.2.3) it necessarily holds that $\widetilde{\lambda}_\alpha < 0$.

The vector equation (11.2.3) supplemented by the constraint (11.2.4) determines the motion of the particle m_α (astronaut) relative to the center of mass of the system of particles m and m_α (satellite–astronaut). In equation (11.2.3), $\mathbf{R}(t)$ is a known function of time, which defines the Keplerian trajectory of the system's center of mass. Equation (11.2.3) itself is a particular case of linear equations that we have already encountered (see [11.2]), namely, the equations that describe the motion in the carrier coordinate system (cf. the 10th essay).

Now let us consider the free motion of the particle m_α relative to the center of mass of the system m, m_α. In this case we must set $\widetilde{\lambda}_\alpha \equiv 0$ in (11.2.3). Then the situation we have to deal with now can be described as follows: An astronaut separates from his spaceship. She/he is not connected to the spaceship by a tether, and carries no individual reactive engine. What will be her/his fate?

3. Free motion of an astronaut relative to his spaceship

For the sake of simplicity we will assume that the orbit of the center of mass of the system is circular and the motion of the astronaut takes place in the plane of this orbit. Since the mass of the astronaut is much smaller than the mass of the spaceship, in what follows we will not distinguish between the center of mass of the spaceship and that of the system spaceship–astronaut. Let us attach to the center of mass O a coordinate system η, ξ, ζ whose axes are permanently directed along the radius vector η of the spaceship, along the transversal direction ξ, and along the normal ζ to the orbital plane. For a circular orbit, in the case of planar motion of the astronaut, the coordinates η and ξ of the astronaut satisfy the equations

$$\eta'' - 2\xi' - 3\eta = 0, \qquad \xi'' + 2\eta' = 0. \tag{11.3.1}$$

These equation can be readily derived as a special case of the vector equation (11.2.3). In (11.3.1) the prime denotes differentiation with respect to the dimensionless time $\tau = \omega_0 t$, where ω_0 is the angular velocity of the motion of the spaceship's center of mass on its orbit. Equations (11.3.1) describe the motion of the astronaut for as long as the tether is not taut. This motion will be referred to here as *free*. To analyze the free motion, let us integrate equations (11.3.1). The solution is obtained with no difficulty:

$$\left.\begin{aligned} \eta &= 2c_1 + c_2 \sin\tau + c_3 \cos\tau, \\ \xi &= c_4 - 3c_1\tau + 2c_2 \cos\tau - 2c_3 \sin\tau, \end{aligned}\right\} \tag{11.3.2}$$

where the arbitrary constants c_1, c_2, c_3, c_4 are expressed in terms of the initial conditions (at $\tau = 0$) of the problem as follows:

$$c_1 = 2\eta_0 + \xi_0', \quad c_2 = \eta_0', \quad c_3 = -3\eta_0 - 2\xi_0', \quad c_4 = \xi_0 - 2\eta_0'. \tag{11.3.3}$$

To get a better representation of the trajectory traced by the astronaut, we note that the coordinates $\eta(\tau)$, $\xi(\tau)$ and the time τ are connected by the following relation (which is an easy consequence of (11.3.2)):

$$\frac{(\eta - 2c_1)^2}{c_2^2 + c_3^2} + \frac{[\xi - (c_4 - 3c_1\tau)]^2}{4(c_2^2 + c_3^2)} = 1. \tag{11.3.4}$$

Here it is assumed that at least one of the constants c_2, c_3 is different from zero – otherwise, as seen from (11.3.2), the astronaut will uniformly move away from the spaceship along the axis ξ, i.e., parallel to the tangent to the orbit (Figure 11.2). The reader should recall that our linear equations (11.3.1) represent only an approximation. Rigorously speaking, both the astronaut and the spaceship move along elliptic orbits that are close to one another, and equations (11.3.1) describe their relative motion only in a small neighborhood of the spaceship, that is, only for as long as the distance between the astronaut and the spaceship is small compared with the dimensions of the spaceship's orbit.

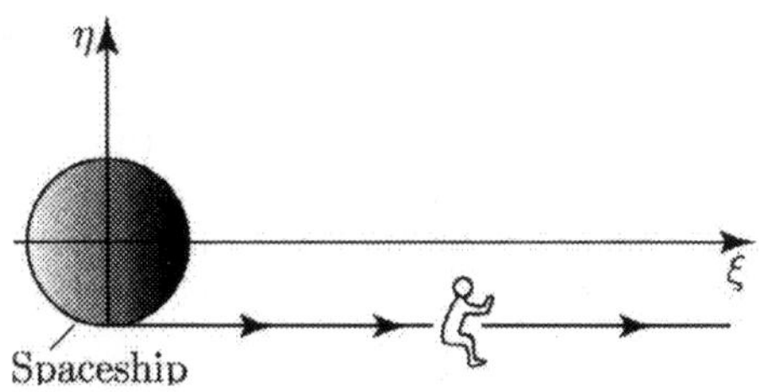

FIGURE 11.2. Uniform motion of the astronaut away from the spaceship

Formulas (11.3.3) show that free uniform motion away from the spaceship is possible only in the absence of an initial velocity directed along the radius vector of the spaceship (this is a necessary, but not sufficient condition; it is also required that $c_3 = 0$).

Returning to the general formula (11.3.4), let us examine first the case of periodic motions of the astronaut. From (11.3.2) and (11.3.4) it follows that such motions may take place only when $c_1 \equiv 0$. Their period coincides with the period of revolution of the spaceship, and the motions themselves take place along an ellipse centered at the point $\eta^* = 0$, $\xi^* = c_4$, whose semi-minor axis is parallel to the η-axis and is equal in magnitude to $\sqrt{c_2^2 + c_3^2}$, and whose semi-major axis is twice as big (Figure 11.3). In particular, it is possible for the astronaut to remain "frozen" in the point $\eta^* = 0$, $\xi^* = \xi_0$ if he was there originally and had no velocity relative to the spaceship. In this last case the astronaut will move around the Earth in the same orbit as the spaceship (but will be located at a different point on the orbit). It is important to note that such a "freezing" and, in general, periodic motions relative to the spaceship are unstable. Once we change the initial conditions so that $c_1 \neq 0$ (even with $|c_1|$ as small as we wish), the periodic motion is destroyed and a new, more general case of motion arises.

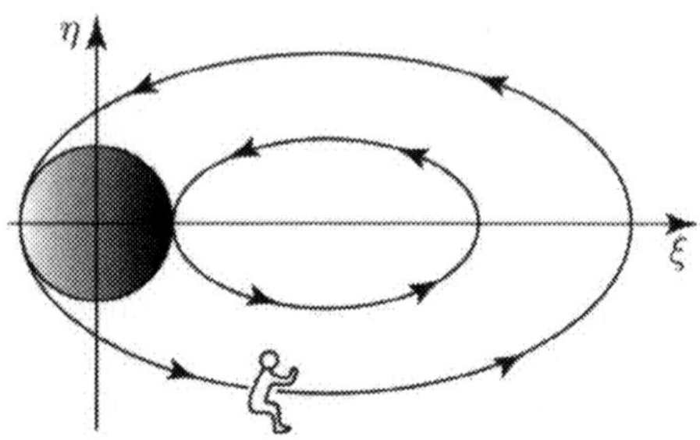

FIGURE 11.3. Periodic motion of the astronaut

In that situation the uniform motion (Figure 11.2) and the periodic motion (Figure 11.3) become superposed, resulting in motion along the trajectory shown in Figure 11.4. From (11.3.4) it is seen that this motion can be represented as the motion on an ellipse with semi-axes $\sqrt{c_2^2 + c_3^2}$ and $2\sqrt{c_2^2 + c_3^2}$, stretched along the ξ-axis. Moreover, the coordinate η^* of the center of this ellipse does not change with time: $\eta^* = 2c_1$, while the second coordinate ξ^* of the center moves uniformly along the ξ-axis: $\xi = c_4 - 3c_1\tau$. Therefore, while the astronaut traverses some arc of the ellipse, the ellipse itself manages to shift along the ξ-axis by some distance.

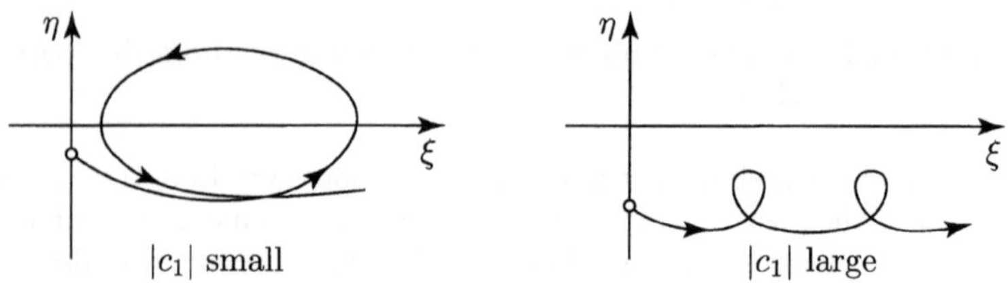

FIGURE 11.4. Sketch of the astronaut's motion relative to the spaceship in the general case

Thus, only in exceptional cases will an astronaut who is not fastened to the spaceship be able to return aboard. In the general case, once he left the spaceship, the astronaut will never be able to return. First he will stay close to the spaceship, but once he moves slightly away, he will gradually remain behind (or outrun) the spaceship. After half a circuit of the spaceship's flight the astronaut will even catch up with the spaceship and may get very close to it. But the little spark of hope will soon die out: the perfidious trajectory of relative motion will take the astronaut into a pirouette, and he will again finds himself moving away from his home spaceship.

From then on the astronaut's fate is horrible: He is doomed to revolve around the Earth forever in his own orbit, loosing the abandoned spaceship from sight. After many, many orbital circuits the astronaut may again get close to the spaceship, but it will be too late...

So, tether the astronaut to the spaceship by a good, decent, reliable cable!

4. Leonov and the lens cap

The preceding analysis of the relative motion applies not only to the investigation of the motion of an astronaut relative to his spaceship, but also to that of the motion of any object relative to any body that moves in orbit. For instance, one can track the motion relative to the astronaut of any object thrown (or lost) by the former.

Such a situation did indeed arise, for example, when the astronaut A. Leonov stepped out of his spaceship in open space. Here is an excerpt from P. Belyaev and A. Leonov's flight report [11.4]:

"... He unscrewed the lens cap of the photo camera attached to the spaceship's outer surface. What to do with it? And the astronaut, swinging his arm, hurled the lens cap toward the Earth. The small object, shining in the Sun, began to distance itself rapidly and soon vanished from sight."

It is instructive to investigate what happened with that lens cap. Pondering over this question, we must necessarily go through several stages.

The first stage does not require mind effort: being used to think in terrestrial categories, we would find nothing surprising in the fact that an object thrown toward the Earth will fall on no other place but on Earth.

The second stage involves some thinking: we recall that the astronaut flies above the Earth at an enormous speed, of about 8,000 m/sec! With what speed did Leonov throw the lens cap? A sufficiently strong throw by a man's arm imparts to a light stone a speed of 15–20 m/sec. For definiteness, let us assume that the initial speed of the lens cap relative to Leonov is equal 10 m/sec. To track the motion of the lens cap relative to the Earth we must geometrically superpose Leonov's velocity relative to the Earth and the velocity of the lens cap relative to Leonov (Figure 11.5). We see that the resultant velocity of the lens cap changes only negligibly in magnitude and direction! This means that the lens cap was simply transferred to an elliptic orbit around the Earth that was very close to Leonov's orbit. Is this in agreement with the assertion that the lens cap, once thrown toward the Earth, distances itself rapidly?

Here we embark on the third, research stage. As persons of reason, we recall that the flight report speaks not about how the lens cap moved relative to the

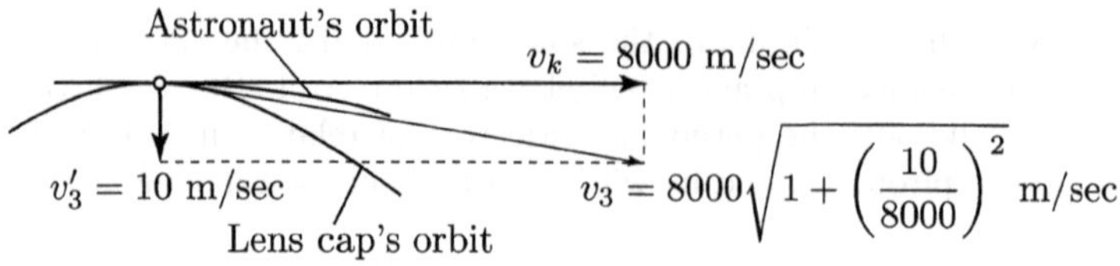

FIGURE 11.5. Motion of the lens cap

Earth, but about how Leonov saw that motion! In other words, we must consider the motion of the lens cap relative to Leonov. And why not, since we are already prepared to do so?! Let us start from the equations (11.3.2) governing the trajectories of relative motion. Since the lens cap was thrown straight toward the Earth, we may take the following initial data: $\eta_0 = \xi_0 = 0$, $\eta_0' \neq 0$, $\xi_0' = 0$. Then, as follows from (11.3.3), $c_1 = c_3 = 0$, $c_2 = \eta_0'$, $c_4 = -2\eta_0'$. Substituting these values of the constants in (11.3.2) and (11.3.4), we discover that the lens cap executes a periodic motion relative to Leonov! This trajectory is shown in Figure 11.6.

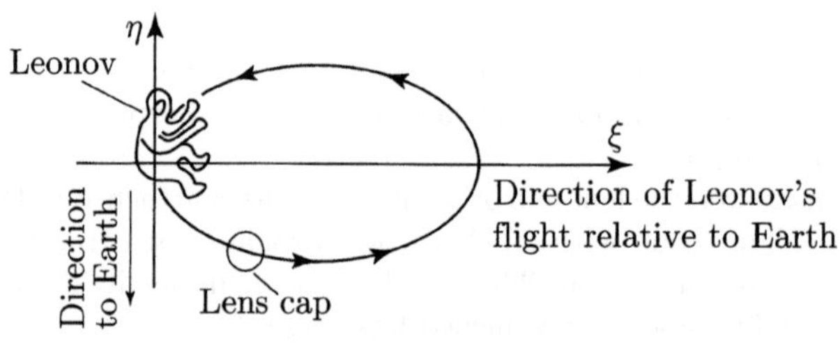

FIGURE 11.6. Periodic trajectory of the lens cap

Should Leonov have proceeded in free flight for a whole lap of orbit, he would have ended up with the lens cap returning right back into his hands, but arriving from the opposite side with respect to the Earth. (Here, of course, we ignore secondary factors – the throw being different from the one we considered, various perturbations that act differently on the astronaut and on the lens cap in their flight, and so on). Examining the figure which shows the trajectory of the lens cap relative to the astronaut, we ask ourselves the following question: could Leonov actually have seen the lens cap moving rapidly away in the direction of the Earth?

It is time for the fourth stage of our analysis – the computational stage. The parametric equation of the trajectory in the situation at hand is given by the formulas

$$\eta = \eta_0' \sin\tau, \quad \xi = 2\eta_0'(\cos\tau - 1) = -4\eta_0' \sin^2(\tau/2). \tag{11.4.1}$$

Here $|\eta_0'| = v/V \approx 10\,\text{m/sec}: 8{,}000\,\text{m/sec} = 0.125 \cdot 10^{-2}$, v is the initial speed of the thrown lens cap relative to Leonov, and V is Leonov's speed relative to Earth. Suppose that Leonov observed the lens cap for a period of $t = 1.5$ minutes. Then $\tau = \omega t = 2\pi t/T = 2\pi 1.5/90$, where $T = 90$ minutes is the period of revolution on

Leonov's orbit. In degrees, $\tau = 360 \cdot 1.5/90 = 6°$. Inserting the numerical values of η_0' and τ in formula (11.4.1) we calculate the coordinates of the lens cap relative to the astronaut: $\eta = -0.131 \cdot 10^{-3}$, $\xi = 0.0137 \cdot 10^{-3}$. To pass to dimensional quantities, we have to multiply these numbers by the distance from the astronaut to the center of the Earth. Taking this distance equal to $R = 6,600$ km, we obtain $\tilde{\eta} = \eta R \approx -865$ m, $\tilde{\xi} = \xi R \approx 90$ m. But the coordinate η [resp., ξ] corresponds to the displacement of the lens cap toward [resp., sideways from] the Earth.

It is doubtful that one can follow just with the naked eye the flight of a small object (the lens cap) for a distance of almost 1 km. Even if we succeed in doing so, it would be hard to notice the shift of the lens cap away from the radial direction, because this shift represents only (approximately) one tenth of the distance traversed by the cap. If one considers that Leonov lost the lens cap from sight earlier (which is far more probable), then the ratio of the transversal shift of the lens cap to the total distance traversed by it will be even smaller. Thus, we conclude that for as long as ones eyes can see, the lens cap seems to fly toward the Earth. After throwing the lens cap toward the Earth, Leonov should indeed have observed its trivial fall to the Earth!

5. Space probe

The free motion of a body relative to another body was considered above under the assumption that the distance between the bodies is small. This allowed us to confine our analysis to the linearized equations.

However, there are problems in which the relative motion must be considered also when the bodies under consideration lie at large distances from one another. Such is, for example, the problem of probing the space near the Sun. Let us imagine that a spaceship is launched from Earth with the mission of penetrating deeply into space and returning to Earth after several months (or years). During its flight this spaceship – an unmanned scientific station ("space probe") – collects and stores information about space and then delivers itself this information to Earth. Can such a space-probe flight trajectory be realized? Clearly, the answer is affirmative. If the probe moves around the Sun in an elliptic orbit with a period commensurable with the period of revolution of the Earth around the Sun, then, naturally, the probe and the Earth will meet periodically. For example, if the probe's period of revolution is 1.5 years, then after three years the probe will return to Earth. It is both interesting and necessary to track the motion of the probe not relative to the Sun, but relative to the Earth.

Such a problem was considered by V. A. Egorov in [11.5]. Figures 11.7 and 11.8, borrowed from [11.5], show trajectories of a space probe in a coordinate system that rotates together with the Earth (Figure 11.7) and the same trajectories in an absolute coordinate system (Figure 11.8). One of these trajectories has a period of motion relative to the Sun of 0.8 years and the minimal distance from the Sun that it reaches is of 108 million km (the maximal distance is obviously identical to the radius of Earth's orbit and equals 149 million km). Two successive

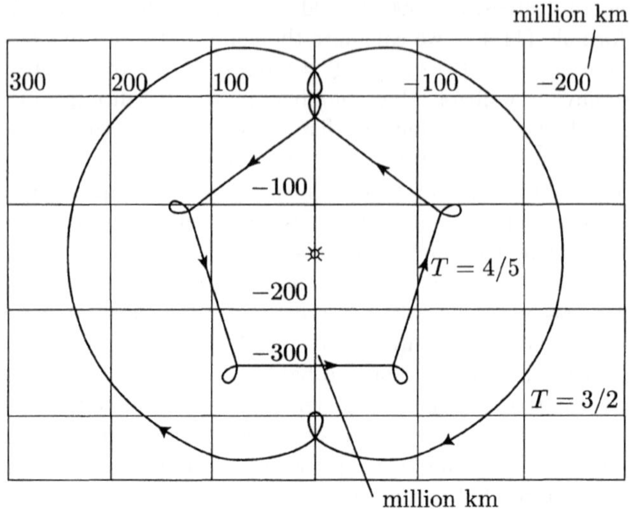

FIGURE 11.7. Two space-probe trajectories relative to the Earth

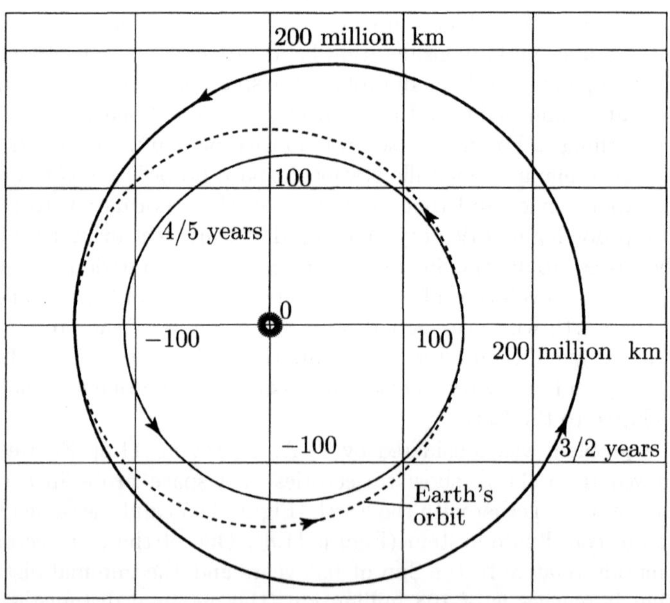

FIGURE 11.8. Two space-probe trajectories relative to the Sun

encounters with Earth are separated by an interval of 4 years; over this period the probe makes 5 revolutions around the Sun. Another trajectory has a period of 1.5 years (of motion relative to the Sun) and the probe encounters Earth after 3 years. For this trajectory the minimal [resp., maximal] distance to the Sun is 149 million km [resp., 242 million km].

For other space-probe problems trajectories with periods of different multiplicity (3/4, 2/3 years or 2, 3 years, and so on) may be suitable.

But let us return to our problem of relative motion of one particle close to another.

6. Boleadoras in space

Finally, let us investigate the dynamics of a link of two bodies launched in orbit [11.1], [11.2]. Such a structure, consisting of two masses connected by an "ideal" string (Figure 11.1), resembles the Argentinian "boleadoras" (or "bolas"). In the Argentinian pampas boleadoras are used to catch running-wild cattle; a somewhat more unusual usage of boleadoras was beautifully described by G. Durrell.

> "All the peons now had their boleadoras out, and I could see the balls gleaming on the ends of the long strings as they whirled them round and round their heads. The rheas turned in a bunch and ran towards us ... I could hear the whine of the boleadoras reach a crescendo ending in a sort of long-drawn 'wheep.' The cord and the balls flew flailing through the air, wound themselves with octopus-like skill round the legs and neck of the flying bird. It ran for two more steps, then the cord tightened and it fell to the ground, legs and wings thrashing." (G. Durrell, *The Drunken Forest*, Rupert Hart-Davis, Soho Square, London, 1956.)

In what follows we will speak about an astronaut connected by a tether to a spaceship (satellite), but one has to keep in mind that our idealized scheme is equally well suitable for the description of the motion of other orbital systems, for example, a two-section satellite whose sections are connected by a cable.

All that we said above about the motion of an astronaut concerned the case of free motion of the astronaut. Our analysis shows that in the general case the astronaut unavoidably distances himself from the spaceship (indeed, the probability of engaging a periodic trajectory is small), and at some moment of time he reaches a point where the tether is fully stretched (let l denote the length of the tether). This is when things become the most interesting. The astronaut's future now depends in essential manner on the elastic properties of the tether. First let us consider an absolutely anelastic tether. This assumption is quite essential: under other conditions the motion of the astronaut is of a completely different nature.

So, we are at the moment when the astronaut reaches a distance to the spaceship equal to the length l of the tether. Let us examine the motion starting from this moment, considering that the initial data of this motion are known (at least from the solution of the preceding problem on the free motion of the astronaut). From general considerations, one can expect that the astronaut will

either continue his free motion, or move along an arc of the circle of radius l around the center of mass of the spaceship. The second type of motion is termed *constrained motion*.

Under what conditions the free motion passes into constrained motion? What is the character of the constrained motion? Can the constrained motion go back into free motion, and under what conditions? Let us attempt to answer these questions.

When the astronaut reaches some point of the sphere of radius l – or of the circle of radius l in the planar case (we will simply say "reaches the sphere"), the component of his velocity in the direction pointing along the taut tether away from the spaceship vanishes completely due to the absolute anelasticity of the tether. The component of the velocity normal to the taut tether is preserved. From equations (11.3.1) one can readily derive the energy integral in the form

$$(\eta')^2 + (\xi')^2 - 3\eta^2 = h = \text{const.} \tag{11.6.1}$$

Let us introduce polar coordinates, as indicated in Figure 11.9:

$$\eta = \rho \sin\varphi, \quad \xi = \rho\cos\varphi. \tag{11.6.2}$$

In these variables the integral (11.6.1) becomes

$$(\rho')^2 + \rho^2(\varphi')^2 - 3\rho^2\sin^2\varphi = h. \tag{11.6.3}$$

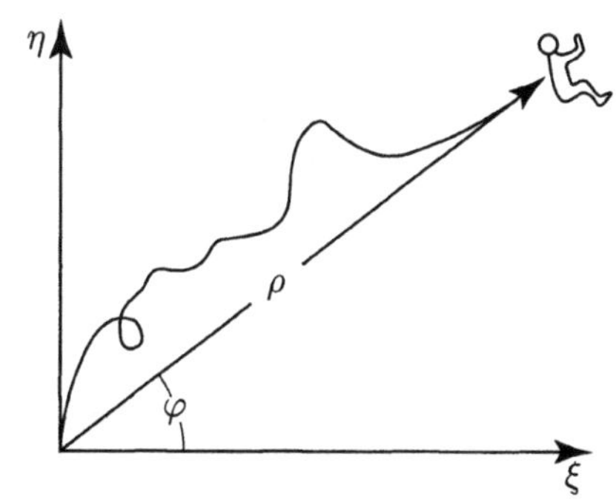

FIGURE 11.9. Polar coordinates

In free motion the energy h of the astronaut is conserved up to the moment τ_s when he reaches the sphere. Hence, at time τ_s we have

$$h_s^{(0)} = (\rho_s')^2 + \rho_s^2(\varphi_s')^2 - 3\rho_s^2\sin^2\varphi_s = (\rho_0')^2 + \rho_0^2(\varphi_0')^2 - 3\rho_0^2\sin^2\varphi_0, \tag{11.6.4}$$

where of course $\rho_s = l$; here the subscript 0 [resp., s] indicates the initial values of variables [resp., the values at time τ_s]. As we mentioned above, at τ_s the radial velocity ρ' undergoes a jump, dropping to zero. However, the transversal velocity $\rho\varphi'$ is fully preserved.[1] This means that the energy also jumps, from the value $h_s^{(0)}$ given by formula (11.6.4), to

$$h_s^{(1)} = \rho_s^2(\varphi_s')^2 - 3\rho_s^2 \sin^2\varphi_s, \quad \rho_s = l.$$

From this moment on the motion takes place on the sphere of radius $\rho = l$. This motion is completely determined if one knows the dependence $\varphi(\tau)$, which in turn is obtained by integrating the equation

$$(\varphi')^2 - 3\sin^2\varphi = h_1, \tag{11.6.5}$$

where $h_1 = h_s^{(1)}/l^2$. Equation (11.6.5) is obtained from (11.6.3) by setting $\rho' = 0$, $\rho = l$, and represents an energy integral similar to that of the motion of a physical pendulum; hence, it can be integrated or analyzed.

However, the motion described by equation (11.6.5) can take place only when the tether is fully stretched, and that is turn can take place only when the second derivative ρ'' is positive. Indeed, since on the sphere one has that $\rho' = 0$, the condition $\rho'' > 0$ means that the radial velocity has the tendency to increase, that is, the astronaut has the tendency to leave the sphere along the radius ρ. But the tether does not let go and "kills" this tendency, preserving a constant (equal to zero) value of the radial velocity.

Now assume that $\rho'' < 0$. Then the tether begins immediately to slacken! Indeed, this condition means that the radial velocity decreases, i.e, becomes smaller than zero, and then the astronaut leaves the sphere for its interior, with nothing to hold him. This show how important is to know the value of φ'' on the sphere. Let us write its expression. On the sphere we have $\rho' = 0$ and $\rho = l$, and so using the equation of motion (11.3.1) we obtain

$$\frac{1}{l}\rho'' = (\varphi')^2 - 2\varphi' + 3\sin^2\varphi.$$

[1]This fact can be established rigorously using Carnot's theorem on the change in the kinetic energy in an inelastic collision. By Carnot's theorem, the kinetic energy lost, ΔT, is equal to the energy corresponding to the loss in velocity, so that

$$2\Delta T = \left[(\rho')^2 + \rho^2(\varphi')^2\right]_1 - \left[(\rho')^2 + \rho^2(\varphi')^2\right]_2 = (V_1 - V_2)^2;$$

here the subscript 1 [resp., 2] indicates the moment before [resp., after] the collision. Clearly, $\rho_1' \neq 0$, $\rho_2' = 0$, $\rho_1 = \rho_2$, $\varphi_1' \neq 0$, and φ_2' is subject to determination; also,

$$\mathbf{V}_1 = \rho_1'\boldsymbol{\rho}_0 + \rho_1\varphi_1'\boldsymbol{\tau}_0, \quad \mathbf{V}_2 = \rho_2\varphi_2'\boldsymbol{\tau}_0 = \rho_1\varphi_2'\boldsymbol{\tau}_0$$

where $\boldsymbol{\rho}_0$ and $\boldsymbol{\tau}_0$ are the radial and the transversal unit vector, respectively. Calculating the square of the difference $V_1 - V_2$ and substituting in the above expression of Carnot's theorem, we convince ourselves that $\varphi_1' = \varphi_2'$.

Hence, if

$$(\varphi')^2 - 2\varphi' + 3\sin^2\varphi > 0, \tag{11.6.6}$$

then the motion is governed by the first integral (11.6.5).

If, however,

$$(\varphi')^2 - 2\varphi' + 3\sin^2\varphi < 0, \tag{11.6.7}$$

then the astronaut "separates" from the sphere and moves toward its interior. Thus, inequality (11.6.7) represents the condition for transition to free motion, while equation (11.6.5) and inequality (11.6.6) define the constrained motion. It is convenient to consider the motion in the phase plane φ, φ' (Figure 11.10). To the constrained motion there correspond the integral curves of equation (11.6.5), shown in this figure. The "separation zones," where condition (11.6.7) is satisfied, are hatched. The free motion in a "separation zone" cannot be depicted in the plane φ, φ', at least for the reason that the phase space of the free motion is four-dimensional.

Let us consider first motions of the astronaut in which he does not enter a "separation zone" and remains all the time on the sphere, so that the tether remains fully stretched. Examining the phase portrait (Figure 11.10) we see that in this case the astronaut's motions are of two types: rotational (continuous motion along a circle without change in direction) and oscillatory (oscillations about the direction of the radius vector of the spaceship). By the definition of the angle φ, the values $\varphi = \pi/2$ and $\varphi = 3\pi/2$ mean that the tether is stretched along the radius vector of the spaceship; moreover, when $\varphi = \pi/2$ [resp., $\varphi = 3\pi/2$] the astronaut is farther from [resp., closer to] the Earth compared to the spaceship. Let us derive the conditions under which the astronaut leaves a regime of rotation or oscillations without separation and goes into free motion.

Note that, generally speaking, the rotational and oscillatory regimes of motion are demarcated by a curve (*separatrix*) which passes through the origin of coordinates, and hence corresponds to the value $h_1 = 0$ in the integral (11.6.5). For $h_1 > 0$ [resp., $h_1 < 0$] the motions are rotational [resp., oscillatory] (for the moment we omit separation). The phase portrait shows that the astronaut will rotate in one direction, without separation, in the following two cases:

(1) $h_1 \geq 4$ in the region $\varphi' > 0$;

(2) $h_1 \leq 0$ in the region $\varphi' < 0$.

The motions corresponding to these cases are schematically represented in Figure 11.11. The condition for oscillatory motion without separation is $h_1 < h^*$, where h_1^* is determined from the condition of tangency of the curves $(\varphi')^2 - 3\sin^2\varphi - h_1 = 0$ and $(\varphi')^2 - 2\varphi' + 3\sin^2\varphi' = 0$. As one can readily verify, at the point of tangency of the phase curve corresponding to the oscillatory motion to the curve bounding the separation zone one has that $\varphi' = 1/2$ and $\sin\varphi = -1/2$, and consequently $h_1^* = -1/2$.

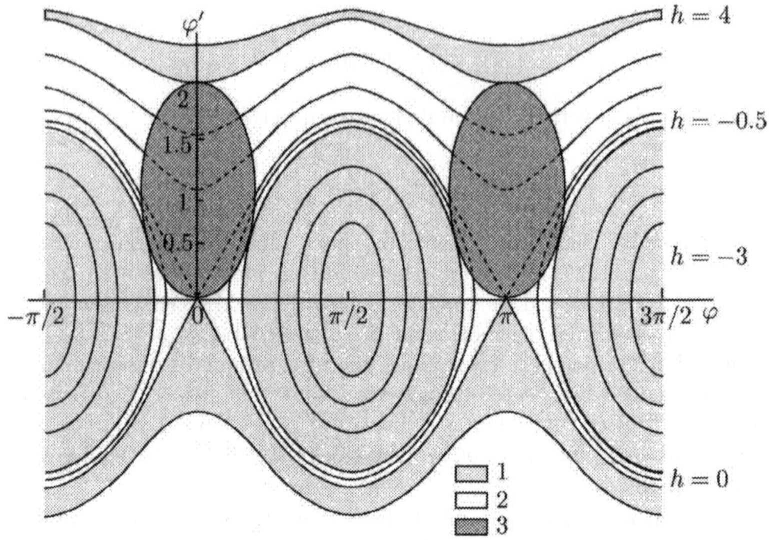

FIGURE 11.10. Phase portrait: 1–zones of nonevolving constrained motion; 2–zones of evolving motion (constrained motion along the integral curves passes into free motion at the boundary of a 3-zone); 3–separation zones

Therefore, if

(3) $h_1 \leq -1/2$,

then the astronaut performs an oscillatory motion with respect to the direction of the spaceship's radius vector, with no separation, i.e., no transition to free motion (Figure 11.12). Obviosuly, there exist a limit amplitude of oscillations without separation. This amplitude is obtained from (11.6.5) by setting $h_1 = h_1^* = -1/2$ and $\varphi' = 0$, which yields $\sin^2 \varphi = 1/6$. Recalling the definition of the angle φ, we conclude that the limit amplitude of the oscillations about the radius vector is

$$a_{\max} \approx 65°55'.$$

"Pure" oscillations with amplitude larger than $a_{\max}$ are not possible: separation and transition of the astronaut to free motion is unavoidable.

In the cases (1)–(3) considered above the motion of our system is analogous to the motion of a rigid dumbbell-shaped satellite around its center of mass (see the sixth essay).

Finally, if

(4) $-1/2 < h_1 < 0$,

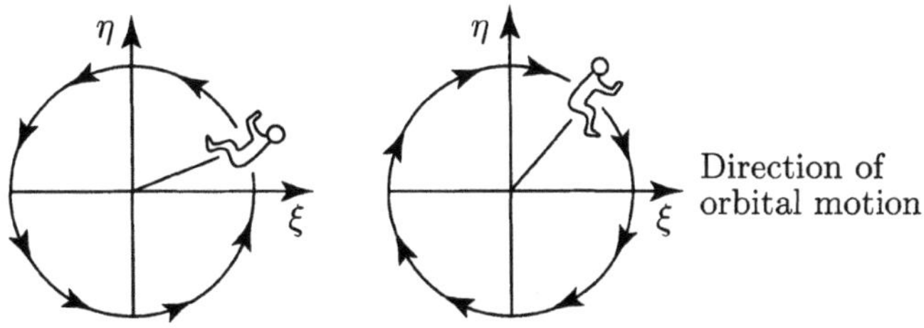

(1) $h_1 \geq 4$, $\varphi' > 0$ (2) $h_1 \geq 0$, $\varphi' < 0$

FIGURE 11.11. Schematic representation of the rotational motion of the astronaut on a taut tether

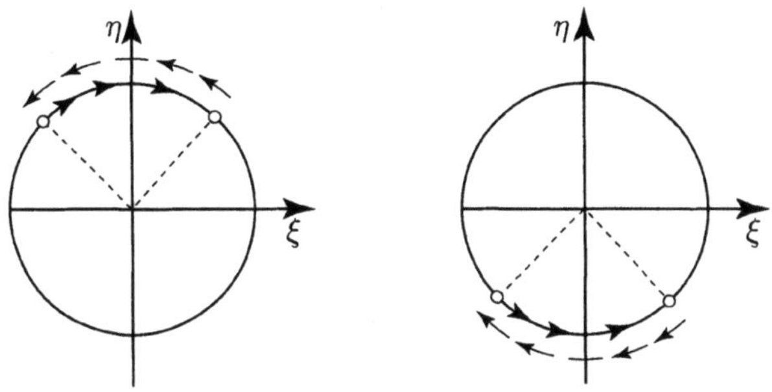

FIGURE 11.12. Schematic representation of the oscillations of the astronaut on a taut tether

then the astronaut executes oscillations with separation to free motion, while if

(5) $0 \leq h_1 \leq 4, \quad \varphi' > 0,$

he executes rotations with separation to free motion.

Regions (4) and (5) form the zone of incomplete mixed motion. Let us examine this zone in more detail.

7. The evolution of mixed motion

One can ask following questions: If the motion on the sphere were to take place along a phase curve corresponding to the value h_1^* of the energy constant in (11.6.5), and then enter the separation zone, then

(a) Would it exit the separation zone again on the sphere?

(b) If yes, then which phase curve will it follow?

The answer to the first question is generally speaking affirmative, because in view of the properties of the free motion discussed above we know that sooner or later the astronaut will again reach the sphere.

The answer to the second question is more complex. We need to track the entire free motion, from separation up to the new moment when it reaches the sphere, and calculate the angular coordinate at the point where it leaves the sphere and the corresponding angular velocity. This calculation associates to a point (φ_1, φ_1') of the phase plane another such point (φ_2, φ_2'). The point (φ_1, φ_1') lies on the boundary of the separation zone; it is the point at which the integral curve along which the motion took place until separation abuts. The point (φ_2, φ_2') is the origin of a new segment of constrained motion; the ensuing motion takes place along the integral curve on which this point lies (provided that (φ_1, φ_1') does not lie in the interior of the separation zone). Thus, in order to understand the nature of the motion it is necessary to construct the so-called *point mapping* on portions of the boundary of the separation zone.

The method of investigation of problems of nonlinear mechanics that uses such constructions is called the *method of point mappings*. This method is convenient for studying motions composed of sections that are described by different systems of differential equations. The point mapping arising in our problem was constructed by numerical methods using computers. Some results of the computations are shown in Figure 11.13, which shows the mapping for one separation zone. For the other zone the picture is similar.

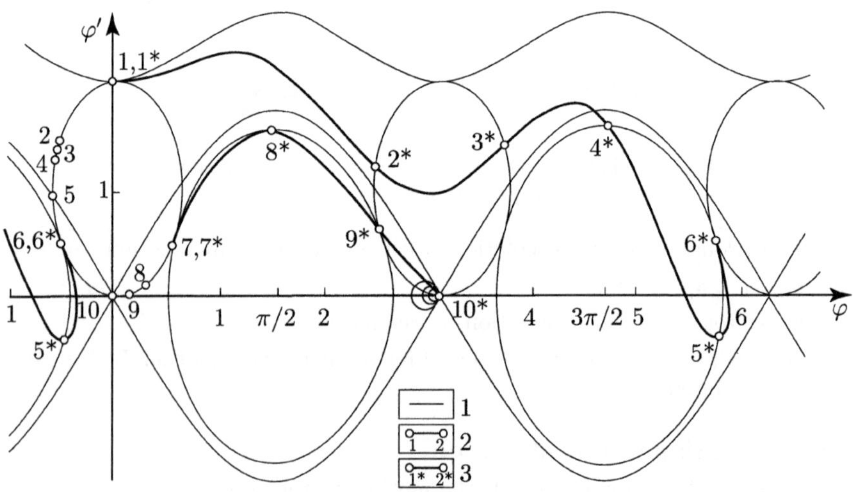

FIGURE 11.13. Point mapping of the boundaries of a separation zone: 1 – phase trajectories and boundaries of the separation zone; 2 – the segments of the boundary of the separation zone on which the mapping acts; 3 – action of the mapping on these segments. The segments $(1.1^* - 7.7^*)$ and $(6.6^* - 10)$ are not mapped

What is most important here is that this point mapping helps us understand the whole picture of the motion. Indeed, if under our mapping $(\varphi_1, \varphi_1') \mapsto (\varphi_2, \varphi_2')$ for any point (φ_1, φ_1') on the boundary of the separation zone, then the subsequent motion becomes clear: the phase point point leaves the point (φ_2, φ_2') and moves along an integral curve till it reaches again the boundary of the separation zone, in a point $(\varphi_1^{(1)}, \varphi_1^{(1)\prime})$, the image of which, $(\varphi_2^{(1)}, \varphi_2^{(1)\prime})$, we already know how to find, and so on. Some complication arises because the initial point or its image under our mapping may lie inside the separation zone. Physically this means that the moment the tether is fully stretched it begins to slacken, so that the trajectory of the motion has only one point on the sphere (rather than a whole segment).

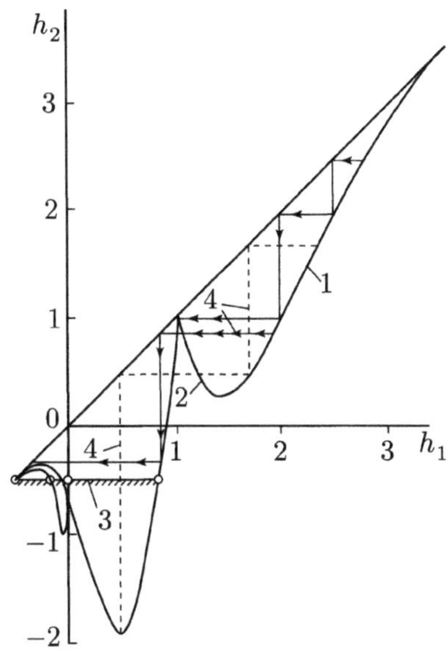

FIGURE 11.14. Energy curve

Hence, to completely analyse the motion we must construct the point mapping on the whole zone, and not only on its boundary. For details the reader is referred to our paper [11.2]. Let us mention here that the aforementioned circumstance does not play an essential role in the evolution of the motion of our system. This evolution is particularly simple to follow on the energy curve, which gives the dependence of the energy h_2 after reaching the constraint sphere (after impact) on the energy h_1 before the constraint sphere is reached. This energy curve $h_2(h_1)$ is shown in Figure 11.14. Let the initial energy equal h_1. Find the corresponding value h_2 on the curve. Now take h_2 as a new initial value h_1 and find the new h_2. Continue in this manner. This step-like process reflects the step-like manner in which energy changes at succesive arrivals at the constraint sphere. Examining Figure 11.14 one readily sees that after the motion reaches the constraint sphere finitely or infinitely many times, it passes into a limit motion whose energy can no longer change. Such a limit motion is for instance the motion with $h = 1$. It has a mixed character, partially constrained and partially free, and is shown in Figure 11.15. The motion reaches this periodic regime in a finite time; however, the set of initial values of the energy that lead to this regime is discrete ("of measure zero" relative to the whole set of all initial values of the energy). Consequently, the probability of occurence of such a regime is very small (equal to zero in a rigorous sense).

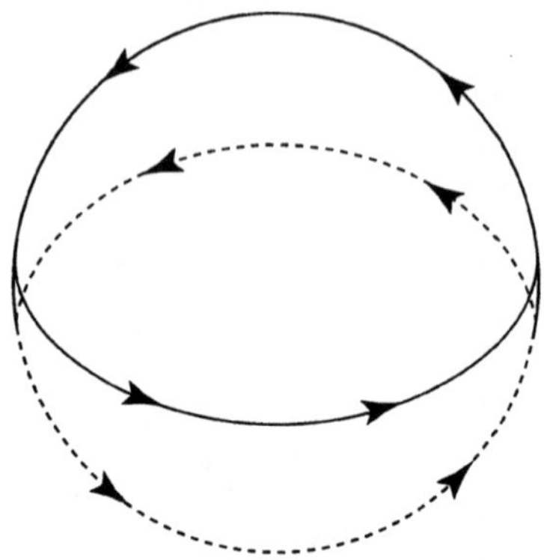

FIGURE 11.15. Periodic motion with $h = 1$

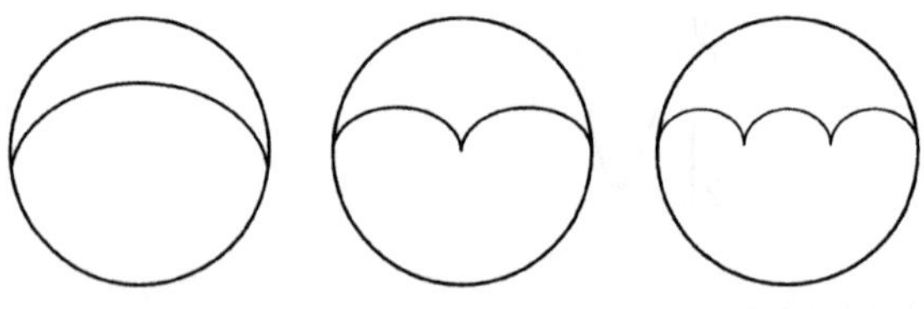

FIGURE 11.16. Periodic regimes with negative energies

A similar character is displayed by an infinite series of periodic limit regimes with small negative values of energy. Some of these regimes are shown in Figure 11.16. They are regimes with very large periods (oscillations with small negative values of the energy take place on segments of phase trajectories close to the separatrix; each "wave" of free motion corresponds to half of the satellite's period of orbital revolution).

But there are also limit regimes, which are reached for dense, and not discrete sets of values of the energy. These are purely constrained motions – oscillations about a relative equilibrium position with values of the energy in the range $-1.85 \leq h \leq -0.5$. The motion "slides" over a finite time interval to oscillations with some fixed energy in the range indicated above; but it is also possible that the motion will tend over an infinite time interval to oscillations with energy $h = -0.5$. Figure 11.17 and Table 11.1 show all possible limit motions.

	The set of initial data leading to motion	Type of motion
	A. Motions that do not reach the circle $\eta^2 + \xi^2 = 1$	
	$2\eta_0 + \xi_0' = 0$, $\lvert c_4\rvert + 2(c_2^2 + c_3^2)^{1/2} \leq 1$ $c_2 = \eta_0'$, $c_3 = -3\eta_0 - 2\xi_0'$, $c_4 = \xi_0 - 2\eta_0'$	Periodic, with the orbital period, on the ellipse $\eta^2/(c_2^2 + c_3^2) + (\xi - c_4)^2/4(c_2^2 + c_3^2) = 1$
	B. Motions that reach the circle $\eta^2 + \xi^2 = 1$	
either $2\eta_0 + \xi_0' \neq 0$, or $\lvert c_4\rvert + 2(c_2^2 + c_3^2)^{1/2} > 1$		I Nonevolving
	I–1° $h = \varphi_0'^2 - 3\sin^2\varphi \geq 4$, $\varphi' > 0$	1° Rotation on the circle $\eta^2 + \xi^2 = 1$ opposite the direction of orbital motion
	I–2° $h \geq 0$, $\varphi' < 0$	2° Rotation on the circle $\eta^2 + \xi^2 = 1$ in the direction of orbital motion
	I–3° $-3 \leq h \leq -0.5$	3° Oscillations on the circle $\eta^2 + \xi^2 = 1$ about the position of stable equilibrium
		II Evolving
	II–1° Set of initial conditions of nonzero measure	1° Limit cycle – oscillations on the circle $\eta^2+\xi^2 = 1$ for $h = -0.5$; infinite transition time
	II–2° Set of initial conditions of nonzero measure	2° Limit oscillations on the circle $\eta^2 + \xi^2 = 1$ for any h in the interval $-1.85 \leq h \leq -0.05$; finite transition time
	II–3° Discrete ∞^1 set of initial conditions	3° Mixed periodic regime with exit from the circle $\eta^2 + \xi^2 = 1$, $h = 1$; finite transition time
	II–4° Discrete ∞^2 set of initial conditions	4° Discrete set of mixed periodic regimes with exit from the circle $\eta^2 + \xi^2 = 1$: $h_1 = -0.00825$, $h_2 = -0.00215$, $h_3 = -0.00095$, ..., $h_\infty = -0$; finite transition time

TABLE 11.1.

Of course, the reader probably realizes that the evolution of motion considered above can only ideally be regarded as relevant for the real picture of the motion of an astronaut in the vicinity of his spaceship. First of all, to analyse the motion of the astronaut there is no need to consider very long time intervals and limit motions. But the same problem is applicable to any two-body system connected by a tether and flying in space, for example, for a satellite composed of two sections. Its relative motion is described in exactly the same manner. Such a two-section design is used, in particular, to enable one to orient the satellite toward the Earth by means of gravitational forces (i.e., by means of the so-called "passive orientation systems" discussed in the sixth essay). For such satellites it is indeed important to have information on the limit regimes of their motion. Specifically, it is desirable that in the limit the satellite will be oriented strictly toward the Earth. (In the scheme considered above we see that if the first time the satellite enters the constrained motion regime one does not achieve a regime of oscillations

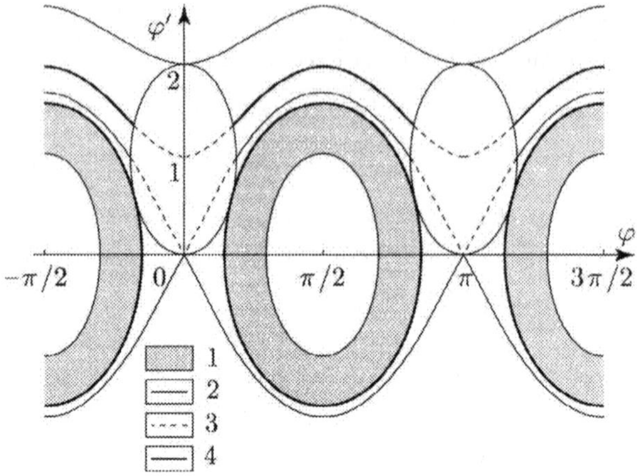

FIGURE 11.17. Limit regimes on the phase plane: 1–the set of limit regimes, bounded by a limit cycle with $h = -0.5$; 2–phase trajectories on the boundary of the separation zone; 3–region of limit motions with $h = -0.00825, h = -0.00095$, and so on; 4–limit cycle with $h = -0.5$ and limit motion with $h = 1$

about the direction to Earth with amplitude smaller than 48° ($h = -1.85$), then the amplitude will never be less than 48°).

An even higher degree of idealization is introduced by our assumption that entering the constrained motion regime is an absolutely anelastic process. In point of fact, this process is partially elastic; moreover, in the "free" motion energy is dissipated due to the bending and torsion of the tether, and so the motion is more complex. But the investigation of more rigorous formulations is the task of special research projects. Our goal here was different, namely, to understand the character of the problem and its place among other problems of the mechanics of space flight, to get a feeling for the problems that arise and possible methods for their investigation.

A two-section satellite whose sections are connected by a sufficiently long tether is a gratifying object to study. The most innocently looking question that still awaits an answer is whether such a system can get itself into a knot.

8. System of linked bodies in space

One can imagine a space station consisting of several identical sections connected by cables so that the station is a three-dimensional configuration. For example, four identical sections connected to one another by cables of the same length form a tetrahedral system of linked bodies (Figure 11.18). Can such a system preserve an invariable configuration in orbit? The answer to this question, posed by E. A. Devyanin, was given by E. T. Novikova. Let us consider this problem.

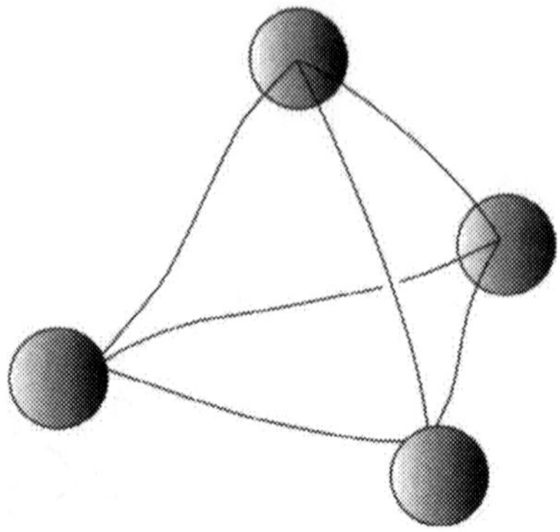

FIGURE 11.18. Tetrahedral system of linked bodies

By analogy to (11.2.3), the equations of dynamics for the present case have the form

$$\ddot{\boldsymbol{\rho}}_i + \frac{\mu}{R^3}\boldsymbol{\rho}_i - \frac{3\mu}{R^5}(\mathbf{R}, \boldsymbol{\rho}_i)\mathbf{R} = \sum_{j=1}^{3} \lambda_{ij}(\boldsymbol{\rho}_i - \boldsymbol{\rho}_j). \tag{11.8.1}$$

Here $\boldsymbol{\rho}_i$ is the radius vector of the i-th section of the space station relative to the station's center of mass, and λ_{ij} is the Lagrange multiplier introduced by the link between the i-th and the j-th sections. In constrained motion (i.e., when the tetraheral configuration is preserved) all λ_{ij} must be negative. The constraints are given by

$$|\boldsymbol{\rho}_i - \boldsymbol{\rho}_j| \leq l.$$

If the configuration is preserved, then the station will unavoidably rotate with an angular velocity $\boldsymbol{\Omega}$ that is constant in magnitude and direction, because the ellipsoid of inertia of the "rigidized" station is a sphere ($A = B = C$).

Then

$$\frac{d\boldsymbol{\rho}_i}{dt} = \boldsymbol{\Omega} \times \boldsymbol{\rho}_i, \qquad \frac{d^2\boldsymbol{\rho}_i}{dt^2} = \boldsymbol{\Omega} \times (\boldsymbol{\Omega} \times \boldsymbol{\rho}_i) = (\boldsymbol{\Omega}, \boldsymbol{\rho}_i)\boldsymbol{\Omega} - \boldsymbol{\Omega}^2\boldsymbol{\rho}_i,$$

and equation (11.8.1) becomes

$$(\boldsymbol{\Omega}, \boldsymbol{\rho}_i)\boldsymbol{\Omega} + \left(\frac{\mu}{R^3} - \boldsymbol{\Omega}^2\right)\boldsymbol{\rho}_i - \frac{3\mu}{R^5}(\mathbf{R}, \boldsymbol{\rho}_i)\mathbf{R} = \sum_{j=1}^{3} \lambda_{ij}(\boldsymbol{\rho}_i - \boldsymbol{\rho}_j). \tag{11.8.2}$$

Taking the scalar product of both sides of this equation by $\boldsymbol{\Omega}$ we obtain

$$\frac{\mu}{R^3}\,\Omega\rho_i^{\boldsymbol{\Omega}} - \frac{3\mu}{R^5}\,(\mathbf{R}, \boldsymbol{\rho}_i)\,(\mathbf{R}, \boldsymbol{\Omega}) = \sum_{j=1}^{3}(\rho_i^{\boldsymbol{\Omega}} - \rho_j^{\boldsymbol{\Omega}}), \tag{11.8.3}$$

where ρ_k^{Ω} denotes the projection of the vector $\boldsymbol{\rho}_k$ on the axis $\boldsymbol{\Omega}$. All these projections ρ_k^{Ω} are constant because the space station as a whole rotates around the $\boldsymbol{\Omega}$ axis. Let $\rho_i^{\Omega} > 0$ be the largest of the projections ρ_k^{Ω}. Then $\rho_i^{\Omega} - \rho_j^{\Omega} > 0$ for all $j \neq i$. Since $\boldsymbol{\Omega}$ is constant and $\mathbf{R}$ sweeps a plane (the orbital plane), there comes a moment of time when $(\mathbf{R}, \boldsymbol{\Omega}) = 0$. At that moment the left-hand side of (11.8.3) is positive, which is possible only when one of the multipliers λ_{ij} is positive (because $\rho_i^{\Omega} - \rho_j^{\Omega} > 0$). But the positivity of λ_{ij} means that the cable connecting sections i and j "slackens", i.e., the motion is no longer constrained. We conclude that an invariable configuration of the three-dimensional multi-body link considered here is not possible.

9. Cloud of particles in orbit and Poincaré's recurrence theorem

One of the most notable problems of the dynamics of relative motion is the problem of a cloud of particles released from a satellite. How does such a cloud behave with the passage of time? In 1963 the American *Westford* project was carried out, in which a "cloud" of metalic needles was released in orbit for radio communication purposes. The same question can be of course posed about a large cluster of satellites, should such a cluster exist.

Let us consider an (indexed) chain of problems, simplified to the maximum, but leading to the correct answer to our question.

Problem 1. *Suppose that at the initial time t_0 a large number of satellites are arranged along the same radius, each having its own angular velocity. What will the configuration of this system of satellites be after time t?*

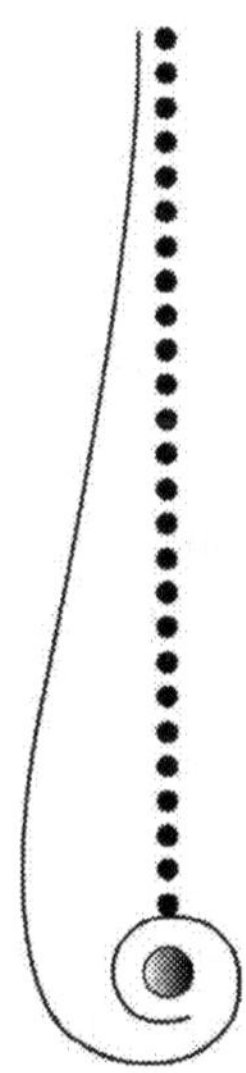

FIGURE 11.19. Configuration of satellites

Solution. Any individual satellite in our system moves on a circular orbit of radius r, along which the polar angle φ changes according to the law $\varphi = \omega(t - t_0)$. By Kepler's third law, $\omega = \sqrt{\mu}/r^{3/2}$. Hence, the configuration of the system after a fixed time $T_* = t - t_0$ is defined by the equation

$$\varphi = \frac{\sqrt{\mu}}{r^{3/2}} T_*,$$

i.e.,

$$r = \frac{c}{\varphi^{3/2}}, \qquad c = \left(\mu T_*^2\right)^{1/3}. \tag{11.9.1}$$

This is the equation of a spiral, shown in Figure 11.19. The larger the time T_*, the more turns are included bewteen two fixed values of the radius, r_1 and r_2.

Incidentally, if the ring $r_1 \leq r \leq r_2$ were from the very beginning filled by particles, except for some radial "gap," then at each moment of time $t \neq t_0$ the gap would look like the spiral (11.9.1). The spiral structure of galaxies is to a considerable extent explained by this simple fact. "Protuberances" or "gaps" that for some or another reason are initially present in a galaxy are twisted with the passage of time into spirals of the type (11.9.1), this being an unavoidable consequence of Kepler's laws.

Problem 2. *Pick any two satellite from the system in Problem* 1. *What is their relative motion?*

Solution. Let us denote by r_1 and ω_1 [resp., r_2 and ω_2] the radius and the frequency of orbital revolution of the first [resp., second] satellite. The relative motion $\boldsymbol{\rho} = \mathbf{r}_1 - \mathbf{r}_2$ is given in polar coordinates ρ, φ by the formulas

$$\rho = \sqrt{r_1^2 + r_2^2 - 2r_1r_2\cos(\omega_1 - \omega_2)(t - t_0)} \tag{11.9.2}$$

and

$$\tan\varphi = \frac{r_2\sin\omega_2(t - t_0) - r_1\sin\omega_1(t - t_0)}{r_2\cos\omega_2(t - t_0) - r_1\cos\omega_1(t - t_0)}. \tag{11.9.3}$$

We see that $\rho(t)$ is a periodic function of period

$$T_\rho = \frac{2\pi}{\omega_1 - \omega_2} = \frac{T_1T_2}{T_2 - T_1}. \tag{11.9.4}$$

The quantity ρ changes periodically between its maximal value ($\rho_{\max} = r_1 + r_2$) and minimal value ($\rho_{\min} = r_2 - r_1$). The period T_ρ is very large for satellites for which the periods T_1 and T_2 are close to one another.

For example, let $T_2 = T_1 + (T_1/n)$, where n is a large number. Then

$$T_\rho = (n + 1)T_1. \tag{11.9.4'}$$

Incidentally, the rate of change of the angle φ is determined by the frequencies ω_1 and ω_2 (and not by their difference). Consequently, the trajectory $\rho(\varphi)$ is a multi-turn "slowly pulsating spiral." After about $(n+1)/2$ revolutions of the first satelite on its orbit, the second satelite finds itself at the maximal distance from the first, and after about $n+1$ revolutions the satellites return to the initial configuration.

Problem 3. *Follow the evolution of the "cloud of satellites" over a long time interval.*

Formulation and solution. To simplify to the extreme the structure of the cloud, we will assume that the particles move along a series of circular orbits with periods T_i, which can be ordered as follows:

$$T_1 = T_0, \quad T_2 = T_0 + \frac{1}{n}T_0, \quad T_3 = T_0 + \frac{2}{n}T_0, \quad \dots \quad , T_{k+1} = T_0 + \frac{k}{n}T_0, \qquad (11.9.5)$$

where n is a suficiently large number (say, positive integer) and $k \geq n$. For each period T_i there is a group of particles, but we will speak of a single particle with the given period.

Let us follow the evolution of such a cloud. From the solution to Problem 1 it follows that the cloud will quite rapidly "spread" over some annular region $r_1^* \leq r \leq r_2^*$. Indeed, over the time mT_0 the first particle makes m revolutions in its orbit, while the $(k+1)$st particle makes $mT_0/T_{k+1} = m/(1+\frac{k}{n})$ revolutions, so that the spiral described in Problem 1 has

$$m - \frac{m}{1+\dfrac{k}{n}} = \frac{m}{\dfrac{n}{k}+1} \approx m$$

turns (provided that $k \gg n$). The period of the satellite that is located the closest to the Earth is about 1.5 hours. This means that after as little as 24 hours the particles of the cloud will disperse over a 16-turn spiral, littering practically the entire space around the Earth in the annulus $r_1^* \leq r \leq r_2^*$. If $k \sim n$, then the number of turns will be $m/2$ (that is, after 24 hours one obtains an 8-turn spiral).

But let us see what happens next.

According to the solution to Problem 2 (formula (11.9.4′)), after the time $\widetilde{T}_1 = (n+1)T_0$ the first and second satellites return to their initial configuration. Over that time interval the third satellite makes $T_1/T_3 = n/(n+2)$ revolutions in its orbit. After the time $\widetilde{T}_2 = (n+2)\widetilde{T}_1$ the configuration of the first two satellites repeats itself (for the $(n+2)$nd time), while the third satellite makes n revolutions in its orbit; therefore, the initial configuration of the first three satellites repeats itself for the first time. Iterating this argument, we see that after the time

$$\widetilde{T}_k = (n+1)(n+2)\cdots(n+k)T_0 \qquad (11.9.6)$$

the initial configuration of all $k+1$ satellites repeats itself, that is, our cloud of particle-satellites regroups itself in the initial position!

Note, however, that the time $\widetilde{T}_k$ is incredibly large. Indeed,

$$\widetilde{T}_k > n^{k-1}\widetilde{T}_1 > 2^{k-1}\widetilde{T}_1. \tag{11.9.7}$$

Suppose the repetition time $\widetilde{T}_1$ of the configuration of two particles is of the order of a day (a lowered estimate, corresponding to $n = 15$). Then for a cloud of merely 120 particles the time after which the system returns to the initial configuration will be manifestly larger than 2^{100} days, which amounts to more than 10^{30} days. For comparison, the age of the Solar system is ("only") about $10^{12} \div 10^{13}$ days.

Our problem is just a pale reflection of a remarkable theorem of dynamics – the "*Poincaré recurrence theorem,*" which can be formulated as follows. *Assume that the motion of a system is*: (a) *conservative, and* (b) *bounded with respect to the phase variables. Then for the majority of initial conditions, after a sufficiently long time the system returns to a position that is arbitrarily close to the initial one.*

Thus, the fact that a system returns arbitrarily close to its initial position is established by Poincaré's theorem under rather general assumptions. It follows that in the problem considered above this fact does not depend at all on the concrete arrangement (11.9.5) of the particles; moreover, it holds in the presence of various perturbing factors acting on the particles (as long as they are conservative!), for arbitrary conservative interactions between the particles, and so on. The only thing that matters is the conservative nature of the system and the natural assumption that the phase space in which the motion takes place is bounded (what flies away to infinity cannot return).

The Poincaré recurrence theorem is one of the few general conclusions on the nature of motion, though the details of the motion "are known to no one" [11.6], even in the most elementary problems (for instance, nonintegrable two-dimensional problems).

This theorem has numerous, at times unexpected concrete formulations. Here is a phenomenon that is mentioned in V. I. Arnold's book [11.6]: if we partition a room by a screen, pump out the air from one of the halves, and after that remove the screen, then "after a sufficiently long time" all the air will gather, by itself, in one of the two halves of the room (keep in mind, however, that the required time is comparable with the age of the Universe).

We will give the formulation and proof of Poincaré's theorem as they appear in Arnold's magnificent book [11.6].

Consider a Hamiltonian system of equations (2.4.4) in the general case where $i = 1, \dots, n$. The $2n$-dimensional space with the coordinates p_i, q_i $(i = 1, \dots, n)$ is called the *phase space* of the system.

Definition. The *phase flow* is the one-parameter group of transformations of the phase space defined by

$$g^t : (p_i(0), q_i(0)) \to (p_i(t), q_i(t)), \tag{11.9.8}$$

where $p_i(t)$, $q_i(t)$ is the solution of Hamilton's system of equations (with initial condition $(p_i(0), q_i(0))$).

Liouville's theorem. The phase flow preserves the phase volume (see Figure 11.20): *for any domain D in phase space,*

$$\text{volume}(g^t D) = \text{volume}(D).$$

We omit the proof, which can be found, for example, in [11.6].

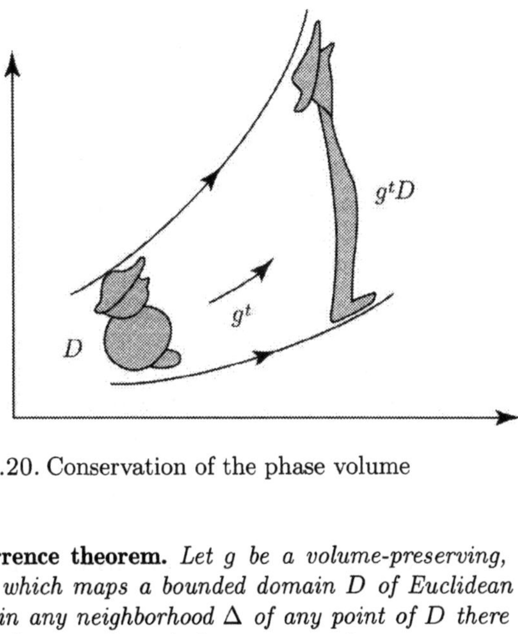

FIGURE 11.20. Conservation of the phase volume

Poincaré's recurrence theorem. *Let g be a volume-preserving, continuous, one-to-one mapping which maps a bounded domain D of Euclidean space into itself: $gD = D$. Then in any neighborhood Δ of any point of D there is a point $x \in \Delta$ which returns to Δ, i.e., $g^n x \in \Delta$ for some $n > 0$.*

Proof. Consider the images of the neighborhood Δ under g (Figure 11.21): Δ, $g\Delta$, $g^2\Delta, \dots, g^n\Delta, \dots$. They all have the same positive volume. If they would not intersect, then the volume of D would be infinite. Hence, there exist $k \geq 0$ and $l \geq 0$ ($k > l$), such that the intersection of $g^k\Delta$ and $g^l\Delta$ is not empty: $g^k\Delta \cap g^l\Delta \neq \emptyset$. Consequently, $g^{k-l}\Delta \cap \Delta \neq \emptyset$. Let $g^{k-l}x = y$, $x \in \Delta$, $y \in \Delta$. Then $x \in \Delta$ and $g^n x \in \Delta$ ($n = k - l$), as claimed. □

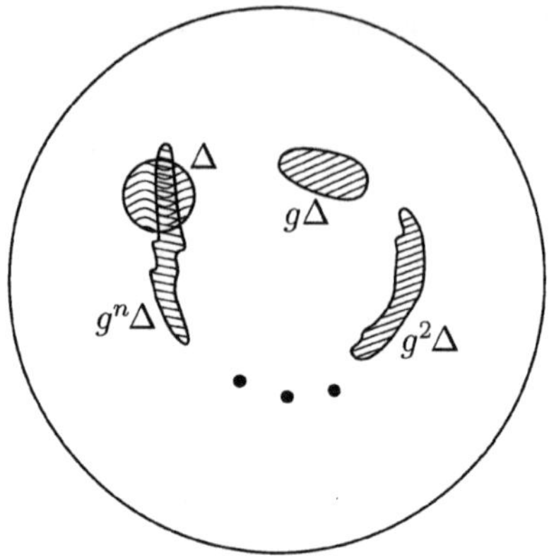

FIGURE 11.21. Regarding the proof of Poincaré's theorem

Poincaré's theorem has a general, abstract, mathematical character. Its applications to mechanics rest on the fact that the mapping g can be given in the form (11.9.8) (which satisfies the assumptions of Poincaré's theorem thanks to Liouville's theorem). An essential requirement in Poincaré's theorem is that the domain D in which the motion takes place be *bounded* (and mapped by g into itself). For one-dimensional systems (for example, a mathematical pendulum) this requirement is satisfied by the domain "inside" the separatrices; for two-dimensional systems with a potential growing at infinity $V = -U$, the requirement is satisfied by the domain "inside" a fixed energy level: $T + V \leq h$; and so on.

Problem. In [11.7] and other papers, T. M. Eneev, N. N. Kozlov , and R. A. Syunyaev modeled the interaction of galaxies as follows. A cloud of mutually nonattracting particles move with initial angular velocities with respect to a mass m_0 ("nucleus of the galaxy"), and each particle is attracted by m_0 according to Newton's law. The particles of the cloud are attracted according to the same law by another mass m_1 ("perturbing mass"). In their turn, the masses m_0 and m_1 attract each other according to Newton's law; hence, by Kepler's laws, m_1 moves with respect to m_0 in a hyperbolic orbit.

Does the system described satisfy the conditions of Poincaré's recurrence theorem?

Answer. No.

The obstruction is the unbounded (by the formulation of the problem) motion of the mass m_1. Therefore, we cannot assert that a galaxy will return, after some time, to the initial state of motion, with all the particles in the cloud moving on circular orbits around the mass m_0.

Calculations carried out by Kozlov, Syunyaev, and Eneev showed that the perturbing mass can "capture" and carry away with it part of the particles; other particles attain escape velocities relative to both masses m_0 and m_1, and are "ejected" from the galaxy. The particles that remain in the vicinity of m_0 have elliptic velocities with respect to m_0, and the cloud of particles itself changes considerably, acquiring a pronounced spiral structure.

Now let us isolate the spiral cloud of particles that *remain* in the vicinity of m_0 from the flying away mass m_1: we assume that m_1 "ceases" completely to attract the particles of the cloud. Then we arrive at a new *initial* problem: Consider a mass m_0 and a spiral cloud of particles in the vicinity of m_0, and assume that these particles are attracted by m_0, but do not attract one another, and that *all* the particles have *elliptic* velocities. *Such a system* already does satisfy the conditions of Poincaré's theorem. Hence, we can assert that regardless of how such a galaxy "deforms," it will eventually return to the original *spiral* structure!

Again, the reader must keep in mind that the time required for this is considerably larger than any reasonable estimate of the physical lifespan of the galaxy.

Additional comments for this translation

The author cannot complain about lack of interest in the present volume *Essays*. In the time that passed from the publication of the first edition of the *Essays* (1972) the book was repeatedly reviewed (by V. I. Arnold and Ya. B. Zeldovich in the Russian journals *Priroda* (No. 10, 1973) and *Uspekhi Fizicheskihk Nauk* (No. 3, 1974)), reprinted (a Polish edition in 1976, a second Russian and a Bulgarian edition in 1977, and a French edition in 1986), and quoted. For instance, V. I. Arnold did include the problem "Leonov and the lens cap," discussed in this essay, in his well-known textbooks [11.6] and [11.8]. The problems treated in this essay continue to be investigated and developed by a number of authors. Some of these investigations will be briefly described below.

A. Development of the theory of impact-free motions.

In 1992 A. P. Ivanov published the paper [11.9], which contains his general theory of systems with unilateral constraints.[2] This allowed one, in particular, to examine from a new position the dynamics two connected bodies, described in Section 7 of the present essay. According to a theorem proved by Ivanov, the *measure* of the set of initial values of the energy that in the evolution process yield the impact-free trajectories shown in figures 11.5 and 11.6 *is different from zero*. In [11.10], Ivanov and A. B. Baziyan found estimates of the domains of attraction of the impact-free motions classified in Table 11.1 and shown in Figure 11.5 ($h = h_0 \sim 1$) and Figure 11.6 ($h = h_n, n = 1, 2, 3 \ldots$). It turned out that the domain of attraction $(\Delta h)_0$ of the solution with $h = h_0$ exists and is of order 10^{-7}; more precisely, $(\Delta h)_0 < 1.8 \cdot 10^{-7}$. In other words, in Figure 11.14, in the point $h_1 \cong 1$, $h_2 \cong 1$ the energy curve is *tangent* to the line $h = h_2$, and the point itself is "semi-stable." All these details could have been discerned by using a good microscope with 10^7 resolution, provided that the curve shown in Figure 11.14 would have been computed with sufficient precision.

[2]See also A. P. Ivanov's recently published book *Dynamics of Systems with Mechanical Impacts*, Mezdunarod. Programma Obraz., Moscow, 1997 (in Russian).

The domains of attraction $(\Delta h)_n$ of the solutions with $h = h_n$, $n = 1, 2, 3, \dots$ obey the estimate $(\Delta h)_n < 2.5 \cdot 10^{-9} n^{-6}$.

The values of the initial data and energies of the mixed impact-free periodic regimes discussed here were sharpened in the papers [11.11], [11.12], and independently in [11.10]. For example, for the two trajectories shown in Figure 11.15, sufficiently accurate values are as follows:

$$\varphi_0^{(1)} = 0.586, \qquad \varphi_0^{(2)} = \pi + \varphi_0^{(1)}; \qquad \varphi_0' = 1.3706;$$

$$T = 2.9028; \qquad h = h_0 = 1.0158;$$

here T is the dimensionless flight duration on an arc of (unconstrained) motion. The exact values of the indicated quantities satisfy the system of equations

$$\tan\frac{T}{2} = \cot\varphi_0 \frac{\varphi_0'}{2\varphi_0' - 3}, \qquad \frac{T}{2} = \frac{1}{3}\cot\varphi_0 \frac{1 - 2\varphi_0'}{2 - \varphi_0'},$$

$$\varphi_0' - 2\varphi_0' + 3\sin^2\varphi_0 = 0, \qquad h = (\varphi_0')^2 - 3\sin^2\varphi_0.$$

For the trajectories with $h = h_n$, $n = 1, 2, 3, \dots$, shown in Figure 11.16, these formulas yield the approximate solution

$$\varphi_0 = (6\pi n)^{-1}, \qquad \varphi_0' = (24\pi^2 n^2)^{-1}; \qquad h_n = -(12\pi^2 n^2)^{-1}; \qquad T = 2\pi n - \varphi_0.$$

This in turn yields the following sharpened values of h_n in Table 11.1 (and in the explanation to Figure 11.17):

$$\begin{array}{llll} h_0 = 1.01580, & h_1 = -0.00844, & h_2 = -0.00211, & h_3 = -0.00094, \\ h_4 = -0.00053, & h_5 = -0.00034, & h_6 = -0.00023, & h_7 = -0.00017, \end{array}$$

and so on. In [11.10] the values of these quantities were computed to 11 significant figures.

B. A system of two connected bodies in orbit as a dynamical billiard.

In [11.11]–[11.13] the problem of the dynamics of a system of linked bodies was considered from a modern position under the assumption of absolutely elastic impact when the constraint set is reached. Specifically, in those papers the structure of the phase space is studied in its whole richness – with a chaotic "sea," islands of regularity, impact-free motions, and so on. The problem is piecewise-integrable and the motion on the time interval bewteen two impacts is described by closed-form formulas. Thanks to the hypotesis that impacts are absolutely elastic, there exists an energy integral whose constant value serves as a parameter of the problem; the phase portrait of the problem depends on the value of this constant. As usual, the phase portrait is constructed at discrete moments of times, namely, in our problem, at the moments of impact on the constraint set. In an "ordinary," kinematic billiard, each trajectory is a broken line composed of linear segments. In the problem at hand, each broken-line trajectory is composed of curvilinear segments whose shape depends on the acting forces; it is terefore natural to refer to such problems as "dynamical billiards" [11.14], [11.15]. In the papers [11.11] and [11.12] the governing force is the gravitational gradient force, while [11.13] also takes into account the force of aerodynamic pressure.

30 years have passed since the publication of the papers [11.1] and [11.2]. Fresh achievements of the theory of dynamical systems allowed one to develop and enrich the old results; this is quite oportune since problems of the dynamics of tethered systems are currently attracting a lot of interest (see , e.g., [1.16]).

Twelfth Essay

Cosmic Pinwheel

> "Look, your worship," said Sancho; "what we see there are not giants but windmills, ..."
>
> Miguel de Cervantes Saavedra, *Don Quixote*

1. The Proton satellites

In the sixth essay of this book we talked about remarkable effects in the motion of a satellite around its own center of mass and in the evolution of the rotation and orientation of the satellite. However, therein we have dealt with phenomena that in some sense are general for artificial as well as natural celestial bodies, phenomena that are the result of the action of the gravitational field.

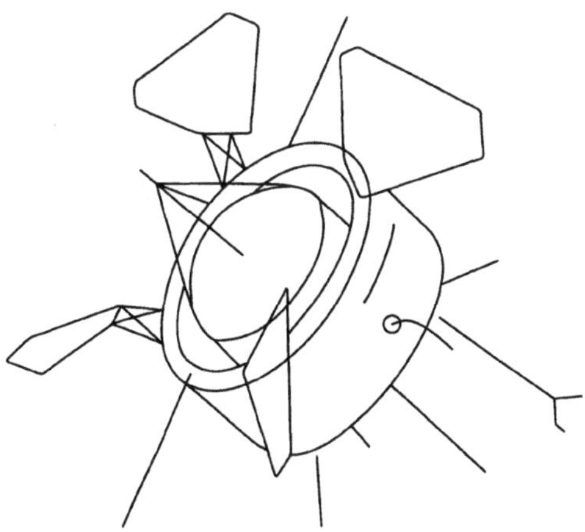

FIGURE 12.1. General view of a Proton satellite

But there are also other forces acting on satellites, for example, aerodynamic ones. Satellites move in the upper, rarefied layers of the terrestrial atmosphere, where the systematic action of very small aerodynamic forces may lead (and indeed does so) to a slow evolution of the satellite's orbit as well as of its rotational motion.

Satellites can have various shapes, at times rather "exotic," and since the aerodynamic properties depend in essential manner on the shape, the evolution of the orientation of a satellite under the action of torques of aerodynamic forces can prove to be quite spectacular, even "exotic."

In the period 1965–1968 Soviet Union launched a series of heavy *Proton* satellites (the heaviest at that time among scientific satellites). A characteristic feature of their construction was that on the satellite there were large blades (panels) carrying solar batteries, arranged skew-symmetrically with respect to the axis of the satellite. A sketch of a Proton satellite is shown in Figure 12.1.

When a satellite moves in orbit it collides with a flow of rarefied medium. Impacting on the battery panels, this flow spins the satellite in the same way in which the wind revolves a toy pinwheel or the sails (arms) of a windmill. The only difference is that the sails of our "windmill" are rigidly attached to the satellite and force it to spin together with them. This results in a number of peculiar phenomena in the evolution of the rotation and orientation of a Proton-type satellite.

2. Here is how all this was discovered

The phenomena alluded to above were discovered during the analysis of experimental data on the rotation and orientation of the Proton-2 satellite.

In the sixth essay we promised to describe what is understood by experimental determination of the orientation of a satellite. It is time to fulfill that promise.

The point is that, as a rule, nothing is known beforehand about the orientation of a satellite and its change with time (we assume that there is no special orientation system on board). But knowing the orientation is necessary: without that it is often impossible to interpret correctly the readings of various scientific instruments on board of the satellite. The job of these instruments is to measure physical characteristics in the region around the Earth and in cosmic space; indeed, the goal of launching a satellite is to enhance our understanding of nature. But what understanding can be achieved if we cannot explain the readings of the instruments? This is why me must know at each moment of time where this or that instrument is "looking," i.e., know the orientation of the satellite in space. To this end some " orientation sensors" are mounted on the satellite. The readings of these sensors are transmitted to Earth. Solar sensors, for instance, give information on whether the sensor's window is illuminated by the Sun or not; magnetic sensors may show how the force lines of the Earth's magnetic field are directed relative to the body of the satellite at a given moment of time, and so on. Our task is, using these data, to determine the orientation of the satellite.

The trouble is, first of all, that usually the readings of the sensors are dispersed so that at each given moment of time there may be readings from no more than one sensor, while at some moments of time there are no readings at all.

Second, the readings of the sensors unavoidably "lie:" they are always affected by errors, though usually small (and, of course, the smaller the better). Indeed, a sensor is a device and its operation as well as the transmission of it readings to Earth and their processing are hindered by thousands of sources of disturbance. As a consequence, what you hold in your hands is alway a bit different than what the sensor wanted to tell you (and again, the smaller the discrepancy, the better).

Third, nothing is known beforehand, even approximately, about the motion of the satellite around its center of mass, and there are no data to initiate an analysis.

Fourth, we still have only poor knowledge of the forces and torques that act on the satellite in space in each concrete case, and in many cases we cannot predict with certainty the evolution of the satellite's orientation.

Fifth, we are confronted with such combinations of motions of the satellite and readings of sensors that from these readings it is in principle impossible to determine its motion ...

So many troubles, yet we must do our work.

Naturally, a serious mathematical processing of the measurement data is needed in order to determine the orientation of a satellite. It is necessary to construct some *model* of the motion of the satellite around its center of mass in order to be able to extract a unique "trajectory" from the uncoordinated readings of the orientation counters. More specifically, we needs to provide a *statistical data-processing algorithm* which would allow us to use in "optimal" manner the readings of the sensors, with all their random errors, to "smooth-out" the influence of these errors, and obtain some "average" trajectory which matches the best the trajectory that the sensors really "wanted to tell" us about. Moreover, if necessary, the data processing must be organized so that in parallel with the satellite's orientation it will also determine the perturbing factors (torques of forces acting on the satellite).

For example, suppose that the motion around the center of mass is described by some vector

$$\mathbf{r}(\mathbf{a}, t), \tag{12.2.1}$$

which defines our model of the satellite's motion. We shall consider that the motion is completely known if the set of constant parameters $\mathbf{a} = \{a_i\}$ $(i = 1, 2 \ldots, m)$ is known. For example, a model of motion is a solution of the differential equations of motion. Suppose that the torques figuring in these equations are written in explicit form, but contain some constant parameters whose values are not known precisely; then the set of parameters a_i consists of the initial data for the integration of the equations of motion plus the constant parameters appearing in the expressions of the torques.

Suppose that at the moments of time t_n $(n = 1, 2, \ldots, N; m < N)$ some measurements are carried out on the satellite, giving a variable ψ_n whose dependence on the satellite's orientation can be calculated, so that the computed values of ψ_n, denoted here by $\widetilde{\psi}_n$, are $\widetilde{\psi}_n(\mathbf{r}(\mathbf{a}, t_n), t_n)$. Consider the differences $\xi_n = \psi_n - \widetilde{\psi}_n$ between the computed and the measured values of the function ψ. We must determine the values of the unknown parameters $\mathbf{a}$ so as to minimize the differences ξ_n according to some criterion. To this end, in mathematical statistics one resorts, for example, to the *least squares method*, in which one is required to minimize with respect to $\mathbf{a}$ the sum

$$\Phi = \sum_{n=1}^{N} \frac{1}{\sigma^2} \xi_n^2, \tag{12.2.2,}$$

where $1/\sigma_n^2$ are some weighting coefficients that depend on the character of the random errors in the measurements ψ_n. The necessary conditions for minimum of the function Φ,

$$\frac{\partial \Phi}{\partial a_1} = 0, \qquad \frac{\partial \Phi}{\partial a_2} = 0, \ldots, \frac{\partial \Phi}{\partial a_m} = 0 \tag{12.2.3}$$

yield m equations for determining the m unknown parameters $a_1, a_2, \ldots, a_m$.

Solving the system of equations (12.2.3) for a_i is a very difficult problem in itself. It is usually approached by the method of successive approximations. To that end it is required to know the zeroth approximation to the values of the parameters a_i. But the only source from which this zeroth approximation can be extracted is the same information on orientation that is provided by the measurements ψ_n. Hence, our algorithm should include as a component an independent algorithm for the determination of the zeroth approximation. That is plenty of work, unthinkable without the use of modern computers.

A whole new direction has crystallized in the theory of motion of artificial satellites [12.1], that of mathematical processing of trajectory measurements with the goal of determining actual orbits and the perturbing factors acting on them. One of the first studies in this direction was carried out by T. M. Eneev, A. K. Platonov, and R. K. Kazakova [12.10]. Numerous papers describe the determination of the density of the upper layers of the terrestrial atmosphere and the harmonics of the potential of the terrestrial gravitational field from trajectory measurements. More recently, studies were made to determine the harmonics of the potential of the Moon's gravitational field from observations of the motion of lunar artificial satellites. One of the first works of this nature is due to E. L. Akim [12.2].

The problem of determining the orientation of a satellite from observations corresponds to a large extent to the problem of orbit determination and is approached by similar methods; however, the former problem has its own specific aspects and difficulties. The determination of the orientation of a satellite from data provided by on-board measurements was carried out for the first time in the world for the third Soviet artificial satellite [12.3]. The method used to determine the orientation of the Proton satellites and its results are described in [12.4], and also in a number of other papers, of which we mention here [12.5]–[12.9].[1]

Let us examine some interesting dynamical effects in the evolution of the orientation of a Proton-type satellite.

[1]In the course of these studies over several years, the author of this book was fortunate to collaborate with a number of researchers. Among those I want to mention here Yu. V. Zonov (determination of the orientation of the third Soviet satellite) [12.3]; V. V. Golubkov and I. G. Khatskevich (elaboration of automated methods for determination of the orientation of satellites in the Proton series and other satellites) [12.3]–[12.8]; E. K. Lavrovskiĭ and S. I. Trushin (determination of the orientation of satellites in the Electron series) [12.5].

3. What was discovered

Figure 12.2 shows the experimentally acquired picture of the time evolution of the rotation and orientation parameters of the Proton-2 satellite [12.4], [12.8].

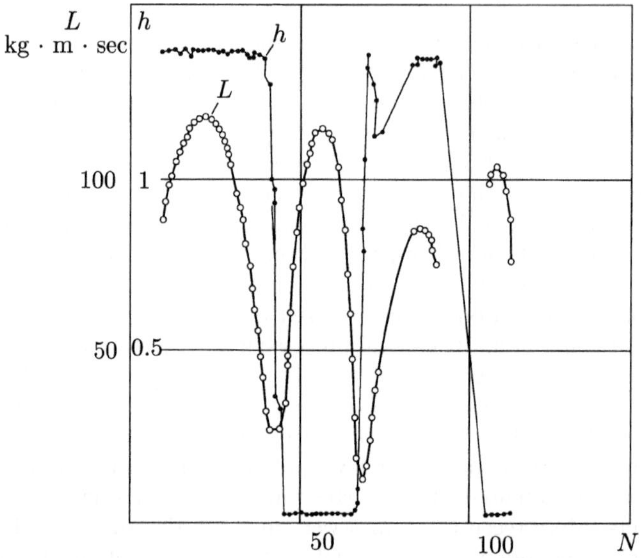

FIGURE 12.2. Evolution of the rotation and orientation parameters of the Proton-2 satellite

The meaning of one of these parameters – the length L of the angular momentum vector – is already familiar to us (see the sixth essay); the other parameter, h, has the following meaning:

$$h = \sin^2 \vartheta (1 + \varepsilon \sin^2 \varphi), \qquad \varepsilon = \frac{B - A}{C - B} \cdot \frac{C}{A}. \tag{12.3.1}$$

Here A, B, C are the satellite's central principal moments of inertia ($C > B > A$), ϑ is the already familiar angle between the longitudinal axis of the satellite (the axis of the moment of inertia C), and the angular momentum vector $\mathbf{L}$, and φ is the angle of proper rotation of the satellite around its longitudinal axis. If no torques of external forces act on the satellite, then in such an unperturbed motion of a triaxial satellite both quantities h and L remain constant. Values $h \sim 0$ correspond to rotation around the longitudinal axis of the satellite ($\vartheta = 0$ or $\vartheta = \pi$), while values $h \sim 1 + \varepsilon$ correspond to rotation in the vicinity of the transversal axis. For a dynamically-symmetric satellite ($B - A = 0$), the change of h from 0 to 1 would simply mean that the rotation regime changes from longitudinal spinning ($\vartheta = 0$)

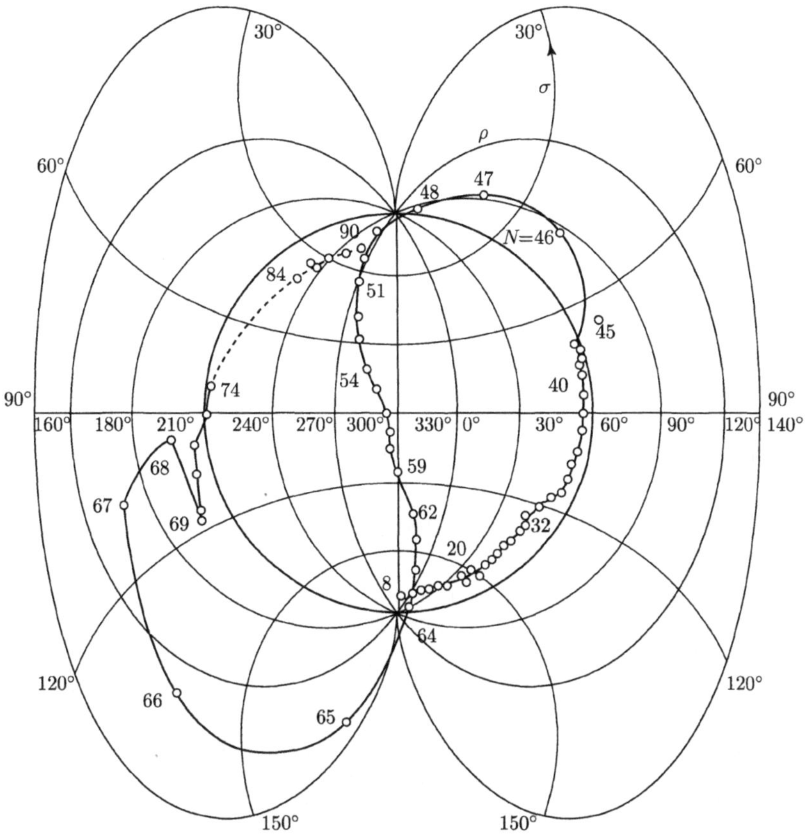

FIGURE 12.3. Trace of the angular momentum vector of the Proton-2 satellite on the celestial sphere, with the number of orbit turns indicated. The points $\rho = 0°$ and $\rho = 180°$ correspond to the North and the South pole of the world, respectively

to topsy-turvy motion ($\vartheta = \pi/2$); thus, the behavior of h matches qualitatively the behavior of the angle ϑ.

From the sixth essay it is clear that for a satellite spinning rapidly in the gravitational field both quantities L and h will remain constant; they remain so in the presence of other influences as well, and if the dissipative forces change, then L and h change in a monotone fashion (for more details see the author's book [12.11]).

Incidentally, in Figure 12.2 we observe a rather unique picture of variation of the parameters L and h: they both oscillate with a large period, and the amplitude of the oscillations is quite large. The quantity L changes by a factor of 5, and the change in h tells us that in its rotation the satellite passes from a regime of axial rotation (spinning) to a regime of topsy-turvy motion. The evolution of the spinning regime is of course slow, with a period of about 55–77 orbital turns of the satellite (in Figure 12.2 on the abscissa axis one indicates the number of orbital turns). The modulation period of the quantity L is one half of the one above; when the satellite spins around its axis, L is maximal (and hence so is the angular velocity of rotation), in the transition regime from spinning to topsy-turvy motion L attains its minimum, in the regime of topsy-turvy motion L has again a maximum, and so on.

We should mention here that the position of the angular momentum vector in space also varies significantly, oscillating within a range of about 180°. Figure 12.3 shows the trace of the angular momentum vector of the Proton-2 satellite on the celestial sphere for the first 90 days of flight; we see that over this period of time Proton-2 passed twice near each of the world poles.

4. Here is how all this is explained

All these oscillations of quantities that at some point were at rest are explained by the so-called *propeller effect* – the action of an incident flow on the asymmetrically mounted solar-battery panels (blades). There is no need to develop here a rigorous theory of such effects, since they are "crude" to the extent that they arise in all reasonably formulated model problems. We shall assume that the satellite is dynamically symmetric, that the straight lines connecting the centers of opposite panels pass through the satellite's center of mass, and that the satellite's body, except for the panels, has no contribution to the aerodynamic torque (this would be the case, for example, for a ball-shaped satellite whose geometric center and center of mass coincide).

Under these assumptions all that is left from the aerodynamic torque is the propelling torque. We will derive a model expression for it based on simple considerations (more rigorous formulations [12.4], [12.9] yield the same qualitative result).

Suppose the longitudinal axis of the satellite coincides with the direction of the velocity $\mathbf{V}$ of the satellite's center of mass. Then due to the presence of the "windmill sails" a constant twisting torque is created, which spins the satellite around its longitudinal axis (when the wind blows frontally, the pinwheel spins). As usual in aerodynamics, this torque is proportional to the dynamic pressure ρV^2, where ρ is the density of the incident flow and V is its speed. Hence, the propelling torque is $m = a_0\rho V^2$; the constant a_0 must have the dimension m^3 (area times the "arm").

It is natural to expect that if the axis of the satellite deviates from the direction of the incident flow then the magnitude of the propelling torque will decrease and even become equal to zero when the axis is perpendicular to the velocity vector $\mathbf{V}$ (when the wind blows sideways, there is no rotation). If the axis of the satellite is opposite in direction to the velocity vector, then the torque must be again equal in magnitude with $a_0 \rho V^2$, but has the opposite sign, since relative to the fixed positive direction of the satellite's longitudinal axis the satellite will rotate in opposite direction (starting from the rest state). The whole dependence of the longitudinal component of the torque on the position of the axis relative to the velocity vector $\mathbf{V}$ is in the simplest way approximated by the following formula (which will be adopted here):

$$M_z = \rho V^2 a_0 \alpha'', \tag{12.4.1}$$

where $\alpha'' = \cos \widehat{(z, \mathbf{V})}$ is the cosine of the angle between the satellite's longitudinal axis z and the velocity $\mathbf{V}$ of its center of mass. Let us direct the other two axes x and y along the straight lines that connect two opposite blades. But the x and y directions are in now way worse than the z direction – the only difference is that two blades work instead of four (hence, a different value of a_0), and the blades are turned in the opposite direction (hence, the torque changes sign). We can therefore write the following expressions for the x- and y-components of the torque:

$$M_x = -\rho V^2 a_1 \alpha, \quad M_y = -\rho V^2 a_1 \alpha', \tag{12.4.2}$$

where $\alpha = \cos \widehat{(x, \mathbf{V})}$, $\alpha' = \cos \widehat{(y, \mathbf{V})}$. If the blades are inclined – as is the case for the Proton-2 satellite – at an angle of 45° to the longitudinal axis, then $a_1 = a_0/2$ (because only two blades out of four operate for different "effective area"[2] of each of the blades). For other inclinations of the blades to the longitudinal axis the relation between a_1 and a_0 will be different because of the different "effective areas," but one can always assume that

$$a_1 < a_0. \tag{12.4.3}$$

Thus, we are faced with the following task: study the evolution of the fast rotation of a dynamically-symmetric satellite subject to the action of the torques (12.4.1) and (12.4.2). This evolution can be considered in the variables previously introduced in the sixth essay: L, ρ, σ (length of the angular momentum vector and the two angles that specify its position relative to the orbital plane) and ϑ (the angle between the angular momentum vector and the longitudinal axis of the satellite). However, here it is more convenient to replace ρ and σ by the angle θ between the angular momentum vector $\mathbf{L}$ and the direction of the velocity at the perigee of the orbit, $\mathbf{V}_\pi$, and the angle λ of rotation of $\mathbf{L}$ around $\mathbf{V}_\pi$, measured from the radius vector of the perigee.

Writing down the evolution equations and carrying out the usual averaging procedure with respect to the fast variable – here the precession angle ψ – we

[2]The area of the projection of the blade on the plane normal to the axis under consideration.

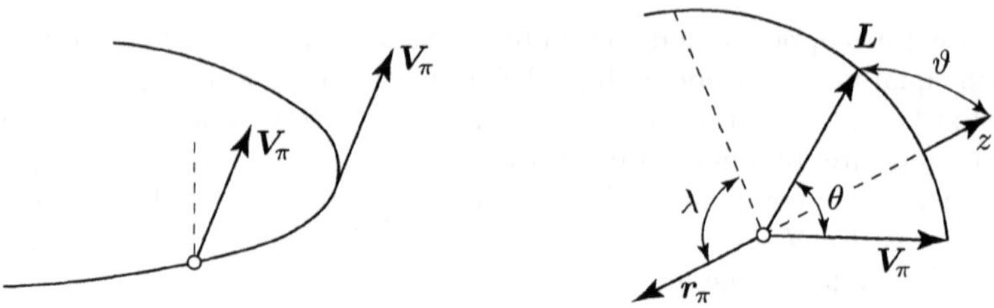

FIGURE 12.4. Variables defining the position of the satellite and of its angular momentum

obtain the following system of evolution equations [12.4]:

$$\left.\begin{aligned}
\frac{d\lambda}{d\nu} &= -\frac{\rho_\pi\sqrt{\mu P}}{L}\left[a_1-(a_0+a_1)\frac{1}{2}\sin^2\vartheta\right]V_2\frac{\sin\lambda}{\sin\theta}\,,\\
\frac{d\theta}{d\nu} &= \frac{\rho_\pi\sqrt{\mu P}}{L}\left[a_1-(a_0+a_1)\frac{1}{2}\sin^2\vartheta\right](V_1\sin\theta+V_2\cos\theta\cos\lambda)\,,\\
\frac{dL}{d\nu} &= \rho_\pi\sqrt{\mu P}\left[-a_1+(a_0+a_1)\cos^2\vartheta\right](V_1\cos\theta-V_2\sin\theta\cos\lambda)\,,\\
\frac{d\vartheta}{d\nu} &= \frac{\rho_\pi\sqrt{\mu P}}{L}(a_0+a_1)\sin\vartheta\cos\vartheta(V_1\cos\theta-V_2\sin\theta\cos\lambda)\,.
\end{aligned}\right\} \quad (12.4.4)$$

Here instead of time we take as independent variable the true anomaly ν. As before, μ is the constant of gravitation, P is the focal parameter of the orbit, and ρ_π is the density of the atmosphere at perigee. If one denotes $\overline{\rho}=\rho/\rho_\pi$, then the functions V_1 and V_2 can be written in the form

$$V_1=\overline{\rho}(\widetilde{h})\,\frac{(e+\cos\nu)\sqrt{1+e^2+2e\cos\nu}}{(1+e\cos\nu)^2}\,,$$

$$V_2=\overline{\rho}(\widetilde{h})\,\frac{\sin\nu\sqrt{1+e^2+2e\cos\nu}}{(1+e\cos\nu)^2}\,,$$

where

$$\widetilde{h}=\frac{Pe}{1+e}\cdot\frac{1-\cos\nu}{1+e\cos\nu}\,.$$

Here e is the eccentricity of the orbit and $\widetilde{h}$ has the meaning of satellite's altitude, measured from the level of the orbit's perigee. Note that we do not actually need the explicit form of the periodic functions V_1 and V_2; we only remark that the right-hand sides of equations (12.4.4) depend explicitly on the "time" ν, which of course complicates an already complicated system of equations. The evolution equations (12.4.4), which describe the influence of the propeller effect on the satellite, are more difficult to analyze than the evolution equations governing motion in a gravity field (see the sixth essay). Indeed, there we had only two equations for the orientation angles of the angular momentum vector, while here we have four. The additional equations describe the variation of the length L of the angular momentum vector and of the angle ϑ between this vector and the satellite's longitudinal axis.

It is therefore even more remarkable that such complicated equations like (12.4.4), with an explicit dependence of the right-hand sides on time, actually admit two exact first integrals. This allows us to carry out an effective qualitative analysis of the solutions of the evolution equations. The first integrals are as follows:

$$L\sin\vartheta\,|\cot\vartheta|^{\alpha} = l, \qquad \alpha = \frac{a_1}{a_1 + a_2} < \frac{1}{2}, \tag{12.4.5}$$

and

$$|\tan\vartheta|^{2\alpha}|\cos\vartheta|\cos^2\rho = c_\rho, \qquad \cos\rho = -\sin\theta\sin\lambda. \tag{12.4.6}$$

Here ρ is the angle between $\mathbf{L}$ and the normal to the orbital plane.

The integral curves (12.4.5) and (12.4.6) are shown in Figure 12.5, (a) and (b). In the actual motion these curves may not be traversed completely. Since the right-hand sides of the differential equations (12.4.4) are explicitly 2π-periodic in ν, their solutions will also contain a periodic component with roughly the same period. As a result, in the planes $\{\rho, \cos\vartheta\}$ and $\{L, \cos\vartheta\}$ (Figure 12.5) the representing point oscillates along an integral curve (12.4.5) and, respectively, (12.4.6), with a period close to the period T_ν of orbital revolution of the satellite; these oscillations have a rather small amplitude, sweeping over the period T_ν a very small arc of the integral curve. Furthermore, in the case of a *circular orbit* the "forward" motion of the representing point along the integral curve during the first half-period of revolution of the satellite is compensated by a similar "backward" drift along the integral curve during the second half-period, and as a result the representing point oscillates about the mean stationary position. In other words, the variables L, ρ, and ϑ will oscillate (with a small amplitude) about their initial values, without changing in the mean. Physically this is readily explained by the fact that all the points of the circular orbit are equivalent; the dynamic pressure exerted by the incident flow is the same at all points, and consequently the perturbations of the motion are compensated in diametrically opposite points of the orbit.

The angular momentum vector preserves its direction in space over an orbital period, and hence changes its direction with respect to the incident flow. The difference between these relative directions is maximal precisely at diametrically opposite points of a circular orbit. In crude terms, at one point of the orbit the "wind blows" in one side of the pinwheel, while in the opposite point it blows in the opposite side.

Not too many possibilities here!

However, on an *elliptic orbit* the situation is completely different. The dynamic pressure at perigee is considerably larger than at apogee and consequently the perturbations of the motion at apogee cannot be compensated by the perturbations at perigee. As a result perturbations gradually accumulate. On the integral

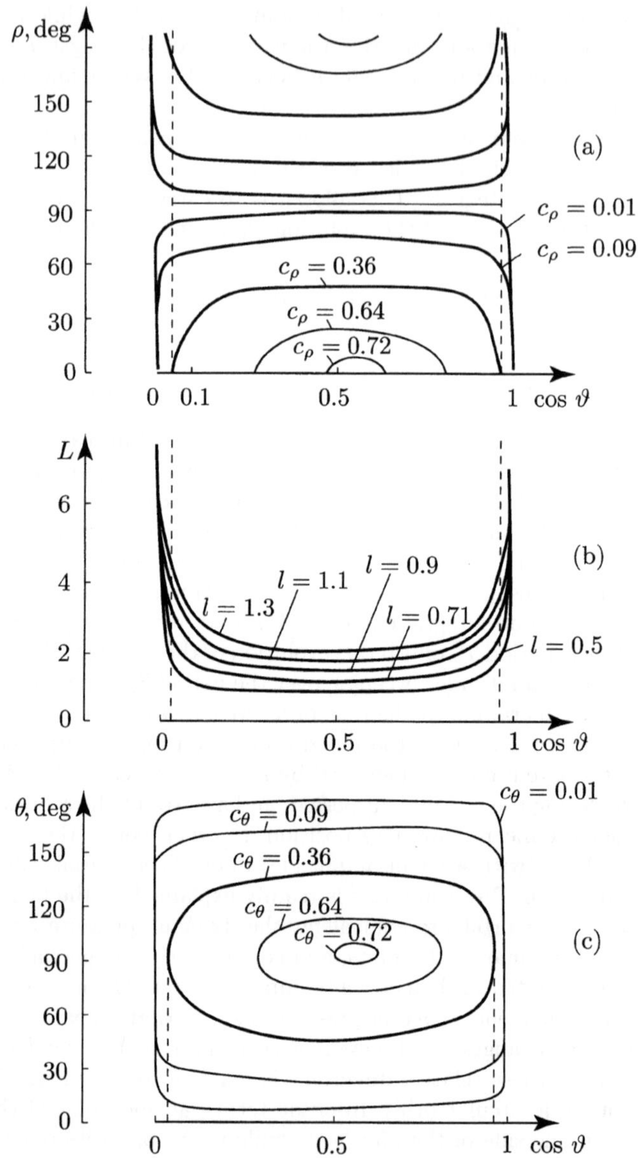

FIGURE 12.5. Integral curves

curves (Figure 12.5) the representing point "advances" more that it "retreats," and after several periods T_ν a substantial drift of the parameters L, ρ, and ϑ is observed.

As a result of this drift of the initial orientation, the incident flow at perigee changes its direction with respect to the satellite, which in the end leads to a drift of the parameters of motion in the opposite direction, and so on. To sum up, the evolution of the rotation and orientation parameters of the satellite in an elliptic orbit has a long-periodic character (with a period several tens of times larger than the period of orbital revolution of the satellite).

As a result of this evolution, on the integral curves shown in Figure 12.5 the representing points sweep a more or less substantial segment of the curve, oscillating along this segment with a long period; these long-period oscillations are in turn a superposition of many, not fully compensated, oscillations of short period (equal to the orbital period). In Figure 12.5 one can distinguish the segments of integral curves traversed in real motion.

An examination of Figure 12.5 (b) reveals that the quantity L may change periodically, with a large period, within rather wide limits, which means that the angular rate of proper precession of the satellite changes. It sometimes grows – when the component of the propelling torque along the vector $\mathbf{L}$ is positive in the mean, and sometimes decreases – when the orientation of the propelling torque with respect to the vector $\mathbf{L}$ changes.

At the same time, the spinning regime of the satellite, characterized by the angle ϑ, also changes. The satellite may switch from a regime of topsy-turvy motion ($\vartheta \approx \pi/2$) to a regime of axial spinning ($\vartheta \approx 0$) and back, in a long-periodic manner. As the integral curves (see Figure 12.5 (b)) make clear, in the regime of axial spinning ($\cos\vartheta \approx 1$) and the regime of topsy-turvy motion ($\cos\vartheta \approx 0$) the magnitude L of the angular momentum vector is maximal, whereas in the transition regime L minimal. This is in full agreement with the picture emerging from experimental data shown in Figure 12.2.

We can extract from this analysis the pure evolution of the motion, neglecting the short-period oscillations. To this end we need to average the evolution equations (12.4.4) one more time, with respect to the fast variable ν. Then

$$I = \frac{1}{2\pi}\int_0^{2\pi} V_1\, d\nu = \text{const} \neq 0, \qquad \text{but} \qquad \frac{1}{2\pi}\int_0^{2\pi} V_2\, d\nu = 0,$$

from which we immediately conclude that within this approximation

$$\lambda = \lambda_0 = \text{const}. \tag{12.4.7}$$

Hence, the vector $\mathbf{L}$ does not rotate around the direction of the velocity $\mathbf{V}_\pi$ at perigee, and moves only in a plane containing $\mathbf{V}_\pi$. In other words, from the two angles that determine the position of $\mathbf{L}$ in space, only one varies, namely θ. In view of (12.4.7), (12.4.6) yields this first integral in the new form

$$|\tan\vartheta|^{2\alpha}|\cos\vartheta|\sin^2\theta = c_\theta. \tag{12.4.8}$$

The integral curves (12.4.8) are depicted in Figure 12.5 (c). They show that if the satellite periodically switches from the regime of topsy-turvy motion $\vartheta \approx \pi/2$ to the regime of axial spinning $\vartheta \approx 0$, then this process is necessarily accompanied by oscillations (with the same period) of the orientation of the satellite's angular momentum vector in a range of almost 180°. This, too, is in full agreement with the experimentally observed motion of the angular momentum vector of the Proton-2 satellite (Figure 12.3). As the reader will recall, the angular momentum vectors "walks" from the world's North pole to the world's South pole and back, while the regime of motion changes back and forth between axial spinning and topsy-turvy motion. Thus, the qualitative agreement between experimental data and the theory developed above is evident.

To require a quantitative agreement in the framework discussed above is meaningless. Indeed, we did not take into account the gravitational torques (for the Proton satellites they are about 5% of the aerodynamic torques), dissipative factors, the aerodynamics of the main body of the satellite itself, the possibility that the blades fall in the "shadow" of the main body, and many other effects. These neglected factors would explain, for instance, the "8-shaped" trace of the angular momentum vector on the celestial sphere (Figure 12.3) – here we only explained the 180° amplitude of this "8-shape."

An accurate formulation of the aerodynamic problem must take into account the character of the interaction between the molecules of the incident air flow and the surface of the satellite. As a matter of fact, using experimental data on the motion of Proton-type satellites around their center of mass one can establish the precise the nature of this interaction [12.9].

However, our goal here was far more modest, namely, to construct a model of the observed phenomena, and this goal has been achieved.

In conclusion, let us remark that the integral curves reveal the existence (in the mean with respect to ν) of a stable stationary point $\theta^* = \pi/2$, $\sin^2\vartheta = 2\alpha$. If the motion reaches these initial conditions (or gets close to them), then it remains there; the magnitude of the corresponding angular momentum vector also remains constant (but arbitrary): $L = L_0$. As we already mentioned, for our model of a Proton satellite one can take $a_0 = 2a_1$; then $\alpha = 1/3$ and the stationary value of the angle ϑ^* is given by the equality

$$\cos\vartheta^* = \frac{1}{\sqrt{3}} \qquad (\vartheta^* \approx 54.7°).$$

For initial data far from the stationary point the motion is strongly modulated (as we already know).

Figure 12.6 shows the dependences $\theta(\nu)$, $\vartheta(\nu)$, and $L/L_0(\nu)$, obtained by numerical integration of the equations (12.4.4). To emphasize the aforementioned effects, a very high value of the propelling torque was chosen; as a result, the short-period oscillations superposed on the long-period evolution are sharply distinguished. As for the long-period evolution, it of course has all the properties investigated above.

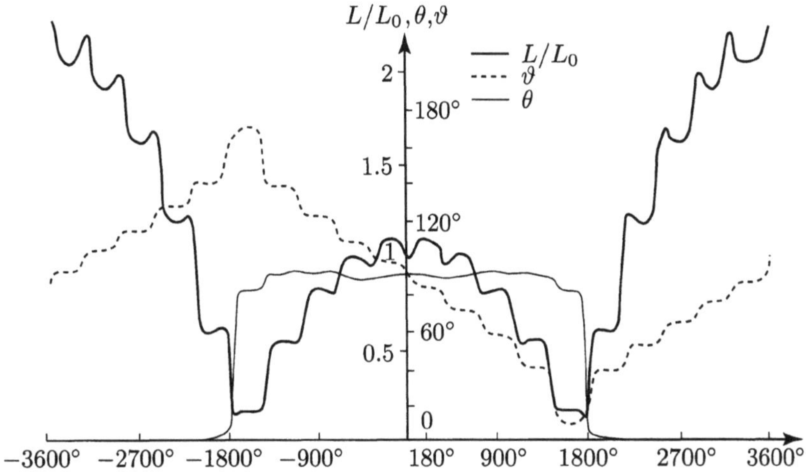

FIGURE 12.6. Graphs of the functions $L/L_0(\nu)$, $\theta(\nu)$, and $\vartheta(\nu)$

We have purposely chosen for this concluding essay a problem that could not even be approached in the old, pre-satellite era mechanics. And with that we say to our reader: "Good-bye!"

Well do I know that I am mortal, a creature of one day. But if my mind follows the winding paths of the stars, Then my feet no longer rest on earth, but standing by Zeus himself I take my fill of ambrosia, the divine dish.

Ptolemy, *Almagest, Book 1*

References

First essay[1]

[1.1] Duboshin, G. N., *Celestial Mechanics. Basic Problems and Methods,* "Nauka," Moscow, 1968. (Russian)

[1.2] Shklovskiĭ, I. S., *Universe. Life. Intelligence,* 4th edition, "Nauka," Moscow, 1976. (Russian); see also Shklovskii, I. S. and Sagan, Carl, *Intelligent Life in the Universe,* Holden-Day, San Francisco, 1966.

[1.3] Berezkin, E. N., *Lectures on Theoretical Mechanics. Part I,* Izdat. MGU, Moscow, 1967. (Russian)

[1.4] Lur′e, A. I., *Analytical Mechanics,* Fizmatgiz, Moscow, 1961. (Russian) [MR **32** #4895]

[1.5] Poincaré, H., *New Methods of Celestial Mechanics. Vol. 1. Periodic and Asymptotic Solutions,* Translated from the French. Revised reprint of the 1967 English translation. With endnotes by V. I. Arnol′d. Edited and with an introduction by Daniel L. Goroff. History of Modern Physics and Astronomy, 13, American Institute of Physics, New York, 1993. [MR **93j:01016a**]

[1.6] Bogolyubov, N. N., and Mitropol′skiĭ, A. Yu., *Asymptotic Methods in the Theory of Non-linear Oscillations,* Fourth edition, revised and augmented, "Nauka", Moscow, 1974. (Russian) [MR **51** #10750]; English transl. of the Second Russian edition, International Monographs on Advanced Mathematics and Physics, Hindustan Publishing Corp., Delhi, Gordon and Breach Science Publishers, New York, 1961. [MR **25** #5242]

[1.7] Volosov, V. M., *Averaging in systems of ordinary differential equations,* Uspekhi Mat. Nauk **17** (1962), no. 6(108), 3–126. (Russian) [MR **26** #3976]

[1.8] Moiseev, N. N., *Asymptotic Methods of Nonlinear Mechanics,* Second edition, "Nauka," Moscow, 1981. (Russian) [MR **83i:70001**]

[1.9] Arnold, V. I., *Proof of a theorem of A. N. Kolmogorov on the preservation of conditionally periodic motions under a small perturbation of the Hamiltonian,* Uspekhi Mat. Nauk **18** (1963), no. 5(113), 13–40. (Russian) [MR **29** #328]; English transl., Russian Math. Surv., **18** (1963), no.5, 9–36.

[1.10] Arnold, V. I., *Small denominators and problems of stability of motion in classical and celestial mechanics,* Uspekhi Mat. Nauk **18** (1963), no. 6(114), 91–192. (Russian) [MR **30** #943]; English transl., Russian Math. Surv., **18** (1963), No. 6, 85–191.

[1]Translator's note. I have tried to find referrences to translations of Russian papers and books whenever I could find them. I have also added Mathematical Reviews numbers whenever available.

[1.11] Demin, V. G., *The Fate of the Solar System,* "Nauka," Moscow, 1969. (Russian)

[1.12] Okhotsimskiĭ, D. E., Eneev, T. M., and Taratynova, G. P., *Determining the lifetime of an artificial Earth satellite and investigating the secular perturbations of its orbit,* Uspekhi Fiz. Nauk **63** (1957), no. 1a; English transl., in: 1958 The Russian literature of satellites. I, pp. 45–70, International Physical Index, Inc., New York. [MR **20** #749]

[1.13] Okhotsimskiĭ, D. E., *Space-Flight Dynamics. Lecture Notes,* Izdat. MGU, Moscow 1968. (Russian)

[1.14] El′yasberg, P. E., *Introduction to the Theory of Flight of Artificial Earth Satellites,* "Nauka", Moscow, 1965. (Russian); English transl., Israel Program for Scientific Translations, Jerusalem, 1967.

[1.15] Balk, M. B., *Elements of Space-Flight Dynamics,* "Nauka", Moscow, 1965. (Russian)

[1.16] Beletsky, V. V., *Orbit of an equatorial Earth satellite,* in: Iskusstvennye Sputniki Zemli, Vol. 13, 1962, pp. 53–60. (Russian); English transl. in: Artificial Earth Satellites, Vol 13, 1963, pp. 56–64.

[1.17] Beletsky, V. V., *Motion of an Artificial Satellite About its Center of Mass,* "Nauka", Moscow 1962 (Russian); English transl., Israel Program for Scientific Translations, Jerusalem, 1966; see also NASA Technical Reports NASA-TT-F-429, TT-67-51366.

[1.18] Beletsky, V. V., *On the libration of a satellite,* in: Iskusstvennye Sputniki Zemli, Vol. 3, 1959, pp. 13–31. (Russian)

[1.18′] Beletsky, V. V., *On the libration of a satellite in an elliptic orbit,* in: Iskusstvennye Sputniki Zemli, Vol. 16, 1963, pp. 45–56. (Russian)

[1.19] Sedov, L. I., *Dynamical effects in the motion of artificial Earth satellites,* in: Iskusstvennye Sputniki Zemli, Vol. 2, 1958, pp. 3–9. (Russian) [MR **23** #B1696]

[1.20] Chernous′sko, F. L., *Resonance phenomena in the motion of a satellite about its center of mass,* Zh. Vychysl. Mat. Mat. Fiz., **3** (1963), no. 3, 528–538. (Russian)

[1.21] Zlatoustov, V. A., Okhotsymkiĭ, D. E., Sarychev, V. A., and Torzhevskiĭ, A. P., *Periodic solutions in the problem of oscillations of a satellite in an elliptic orbit,* Kosmicheskie Issled., **2** (1964), no. 5, 657–666. (Russian); English transl., Cosmic Research **2** (1964), no. 5, 569–576.

[1.22] Beletsky, V. V. and Lavrovskiĭ, E. K., *On the theory of resonance rotation of Mercury,* Astron. Zh., **52** (1975), no. 6, 1299–1308. (Russian); English transl., Soviet Astron. **19** (1976), no. 6, 776–781.

[1.23] Sarychev, V. A., *Orientation of artificial satellites,* in: Itogi Nauki Tekh. Ser. Issled. Kosmich. Prostr., Inst. Nauchn. i Tekhn. Inform. (VINITI), Moscow, Vol. 11, 1978. (Russian)

[1.24] Sarychev, V. A., Sazonov, V. V., and Zlatoustov, V. A., *Periodic oscillations of a satellite in the plane of an elliptic orbit,* Kosmicheskie Issled.,

15 (1977), no. 6, 809-834. (Russian); English transl., Cosmic Research **15** (1978), no. 6, 698–719.

[1.25] Sarychev, V. A., Sazonov, V. V., and Zlatoustov, V. A., *Periodic rotation of a satellite in the plane of an elliptical orbit,* Kosmicheskie Issled., **17** (1979), no. 2, 190–207. (Russian); English transl., Cosmic Research **17** (1979), no. 2, 159–173.

[1.26] Sarychev, V. A., Sazonov, V. V., and Zlatoustov, V. A., *Asymmetric periodic oscillations of a satellite in the plane of an elliptical orbit,* Kosmicheskie Issled., **18** (1980), no. 1, 3–10. (Russian); English transl., Cosmic Research **18** (1980), no. 1, 1–6.

[1.27] Burov, A. A., *Nonintegrability of the equation of plane oscillations of a satellite in an elliptic orbit,* Vestnik Moskov. Univ. Ser. I Mat. Mekh., (1984), no. 1, 71–73. (Russian) [MR **85a:70053**]

[1.28] Wisdom, J., Peale, S. J., and Mignard, F., *The chaotic rotation of Hyperion,* Icarus **58** (1984), no. 2, 137–152.

[1.29] Tong, Xiaohua and Rimrott, Fred P. J., *Numerical studies of chaotic planar motion of satellites in an elliptic orbit,* Chaos Solitons Fractals **1** (1991), no. 2, 179–186.

[1.30] Maciejewski, A. J., *Non-integrability of the planar oscillations of a satellite,* Acta Astronomica **45** (1995), no. 1, 327–344.

[1.31] Gozdziewski, K., *Rotational dynamics of Janus and Epimetheus,* in: Dynamics and Astrometry of Natural and Artificial Celestial Bodies, Proc. IAU Colloquium No. 165 (Poznan, Poland, 1996), pp. 269–274, Kluwer Acad. Publ., 1997.

[1.32] Beletsky, V. V., *Reguläre und Chaotische Bewegung Starrer Körper,* Teubner Studienbücher Mechanik, B. G. Teubner, Stuttgart, 1995. [MR **99m: 70004**]

[1.33] Beletsky, V. V., Pivovarov, M. L., and Starostin, E. L., *Regular and chaotic motions in the applied dynamics of a rigid body,* Chaos **6** (1996), no. 2, 155–166.

[1.34] Bruno, A. D., *Power Geometry in Algebraic and Differential Equations,* "Nauka," Moscow, 1998. (Russian)

[1.35] Bruno, A. D. and Varin, V. P., *The limit problems for the equation of oscillations of a satellite,* Celestial Mech. Dynam. Astronom., **67** (1997), no. 1, 1–40. [MR **99f:70029**]

[1.36] Bruno, A. and Varin, V., *Singularities of oscillations of a satellite of highly eccentric elliptic orbits,* in: Proc. Second World Congress of Nonlinear Analysts, Part 4 (Athens, 1996). Nonlinear Anal., **30** (1997), no. 4, 2541–2546.

[1.37] Bruno, A. D. and Petrovich, V. Yu., *Computation of the periodic oscillations of a satellite,* Mat. Model., **9** (1997), no. 6, 82–94. (Russian) [MR **98h:70025**]

[1.38] Sadov, S. Yu., *Normal form of the equation of the oscillations of a satellite in the singular case,* Mat. Zametki **58** (1995), no. 5, 785–789. (Russian) [MR **98b:34053**]

[1.39] Sadov, S. Yu., *Singular normal form for a quasilinear ordinary differential equation,* in: Proc. Second World Congress of Nonlinear Analysts, Part 8 (Athens, 1996). Nonlinear Anal., **30** (1997), no. 8, 4973–4978.

[1.40] Sadov, S., *Functions that determine stability of rational rotations of a near symmetric satellite. Simplification of systems of algebraic and differential equations with applications,* Math. Comput. Simulation **45** (1998), no. 5-6, 465–484. [MR **99a:70042**]

[1.41] Beletsky, V. V. and Shlyakhtin, A. N., *Extremal properties of resonance motions,* Dokl. Akad. Nauk SSSR **231** (1976), no. 4, 829–832. (Russian)

Second essay

[2.1] Aksenov, E. P., Grebennikov, E. A., and Demin, V. G., *General solution of the problem of the motion of an artificial satellite in the normal field of attraction of Earth,* in: Iskusstvennye Sputniki Zemli, Vol. 8, 1961, pp. 64–71; English transl. in: Artificial Earth Satellites, Vol. 7–8, 1962, pp. 225–232. (Russian)

[2.2] Brouwer, D. and Clemence, G. M., *Methods of Celestial Mechanics,* Academic Press, New York-London, 1961. [MR **23** #B1544]

[2.3] Demin, V. G., *Motion of an Artificial Satellite in a Noncentral Gravitational Field,* "Nauka," Moscow, 1968. (Russian); English transl., *Motion of an Artificial Satellite in an Eccentric Gravitation Field*, NASA Reports, 1970.

[2.4] Duboshin, G. N., *Celestial Mechanics. Analytic and Qualitative Methods,* "Nauka," Moscow, 1964. (Russian)

[2.5] Kislik, M. D., *Motion of an artificial satellite in the normal gravitational field of Earth,* in: Iskusstvennye Sputniki Zemli, Vol. 4, 1960, pp. 3–17. (Russian)

[2.6] Vinti, J. P., *Theory of an intermmediary orbit for satellite astronomy,* J., Res. National Bureau Standards **65B** (1959), no. 3, 169–201.

[2.7] Alekseev, V. M., *A generalized three-dimensional problem of two fixed centers of gravitation. A classification of motions,* Byull. Inst. Teoret. Astronom., **10** (1965), 241–271. (Russian) [MR **32** #3819]

[2.8] Lur′e, A. I., *Analytical Mechanics,* Fizmatgiz, Moscow, 1961. (Russian) [MR **32** #4895]

[2.9] Duboshin, G. N., *Celestial Mechanics. Basic Problems and Methods,* "Nauka," Moscow, 1968. (Russian)

Third essay

[3.1] Beletsky, V. V., *On trajectories of space flights with a constant reactive acceleration vector,* Kosmicheskie Issled., **2** (1964), no. 3, 408–413. (Russian); English transl., Cosmic Research **2** (1964), no. 3, 346–350.

[3.2] Beletsky, V. V., *Description of planar space-flight trajectories with constant reactive acceleration vector,* in: Proc. Conference on General and Applied Problems of Celestial Mechanics, "Nauka," Moscow, 1967. (Russian)

[3.3] Niţă, M. M., *Über die Bewegung der Rakete in einen Zenträlkraftfeld,* Rev. Méc. Appl., **6** (1961), no. 2, 171–190. [MR **55** #5787]

[3.4] Demin, V. G., *On a method for investigating the motion of a spacecraft in the sphere of action of a planet,* Trudy Univ. Druzhby Narodov Patrisa Lumumby, Ser. Teor. Mekh., **17** (1966), no. 4. (Russian)

[3.5] Kunitsyn, A. L., *On the motion of a rocket in a central force field with a constant reactive acceleration vector,* Kosmicheskie Issled., **4** (1966), no. 2, 324–327. (Russian); English transl., Cosmic Research **4** (1966), no. 2, 295–298.

[3.6] Kopnin, Yu. M., *Evolution of the orbit of a satellite under the action of a perturbing force of constant magnitude and direction,* Inzhenernyĭ Zh., **5** (1965), no. 6, 1003–1009. (Russian)

[3.7] Chernikov, Yu. A., *Evolution of finite planetocentric orbits of small bodies under the influence of direct radiation pressure,* Kosmicheskie Issled., **6** (1968), no. 6, 812–824. (Russian)

[3.8] Lidov, M. L., *Secular effects in the evolution of orbits due to radiation pressure,* Kosmicheskie Issled., **7** (1969), no. 4, 467–484. (Russian)

[3.9] Marshak, S. Ya., *The trunk. Norwegian folktale,* in: Collected Works, Vol. 1, 1957. (Russian)

Fourth essay

[4.1] Duboshin, G. N., *Celestial Mechanics. Basic Problems and Methods,* "Nauka," Moscow, 1968. (Russian)

[4.2] Moiseev, N. D., Trudy Gos. Astron. Inst. P. K. Shternberg **15** (1945). (Russian)

[4.3] Lidov, M. L., *On the approximate analysis of the evolution of orbits of artificial satellites,* in: Problems of the Motion of Artificial Spacecraft, Izdat. Akad. Nauk SSSR, Moscow, 1963. (Russian)

[4.4] Lidov, M. L., *Evolution of orbits of artificial satellites of planets under the action of the gravitational perturbations of external bodies,* in: Iskusstvennye Sputniki Zemli, Vol. 8, 1961, pp. 1–45. (Russian); English transl., in: Artificial Earth Satellites Vol. 7–8, 1962, pp. 168–207.

[4.5] Demin, V. G., *The Fate of the Solar System,* "Nauka," Moscow, 1969. (Russian)

[4.6] Chebotarev, G. A., *Analytical and Numerical Methods of Celestial Mechanics,* "Nauka," Moscow, 1965. (Russian); English transl., Modern Analytic and Computational Methods in Science and Mathematics, 9. American Elsevier Publishing, New York, 1967.

[4.7] Kislik, M. D., *Spheres of influence of large planets and of the Moon,* Kosmicheskie Issled., **2** (1964), no. 6, 853–858. (Russian); English transl., Cosmic Research **2** (1964), no. 6, 746–750.

[4.8] Arnold, V. I., *Proof of a theorem of A. N. Kolmogorov on the preservation of conditionally periodic motions under a small perturbation of the Hamiltonian,* Uspekhi Mat. Nauk **18** (1963), no. 5 (113), 13–40. (Russian) [MR **29** #328]

[4.9] Arnold, V. I., *Small denominators and problems of stability of motion in classical and celestial mechanics,* Uspekhi Mat. Nauk **18** (1963), no. 6 (114), 91–192. (Russian) [MR **30** #943]

[4.10] Molchanov, A. M., *The resonant structure of the solar system. The law of planetary distances,* Icarus – Internat. J. Solar System **8** (1968), no. 2, 203-215.

[4.11] Bruno, A. D., *Instability in Hamiltonian systems and the distribution of asteroids,* Mat. Sb. (N. S.), **83 (125)** (1970), no. 2 (10), 273–312. (Russian) [MR **43** #625]

[4.12] Brouwer, D., *The problem of Kirkwood Gaps in the asteroid belt,* Astron. J., **68** (1963), no. 3, 152–159.

[4.13] Rzhiga, O. N., *Results of radar studies of planets,* Kosmicheskie Issled., **7** (1969), no. 1, 84–91 (Russian)

[4.14] Chetaev, N. G., *Stability of Motion. Works on Analytical Mechanics,* Izdat. Akad. Nauk SSSR, Moscow, 1962. (Russian); see also *Stability of Motion,*, Fourth edition. With comments by V. V. Rumyantsev, "Nauka", Moscow, 1990. [MR **92j:34095**]

[4.15] Ishlinskiĭ, A. Yu., *Mechanics,* in the book *October and the Scientific Progress,* Vol.1, Izdat. APN, Moscow, 1967. (Russian)

[4.16] Lichtenberg, A. J. and Lieberman, M. A., *Regular and Stochastic Motion,* Springer Verlag, New York-Berlin, 1983. [MR **85g:58038**] (see also [MR **93c:58071**])

[4.17] Thompson, J. M. T. and Stewart, H. B., *Nonlinear Dynamics and Chaos. Geometrical Methods for Engineers and Scientists,* John Wiley & Sons, Ltd., Chichester, 1986. [MR **87m:58098**]

Fifth essay

[5.1] Egorov, V. A., *Certain problems of moon flight dynamics,* Uspekhi. Fiz. Nauk **63** (1957), no. 1a. (Russian)

[5.2] Egorov, V. A., *The Three-Dimensional Problem of Flight to the Moon,* "Nauka," Moscow, 1965. (Russian); see also *Three-dimensional Lunar Trajectories*, translated from Russian by Ron Harding, Israel Program for Scientific Translations, Jerusalem, 1969.

[5.3] Levantovskiĭ, V. I., *On a Rocket to the Moon,* "Fizmatgiz," Moscow, 1960. (Russian)

[5.4] Leontovich, A. M., *On the stability of Lagrange's periodic solutions of the restricted three-body problem,* Dokl. Akad. Nauk SSSR **143** (1962), 525–528. (Russian) [MR **24** #B2462]; English transl., Sov. Math. Dokl., (1962) no. 3, 425–429.

[5.5] Deprit, A. and Deprit-Bartholome, A., *Stability of the triangular Lagrangian points,* Astron. J., **72** (1967), no. 2, 173–179.

[5.6] Markeev, A. P., *On the stability of the triangular libration points in the circular restricted three-body problem,* Prikl. Mat. Mekh., **33** (1969), no. 1. (Russian); English transl., J. Appl. Math. Mech. **33** (1969) 105–110. [MR **43** #4311]

[5.7] Lidov, M. L., Okhotsimskiĭ, D. E., and Teslenko, N. M., *Investigation of a class of trajectories of the restricted three-body problem,* Kosmicheskie Issled., **2** (1964), no. 6, 843–852. (Russian); English transl., Cosmic Research **2** (1964), no. 6, 737–745.

[5.8] Levantovskiĭ, V. I., *Mechanics of Space Flight: An Elementary Exposition,* "Nauka," Moscow, 1970. (Russian)

[5.9] Sedov, L. I., *Orbits of space rockets towards the Moon,* in: Iskusstvennye Sputniki Zemli, Vol. 5, 1960, pp. 3–15. (Russian)

[5.10] Kozlov, N. N., Syunyaev, R. A., and Eneev, T. M., *Tidal interaction of galaxies,* Zemlya i Vselennaya, no. 6, 1974. (Russian)

[5.11] Kozlov, N. N., Syunyaev, R. A., and Eneev, T. M., *Tidal interaction of galaxies,* Preprint no. 76, Institute of Applied Mathematics, Academy of Sciences of USSR, 1971. (Russian)

[5.12] Kozlov, N. N., Syunyaev, R. A., and Eneev, T. M., *Tidal interaction of galaxies,* Dokl. Akad. Nauk SSSR **204** (1972), no. 3, 579–582. (Russian)

[5.13] Eneev, T. M., Kozlov, N. N., and Sunjaev, R. A., *Tidal interaction of galaxies,* Astronomy and Astrophys., **22** (1973), no. 1, 41–60.

[5.14] Demin, V. G., *The Fate of the Solar System,* "Nauka," Moscow, 1969. (Russian)

[5.15] Bruno, A. D., *The Restricted Three-Body Problem: Plane Periodic Orbits,* "Nauka," Moscow, 1990. (Russian) [MR **93g:70009**]; English transl., de Gruyter Expositions in Mathematics, 17. Walter de Gruyter & Co., Berlin, 1994. [MR **95g:70007**]

[5.16] Hénon, M., *Sur les orbites interplanétaires qui rencontre deus fois la terre,* Bull. Astron. Ser. 3, **3** (1968), no. 3, 377–402.

[5.16] Bruno, A. D., *On periodic flybys of the Moon,* Celestial Mech., **24** (1981), no. 3, 255–268. [MR **82i:70011**]

Sixth essay

[6.1] Beletsky, V. V., *Motion of an Artificial Satellite About its Center of Mass,* "Nauka", Moscow 1962 (Russian); English transl., Israel Program for Scientific Translations, Jerusalem, 1966; see also, NASA Technical Reports NASA-TT-F-429, TT-67-51366.

[6.2] Goldreich, P. and Peale. S. J., *The dynamics of planetary rotations,* Annual Review of Astronomy and Astrophysics, Vol. 6, pp. 287–320, Palo Alto, CA, 1968.

[6.2′] Express Information, *Astronautics and Astrodynamics,* no. 12, 1972. (Russian)

[6.3] Beletsky, V. V., *On the libration of a satellite,* in: Iskusstvennye Sputniki Zemli, Vol. 3, 1959, pp. 13–31. (Russian)

[6.4] Rumyantsev, V.V., *On the stability of stationary motions of satellites,* Vychisl. Ttsentr Akad. Nauk SSSR, Moscow, 1967. (Russian).

[6.5] Stümpke, H., *Bau und Leben der Rhinogradentia,* Gustav Fischer Verlag, Stuttgart, 1967.[2]

[6.6] Stümpke, H., *Body Structure and Life of Rhinogradentia,* Nauka i Zhizhn′ **4**, 1963. (Russian)

[6.7] DeBra, D. B., *Operation principles of passive stabilization systems and fundamental directions of research,* in: Recent Developments in Space Flight Mechanics, American Association for the Advancement of Science and American Astronautical Society, Special Astronautics Symposium, Berkeley, Calif., December 29, 1965, Proceedings, edited by P. B. Richards, American Astronautical Society, Vol. 9, 1966.

[6.8] *Torques and attitude sensing in earth satellites*, edited by S. Fred Singer, New York, Academic Press, 1964.

[6.9] Sarychev, V. A., *Investigation of the dynamics of a gravitational-stabilization system,* in: Iskusstvennye Sputniki Zemli, Vol. 16, 1963, pp. 10–33. (Russian); English transl., in: Artificial Earth Satellites, Vol. 16, 1964, pp. 6–29.

[6.10] Breakwell, J. V. and Pringle, R., *Nonlinear resonances affecting gravity-gradient stability,* in: Proc. XV IAF International Congress, Athens, 1965, Vol. 6, p. 305.

[6.11] Duboshin, G. N., *Celestial Mechanics. Basic Problems and Methods,* "Nauka," Moscow, 1968. (Russian)

[6.12] Kane, T. R., *Attitude stability of Earth-pointing satellites,* AIAA J., **3** (1965), no. 4, 726–731.

[6.13] Khazin, L. G., *On the stability of Hamiltonian systems in the presence of resonances,* Prikl. Mat. Mekh., **35** (1971), no. 3, 423–431. (Russian); English transl., J. Appl. Math. Mech., **35** (1971), no. 3, 384-391.

[6.14] Beletsky, V. V., *On Cassinis' laws,* Preprint no. 79, Institute of Applied Mathematics, Academy of Sciences of USSR, 1971. (Russian)

[6.15] Beletsky, V. V., *Evolution of the rotation of a dynamically symmetric satellite,* Kosmicheskie Issled., **1** (1963), no. 3, p. 339–386. (Russian); English transl., Cosmic Research **1** (1963), no. 3, 279–319.

---

[2]The existence of literature on *Rhinogradentia* is not a sufficient condition for its existence.

[6.16] Beletsky, V. V., *Motion of an artificial Earth satellite about its center of mass,* in: Iskusstvennye Sputniki Zemli, Vol. 1, 1958, pp. 25–43. (Russian)

[6.17] Chernous′sko, F. L., *On the motion of a satellite about its center of mass under the action of gravitational moments,* Prikl. Mat. Mekh., **27** (1963), no. 3, 474–483. (Russian); English transl., J. Appl. Math. Mech., **27** (1963), no. 3, 708–722. [MR **28** #1968]

[6.18] Ishlinskiĭ, A. Yu., *Mechanics,* in the book *October and the Scientific Progress,* Vol. 1, Izdat. APN, Moscow, 1967. (Russian)

[6.19] Beletsky, V. V., *Classification of the motions of an artificial Earth satellite about its center of mass,* in: Iskusstvennye Sputniki Zemli, Vol. 6, 1961, pp. 11–32. (Russian)

[6.20] Trushin, S. I., *Accounting for the motion of the perigee of the orbit in the problem of rotation and orientation of artificial Earth satellites,* Kosmicheskie Issled., **8** (1970), no. 4, 628–629. (Russian)

[6.21] Holland, R. L., and Sperling, H. J., *A first-order theory for the rotational motion of a triaxial rigid body orbiting an oblate planet,* Astron. J., **74** (1969), no. 3, 490–496.

[6.22] Chernous′sko, F. L., *Resonance phenomena in the motion of a satellite about its center of mass,* Zh. Vychsil. Mat. Mat. Fiz., **3** (1963), no. 3, 528–538. 528. (Russian)

[6.23] Lutze, F. H. and Abbitt, M. W., *Rotation locks for near-symmetric satellites,* Celestial Mech., **1** (1969), no. 1, 31–35.

[6.24a] Torzhevskiĭ, A. P. *Fast rotation of an artificial satellite around its center of mass in a resonance regime,* Kosmicheskie Issled., **6** (1968), no. 1, 58–70. (Russian)

[6.24b] Torzhevskiĭ, A. P. *The motion of an artificial satellite around its center of mass and resonances,* Astronautica Acta, **14** (1969), no. 3, 241–259. (Russian)

[6.25] Markeev, A. P., *Investigation of the stability of motion in some problems of celestial mechanics,* Preprint, Institute of Applied Mathematics, Academy of Sciences of USSR, 1970. (Russian)

[6.26] Rzhiga, O. N., *Results of radar probing of planets,* Kosmicheskie Issled., **7** (1969), no. 1, 84–91. (Russian)

[6.27] Pettengill, G. H. and Dyce, R. B., *A radar determination of the rotation of the planet Mercury,* Nature **206** (1965), p. 1240.

[6.28] Dyce, R. B., Pettengill, G. H., and Shapiro, I. I., *Radar determination of the rotations of Venus and Mercury,* Astron. J., **72** (1967), p. 351–359.

[6.29] Colombo, C., *Rotational period of the planet Mercury,* Nature **208** (1965), p. 575.

[6.30] Konashenok, V. N. and Kondrat′ev, K. Ya., *What is New About Venus and Mars,* "Gidrometeoizdat," Leningrad, 1970. (Russian)

[6.31] Kotel′nikov, V. A. et al, *Radar observations of Venus in the Soviet Union in 1964,* Dokl. Akad. Nauk SSSR **163** (1965), no. 1, 50–53. (Russian)

[6.32] Carpenter, R. L., *Study of Venus by cw radar – 1964 results,* Astron. J., **71** (1966), 142–152.

[6.33] Goldstein, R. M., *Moon and Planets,* North Holland, Amsterdam, 1967.

[6.34] Okhotsimskiĭ, D. E. and Sarychev, V. A., *Gravitational stabilization system for artificial satellites,* in: Iskusstvennye Sputniki Zemli, Vol. 16, 1963, pp. 5–9. (Russian); English transl., in: Artificial Earth, Vol. 16, 1963, pp. 1–5.

[6.35] Beletsky, V. V., *Resonance phenomena in the rotational motions of artificial and natural celestial bodies,* Preprint no. 10, Institute of Applied Mathematics, Academy of Sciences of USSR, 1975. (Russian); also, *Resonance phenomena at rotations of artificial and natural celestial bodies,* in: Satellite Dynamics Symposium, Sao Paolo, Brasil, 1974, pp. 192–232 Springer Verlag, Berlin, 1975.

[6.36] *Handbook of Celestial Mechanics and Astrodynamics,* G. N. Duboshin, ed., "Nauka," Moscow, 1971. (Russian)

[6.37] Tisserand, F., *Traité de Mécanique Céleste,* Tome II, Gauthier-Villars, Paris, 1891.

[6.38] Lagrange, J. L., *Oeuvres,* Tome V, Gauthier-Villars, Paris, 1870.

[6.39] Laplace, P. S., *Traité de Mécanique Céleste,* Tome II, Duprat, Paris, 1799. English transl. by Nataniel Bodwich *Mécanique Céleste by the Marquis de Laplace,* with a Commentary, 4 vols., Boston, 1829–1839.

[6.40] Khabibullin, Sh. T., *A system of gravitational stabilization of artificial satellites,* Trudy Kazansk. Gor. Astron. Observat., **34**, Kazan′, 1966. (Russian)

[6.41] Kulikov, K. A. and Gurevich, V. B., *Foundations of Lunar Astrometry,* "Nauka," Moscow, 1972. (Russian)

[6.42] Beletsky, V. V., *Resonance rotation of celestial bodies and Cassini's laws,* Celestial Mech., **6** (1972), no. 2, 356–378.

[6.43] Lidov, M. L. and Neĭshtadt, A. I., *Canonical transformation methods in problems of rotation of celestial bodies and Cassini's laws,* Preprint no. 9, Institute of Applied Mathematics, Academy of Sciences of USSR, 1973. (Russian); see also *The method of canonical transformations in problems of the rotation of celestial bodies and Cassini laws*, in: Determination of the Motion of Spacecraft, pp. 74–106, "Nauka", Moscow, 1975. (Russian) [MR **57 # 4731**]

[6.44] Colombo, G., *Cassini's second and third laws,* Astron. J., **71** (1966), no. 9, 891–896.

[6.45] Peale, S., *Generalized Cassini's laws,* Astron. J., **74** (1969), no. 3, 483–489.

[6.46] Holland, R. L. and Sperling, H. J., *A first-order theory for the rotational motion of a triaxial body orbiting an oblate primary,* Astron. J. **74** (1969), no. 3, 490–496.

[6.47] Johnson, D. B., *Precession rate matching a space station in orbit about an oblate planet,* J. Spacecraft and Rockets **10** (1973), no. 7, 467–470.

[6.48] Hara, M., *Effect of magnetic and gravitational torques on spinning satellite attitude,* AIAA J., **11** (1973), no. 12, 1737–1742.

[6.49] Beletsky, V. V. and Trushin, S. I., *Resonances in the rotation of celestial bodies and the generalized Cassini's laws,* in: Rigid Body Mechanics, Vol. 6, "Naukova Dumka," Kiev, 1974. (Russian)

[6.50] Beletsky, V. V. and Trushin, S. I., *Stability of generalized Cassini's laws,* in: Rigid Body Mechanics, Vol. 6, "Naukova Dumka," Kiev, 1974. (Russian)

[6.51] Beletsky, V. V., *Motion of a Satellite Around its Center of Mass in a Gravitational Field,* Izdat. Mosk. Gos. Univ., Moscow, 1975. (Russian)

[6.52] Shapiro, I. I., *Radar astronomy, general relativity and celestial mechanics,* in: Modern Questions of Celestial Mechanics, Bressanone, May 1967, Cremonese, Roma, 1968.

[6.53] Khentov, A. A., *Influence of the magnetic and gravitational fields of the Earth on the oscillations of a satellite about its center of mass,* Kosmicheskie Issled., **5** (1967), no. 4, 554–572. (Russian); English transl., Cosmic Research **5** (1967), no. 4, 479–495.

[6.54] Khentov, A. A., *Passive stabilization of artificial satellites with respect to the Earth's magnetic field,* Kosmicheskie Issled., **5** (1967), no. 4, 540–553. (Russian); English transl., Cosmic Research **5** (1967), no. 4, 466–478.

[6.55] Beletsky, V. V. and Khentov, A. A., *Magnetic-gravitational stabilization of a satellite,* Izv. Akad. Nauk SSSR, Ser. Mekh. Tverdogo Tela (1973), no. 4, 22–32. (Russian)

[6.56] Chetaev, N. G., *Stability of Motion. Works on Analytical Mechanics,* Izdat. Akad. Nauk SSSR, Moscow, 1962. (Russian); see also *Stability of Motion,* Fourth edition. With comments by V. V. Rumyantsev, "Nauka", Moscow, 1990. [MR **92j:34095**]

[6.57] Beletsky, V. V., Levin, E. M. and Pogorelov, D. Yu., *On the problem of the resonance rotation of Venus,* Astronom. Zh. **57** (1980), no. 1, 158–167. (Russian); English transl., Soviet Astron., **24** (1980), no. 1, 94–99. [MR **81h:70004**]

[6.58] Beletsky, V. V., Levin, E. M., and Pogorelov, D. Yu., *On the problem of resonant rotation of Venus. II,* Astronom. Zh. **58** (1981), no. 1, 198–207. (Russian); English transl., Soviet Astron., **25** (1981), Jan.-Feb., 110-115.

[6.59] Beletsky, V. V., *Tidal evolution of inclinations and rotations of celestial bodies,* Celestial Mech., **23** (1981), no. 3, 371–382.

[6.60] Kozlov, N. N. and Eneev, T. M., *Numerical modeling of the process of formation of planets from a protoplanetary cloud,* Preprint no. 134, Institute of Applied Mathematics, Academy of Sciences of USSR, 1977. (Russian)

[6.61] Khentov, A. A., *Dynamics of the rotation of Venus with allowance for the superrotation of its atmosphere,* Astronom. Zh. **69** (1992), no. 3, 655–659. (Russian); English transl., Soviet Astron., **36** (1992), no. 3, 333–335.

[6.62] Bills, B. G., Kiefer, W. S., and Jones, R. L., *Venus gravity: a harmonic analysis,* J. Geophys. Res. **92** (1987), no. B10, 10335–10351.

[6.63] Beletsky, V. V., Pivovarov, M. L., and Starostin, E. L. *Regular and chaotic motions in applied dynamics of a rigid body,* Chaos **6** (1996), no. 2, 155–166.

[6.64] Beletsky, V. V., *Reguläre und Chaotische Bewegung Starrer Körper,* Teubner Studienbücher Mechanik, B. G. Teubner, Stuttgart, 1995. [MR **99m: 70004**]

[6.65] Beletsky, V. V., Pivovarov, M. L., and Starostin, E. L. *Regular and chaotic motions in the problem of the rotation of a celestial body in the gravitational field of two centers,* Preprint no. 128, Keldysh Institute of Applied Mathematics, Academy of Sciences of USSR, 1990. (Russian)

[6.66] Shapiro, I. I., Campbell, D. B., and De Campli, W. M., *Nonresonance rotation of Venus,* Astrophys. J. Lett., **230** (1979), no. 2.

[6.67] Khentov, A. A., *Dynamics of the formation of resonant rotations of natural celestial bodies,* Astronom. Zh. **59** (1982), no. 4, 769–778. (Russian); English transl., Soviet Astron., **26** (1982), 468–473.

[6.68] Khentov, A. A., *On the problem of the interpretation of the observed rotation of Venus,* Astronom. Zh. **66** (1989), no. 1, 202–204. (Russian); English transl., Soviet Astron., **33** (1989), no. 1, 105–106.

[6.69] Peale, S. J., *Some unsolved problems in the evolutionary dynamics in the solar system,* Celestial Mech. Dyn. Astron., **46** (1989), no. 3, 253–275.

[6.70] Beletsky, V. V. and Khentov, A. A., *Resonant Rotational Motions of Celestial Bodies,* Nizhegorod. Gumanit. Tsentr, Nizhniĭ Novgorod, 1995. (Russian)

[6.71] Beletsky, V. V. and Ponomareva, O. N., *A parametric analysis of relative equilibrium stability in a gravitational field,* Kosmicheskie Issled., **28** (1990), no. 5, 664–675. (Russian); English transl., Cosmic Research, **28** (1991), no. 5, 573–582.

[6.72] Beletsky, V. V. and Levin, E. M., *Dynamics of Space Tether Systems,* "Nauka," Moscow, 1990. (Russian); English transl., Advances in the Astronautical Sciences, Vol. 83, Univelt Inc. Publ., San Diego, 1993.

[6.73] Keldysh, M. V., *Collected Works. Rocket Technology and Astronautics,* "Nauka," Moscow, 1988. (Russian)

Seventh essay

[7.1] Grodzovskiĭ, G. L., Ivanov, Yu. N., and Tokarev, V. V., *Mechanics of Low-Thrust Space Flight,* "Nauka," Moscow, 1966. (Russian); English transl., Israel Program for Scientific Translations, Jerusalem, 1969.

[7.2] Beletsky, V. V. and Egorov, V. A., *Acceleration of a spacecraft in the sphere of action of a planet,* Kosmicheskie Issled., **2** (1964), no. 3, 392–407. (Russian); English transl., Cosmic Research **2** (1964), no. 3, 331–345.

[7.3] Okhotsimskiĭ, D. E., *Investigation of the motion in a central field under the action of a constant tangential acceleration,* Kosmicheskie Issled., **2** (1964), no. 6, 817–842. (Russian); Cosmic Research **2** (1964), no. 6, 715–736.

[7.4] Evtushenko, Yu. G., *Investigation of the motion in a central field under the action of a constant tangential acceleration,* Prikl. Mat. Mekh., **30** (1966), no. 3. (Russian)

[7.5] Efimov, G. B., *Optimal acceleration in a central field to escape velocity,* Kosmicheskie Issled., **7** (1970), no. 1, 26–47. (Russian)

Eighth essay

[8.1] Clarke, A. C., *The wind from the Sun,* in: *The Wind from the Sun; Stories of the Space Age,* Harcourt Brace Jovanovich, New York, 1972.

[8.2] Grodzovskiĭ, G. L., Ivanov, Yu. N., and Tokarev, V. V., *Mechanics of Space Flight. Optimization Problems,* "Nauka," Moscow, 1975. (Russian)

[8.3] Skoptsov, A. P., *Variational problem on the exit of a solar-sail powered spacecraft from the sphere of attraction of a planet,* in: Problems of the Mechanics of Controlled Motion, Vol. 1, Perm′, 1972. (Russian)

[8.4] Beletsky, V. V., *On trajectories of space flights with a constant reactive acceleration vector,* Kosmicheskie Issled., **2** (1964), no. 3, 408–413. (Russian); English transl., Cosmic Research **2** (1964), no. 3, 346–351.

[8.5] Beletsky, V. V., *Description of planar trajectories of space flights with a constant reactive acceleration vector,* in: Proc. Conf. on General and Applied Problems of Celestial Mechanics, Moscow, 1964. (Russian)

Ninth essay

[9.1] Beletsky, V. V. and Giverts, M. E., *Motion of a pulsating system in a gravitational field,* Kosmicheskie Issled., **5** (1967), no. 6, 304–308. (Russian)

[9.2] Beletsky, V. V., *Gravity flyer,* Tekhnika–Molodezhi, (1970), no. 3. (Russian)

[9.3] Beletsky, V. V., *Motion of an Artificial Satellite About its Center of Mass,* "Nauka", Moscow 1962 (Russian); English transl., Israel Program for Scientific Translations, Jerusalem, 1966; see also NASA Technical Reports NASA-TT-F-429, TT-67-51366.

[9.4] Donov, A. E., *Flight theory of a gravity flyer,* Kosmicheskie Issled., **9** (1971), no. 3, 392–396. (Russian); English transl., Cosmic Research **9** (1971), no. 3, 360–363.

[9.5] Giverts, M. E., *On a particular case of optimization of the control of a gravity flyer,* Kosmicheskie Issled., **10** (1972), no. 2, 297–299. (Russian); English transl., Cosmic Research **10** (1972), no. 2, 269–271.

[9.6] Markov, V. E., *Countering the action of external perturbations on a spacecraft by modifying its mass geometry,* Izv. Akad. Nauk SSSR. Mekh. Tverdogo Tela (1974), no. 5. (Russian)

[9.7] Kholshevnikov, K. V. and Markov, V. E., *Stabilization of the altitude of the pericenter,* Kosmicheskie Issled., **13** (1975), no. 2, 181–184. (Russian); English transl., Cosmic Research **13** (1975), no. 2, 158–161.

[9.8] Schaefer, J. F., *Artificial satellite orbit shifting without mass expulsion utilizing gravity gradient,* in: Proc. 17th Astr. Congress, Madrid, 1966; Proc. 4 Astrodyn. Guidance and Control, Paris, 1967.

Tenth essay

[10.1] Beletsky, V. V. and Egorov, V. A., *Interplanetary flights with a constant-power engine,* Kosmicheskie Issled., **2** (1964), no. 3, 360–391. (Russian); English transl., Cosmic Research **2** (1964), no. 3, 303–330.

[10.2] Beletsky, V. V. and Egorov, V. A., and Ershov, V. G., *Analysis of trajectories of interplanetary flights with constant-power engines,* Kosmicheskie Issled., **3** (1965), no. 4, 507–522. (Russian); English transl., Cosmic Research **3** (1965), no. 4, 397–413.

[10.3] Beletsky, V. V., Golubkov, V. V., Egorov, V. A., and Ershov, V. G., *Investigation of low-thrust flight trajectories,* in: Abstracts of the Second All-Union Congress on Theoretical and Applied Mechanics, Izdat. Akad. Nauk SSSR, Moscow, 1964. (Russian)

[10.4] Grodzovskiĭ, G. L., Ivanov, Yu. N., and Tokarev, V. V., *Mechanics of Low-Thrust Space Flight,* "Nauka," Moscow, 1966. (Russian); English transl., Israel Program for Scientific Translations, Jerusalem, 1969.

[10.5] Pontryagin, L. S., Boltyanskiĭ, V. G., Gamkrelidze, R. V., and Mishchenko, E. F., *The Mathematical Theory of Optimal Processes,* 4th edition, "Nauka," Moscow, 1969. (Russian) [MR **84i:93008**]; English transl., A Pergamon Press Book. The Macmillan Co., New York 1964. [MR **32** #3896]

[10.6] Boltyanskiĭ, V. G., *Mathematical Methods of Optimal Control,* Second revised and supplemented edition, "Nauka", Moscow, 1969. (Russian) [MR **50** #5568]; English transl., Holt, Rinehart and Winston, Inc., New York-Montreal, Que.-London, 1971. [MR **50** #5567]

[10.7] Melbourne, W. G. and Sauer, C. G., *Optimum interplanetary rendezvous with power-limited vehicles,* AIAA J., **1** (1963), no. 1, 54–60.

[10.8] Platonov, A. K., *Optimal properties of correcting maneuvers using thrust-limited engines,* Kosmicheskie Issled., **5** (1967), no. 2, 170–175. (Russian); English transl., Cosmic Research **5** (1967), no. 2, 147–151.

[10.9] Grodzovskiĭ, G. L., Okhotsimskiĭ, D. E., Beletsky, V. V., Ivanov, Yu. N., Kur′yanov, A. I., Platonov, A. K., Sarychev, V. A., Tokarev, V. V., and Yaroshevskiĭ, V. A., *Mechanics of space flight,* in: Mechanics in the USSR Over the Last 50 Years, Vol. 1, "Nauka,", Moscow, 1968. (Russian)

Eleventh essay

[11.1] Beletsky, V. V. and Novikova, E. T., *On the relative motion of two connected bodies in orbit,* Kosmicheskie Issled., **7** (1969), no. 3, 377–384. (Russian)

[11.2] Beletsky, V. V., *On the relative motion of two connected bodies in orbit. II,* Kosmicheskie Issled., **7** (1969), no. 6, 827–840. (Russian)

[11.3] Beletsky, V. V. and Egorov, V. A., *Interplanetary flights with a constant-power engine,* Kosmicheskie Issled., **2** (1964), no. 3, 360–391. (Russian); English transl., Cosmic Research **2** (1964), no. 3, 303–330.

[11.4] Belyaev, P. I. and Leonov, A. A., *In open space,* "Pravda" newspaper, April 14, 1965. (Russian)

[11.5] Egorov, V. A., *Some problems of the optimization of space probe trajectories in interplanetary space,* Avtomat. i Telemekh., (1970), no. 5, 5–24. (Russian)

[11.6] Arnold, V. I., *Mathematical Methods of Classical Mechanics,* 3rd edition, "Nauka", Moscow, 1989. (Russian) [MR **93c:70001**]; English transl., corrected reprint of the second (1989) edition, Graduate Texts in Mathematics, 60, Springer-Verlag, New York, 199?. [MR **96c:70001**]

[11.7] Eneev, T. M., Kozlov, N. N., and Syunyaev, R. A., *Tidal interaction of galaxies,* Astron. and Astrophys., **22** (1973), no. 1, 41–60.

[11.8] Arnold, V. I., *Ordinary Differential Equations,* 3rd edition, "Nauka", Moscow, 1984. (Russian) [MR **86i:34001**]; English transl., Springer Textbook, Springer-Verlag, Berlin, 1992. [MR **93b:34001**]

[11.9] Ivanov, A. P., *Shockless motions in systems with unilateral constraints,* Prikl. Mat. Mekh., **56** (1992), no. 1, 3–15. (Russian) [MR **93d: 70024**]; English transl., J. Appl. Math. Mech., **56** (1992), no. 1, 1–12.

[11.10] Ivanov, A. P. and Baziyan, A. B., *Investigation of asymptotic motions of two connected bodies in a circular orbit,* Kosmicheskie Issled., **33** (1995), no. 5. (Russian)

[11.11] Beletsky, V. V. and Pankova, D. V., *Connected bodies in orbit as a dynamic billiard,* Preprint no. 7, Keldysh Institute of Applied Mathematics, 1995. (Russian)

[11.12] Beletsky, V. V. and Pankova, D. V., *Connected bodies in orbit as a dynamic billiard,* Regul. Khaoticheskaya Din., **1** (1996), no. 1, 87–103. [MR **99f:70034**]

[11.13] Beletsky, V. V. and Pankova, D. V., *Influence of aerodynamics on the relative motion of two connected bodies in orbit. Part II. Chaotic motions,* Preprint no. 40, Keldysh Institute of Applied Mathematics, 1996. (Russian)

[11.14] Beletsky, V. V., Kasatkin, G. V., and Starostin, E. L., *The pendulum as a dynamical billiard,* Chaos Solitons Fractals **7** (1996), no. 8, 1145–1178. [MR **97e:58147**]

[11.15] Beletsky, V. V., *Reguläre und Chaotische Bewegung Starrer Körper,* Teubner Studienbücher Mechanik, B. G. Teubner, Stuttgart, 1995. [MR **99m: 70004**]

[11.16] Beletsky, V. V. and Levin, E. M., *Dynamics of Space Tether Systems,* "Nauka," Moscow, 1990. (Russian); English transl., Advances in the Astronautical Sciences, Vol. 83, Univelt Inc. Publ., San Diego, 1993.

Twelfth essay

[12.1] Akim, E. L. and Eneev, T. M., *Determination of motion parameters of a space vehicle from orbital data,* Kosmicheskie Issled., **1** (1963), no. 1, 5–50. (Russian); English transl., Cosmic Research **1** (1963), no. 1, 2–40.

[12.2] Akim, E. L., *Determination of the gravitational field of the Moon from the motion of the lunar orbiter Luna-10,* Kosmicheskie Issled., **4** (1966), no. 6, 823–826. (Russian); English transl., Cosmic Research **4** (1966), no. 6, 712–715.

[12.3] Beletsky, V. V. and Zonov, Yu. V., *Rotation and orientation of the third Soviet satellite,* in: Iskusstvennye Sputniki Zemli, Vol. 7, 1961, 32–55. (Russian) English transl., in: Artificial Earth Satellites Vol. 7–8, 1962, pp. 29–52.

[12.4] Beletsky, V. V., Golubkov, V. V., Stepanova, E. A. and Khatskevich, I. G., *Determination of the orientation of artificial satellites from measurement data. Part I. Methods,* 1967; *Part II. Results. Analysis of motion,* 1968, Preprints, Institute of Applied Mathematics, Academy of Science of USSR, Moscow. (Russian)

[12.5] Beletsky, V. V., Golubkov, V. V., Lavrovskiĭ, E. K., Trushin, S. I., and Khatskevich, I. G., *Determination of the orientation and rotation of artificial satellites from measurement data,* Kosmicheskie Issled., **5** (1967), no. 5, 686–702. (Russian); English transl., Cosmic Research **5** (1967), no. 5, 585–598.

[12.6] Beletsky, V. V., Golubkov, V. V., and Khatskevich, I. G., *Rotation and orientation analysis of the Proton 2 satellite,* in: Proc. XVIII IAF International Astronautical Congress, Belgrad, 1967, Pergamon Press, 1968.

[12.7] Golubkov, V. V. and Khatskevich, I. G., *Determination of the orientation of artificial Earth satellites from a given system of measurements,* Kosmicheskie Issled., **7** (1969), no. 4, 510–521. (Russian)

[12.8] Beletsky, V. V., Golubkov, V. V., Stepanova, E. A., and Khatskevich, I. G., *Results of the determination of the orientation of the Proton 2 satellite and description of its motion realtive to the center of mass,* Kosmicheskie Issled., **7** (1969), no. 4, 522-533. (Russian)

[12.9] Beletsky, V. V., *Estimate of the nature of the interaction between an aerodynamic flow and a satellite on the basis of an analysis of the motion of the Proton 2 satellite relative to the center of mass,* Kosmicheskie Issled., **8** (1970), no. 2, 206–217. (Russian)

[12.10] Eneev, T. M., Platonov, A. K., and Kazakova, R. K., *Determination of the orbital parameters of an artificial satellite from terrestrial measurements,* in: Iskusstvennye Sputniki Zemli, Vol. 4, 1960, pp. 43–55. (Russian)

[12.11] Beletsky, V. V., *Motion of an Artificial Satellite About its Center of Mass,* "Nauka", Moscow 1962 (Russian); English transl., Israel Program for Scientific Translations, Jerusalem, 1966; see also NASA Technical Reports NASA-TT-F-429, TT-67-51366.

Subject Index

Author Index

9 783764 358662

MIX
Papier aus verantwortungsvollen Quellen
Paper from responsible sources
FSC® C105338

If you have any concerns about our products, you can contact us on
ProductSafety@springernature.com

In case Publisher is established outside the EU, the EU authorized representative is:
Springer Nature Customer Service Center GmbH
Europaplatz 3, 69115 Heidelberg, Germany

Printed by Libri Plureos GmbH
in Hamburg, Germany